Practical Carpentry

BEING A COMPLETE, UP-TO-DATE EXPLANATION OF MODERN CAR-
PENTRY AND AN ENCYCLOPEDIA ON THE MODERN METHODS
USED IN THE ERECTION OF BUILDINGS, FROM THE
LAYING OF THE FOUNDATION TO THE
DELIVERY OF THE BUILDING
TO THE PAINTER

IN TWO VOLUMES

EDITED UNDER THE SUPERVISION OF

WILLIAM A. RADFORD

EDITOR-IN-CHIEF OF THE "AMERICAN CARPENTER AND BUILDER," PRESIDENT OF
"THE RADFORD ARCHITECTURAL CO.," AUTHOR OF "THE STEEL SQUARE
AND ITS USES" AND THE BEST AUTHORITY IN THE COUNTRY ON
ALL THINGS PERTAINING TO THE BUILDING TRADE

ASSISTED BY

ALFRED W. WOODS AND WILLIAM REUTHER

VOLUME I.

Books for Business
New York - Hong Kong

Practical Carpentry:
A Complete Up-To-Date Explanation of
Modern Carpentry

Edited Under the Supervision of
William A. Radford

ISBN: 0-89499-175-2

Reprinted from the 1907 edition

Books for Business
New York _ Hong Kong
http://www.BusinessBooksInternational.com

CONTENTS

Part I

Part II

Part III

———

Part IV

———

Part V

Part VI

PREFACE

In getting out the present work it has been my constant aim to take up the subject of building construction in a systematic and concise way. With this object in view, I have first treated the subject of geometry in so far as it proved a help in carpentry. Every example given involves some fundamental principle of carpentry and if thoroughly mastered will prove of inestimable value later on in the work.

One chapter is devoted to the use of the steel square in carpentry in which a number of suggestions are given, especially along the line of roof framing which will help to simplify what was formerly a troublesome part of carpenter work. In treating the subject of house framing, both good and faulty methods of construction are given so as to more clearly bring out the errors most common in the trade. Various ways of constructing cornice, sills and porches are illustrated. Framing studding and joist bearer and cutting window openings are also described in a comprehensive manner.

In the chapter of roof construction it has been thought best to start with the simplest shed roof and work into the more complicated roofs, covering in the chapter all the principles involved in their construction.

The department of questions and answers is one of the most useful, interesting and instructive parts of this work. They are questions which have arisen in the daily work of practical carpenters all over the country and in answering them my aim has been to make them as clear and concise as possible so as to be understood by everyone.

Original illustrative diagrams, over two hundred in number, are contained in the work, including many full-page illustrations which give clearer and better ideas than could be given in any other way.

Great care has been taken to secure absolute accuracy in all of this work so as to make it reliable in every detail.

All the rules and examples are placed under appropriate headings, with index commencement words printed in bold type, so that the eye of the reader can catch the particular information wanted at a glance.

We are fortunate in being able to present a number of full-page details, showing the construction of cornices, porches, stairs, etc., which were prepared for this work by G. W. Ashby.

It has not been thought necessary to give any explanation of these details, as they are self-explanatory. They show the construction of each part completely and complete dimensions are given in all cases.

In Volume II the subject is taken up where it was left off in this volume, treating the subject of stair building, shingling, window construction, fireplace construction and the various kinds of mouldings.

WILLIAM A. RADFORD,
Chicago, Ill.

Practical Carpentry

Part I

GEOMETRY. Magnitude—Solid—Surface—Plane—Angle—
Quantity of an angle — Right angle — Acute angle — Obtuse
angle — Parallelogram — Rectangle — Diagonal — Polygon —
Circle — Radius of a circle — Chord of an arc — Tangent —
Altitude — Inscribed polygon — Problems in Geometry — To
bisect a given angle — To erect a perpendicular — To bisect
a given straight line—To describe a square equal to two given
squares — Inscribing a pentagon in a circle — To describe the
circumference of a circle through three given points, Etc.

While it is not essential for a man to have a
thorough knowledge of geometry in order to be
a good carpenter, yet none will deny that with a
knowledge of the science they would be more
competent workmen and have a larger field to
work in.

With this object in view we have thought it
best to enter into some of the details of geometry
and explain some of the more simple terms and
problems, with the hope that it will give the
student a desire for a more thorough knowledge
of the subject.

The terms and definitions used in geometry
are as follows:—

Geometry is the science of magnitude.

13

Magnitude has three dimensions: length, breadth and thickness.

A **Point** has position but not magnitude. Practically, it is represented by the smallest visible mark or dot; but geometrically understood, it occupies no space. The extremities or ends of lines are points; and when two or more lines cross one another, the places that mark their intersection are also points.

A **Line** has length, without breadth or thickness, and, consequently, a true geometrical line cannot be exhibited; for however fine a line may be drawn it will always occupy a certain extent of space.

A **Surface** has length and breadth, but no thickness. For instance, a shadow gives a very good representation of a surface—its length and breadth can be measured; but it has no depth or substance. The quantity of space contained in any plane surface is called its area.

A **Plane** is a flat surface, which will coincide with a straight line in every direction.

A **Curved or uneven surface** is one that will not coincide with a straight line in all directions. By the term surface is generally understood the outside of any body or object.

A **Solid** is anyth ng which has length, breadth, and thickness; consequently the term may be applied to any visible object containing substance, but practically it is understood to signify the solid contents or measurements contained within the different surfaces of which any body is formed.

Lines may be drawn in any direction, and are termed straight, curved, mixed, concave or convex lines, according as they correspond to the following definitions:—

A **Straight Line** is one every part of which lies in the same direction between its extremities, and is, of course, the shortest distance between two points.

A **Curved Line** is such that it does not lie in a straight direction between its extremities, but is continually changing by inflection. It may be either regular or irregular.

A **Mixed Line** is composed of straight and curved lines, connected in any form.

A **Concave or Convex Line** is such that it cannot be cut by a straight line in more than two points; the concave or hollow side is turned towards the straight line, while the convex or swelling side looks away from it. For instance, the inside of a basin is concave; the outside of a ball is convex.

Parallel Straight Lines have no inclination, but are everywhere at an equal distance from each other; consequently they can never meet, though produced or continued to infinity in either or both directions.

Oblique or Converging Lines are straight lines which, if continued, being in the same plane, change their distance so as to meet or intersect each other.

An Angle is formed by the inclination of two lines meeting in a point; the lines thus forming the angle are called the sides, and the point where the lines meet is called the vertex or angular point.

When an angle is expressed by three letters, as A B C, Fig. 1, the middle letter B should always denote the angular point.

The Quantity of an Angle is estimated by the arc of any circle contained between the two sides or lines forming the angle; the junction of the two lines, or vertex of the angle, being the center from which the arc is described. As the circumferences of all circles are proportional to their

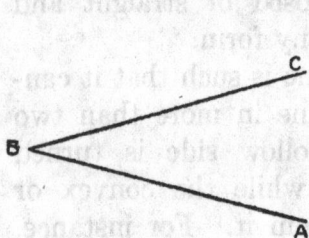

FIG.1.   FIG.2.

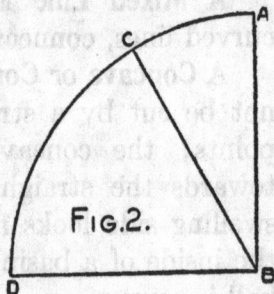

diameters, the arcs of similar sectors also bear the same proportion to their respective circumferences; and, consequently, are proportional to their diameters and, of course, also their radii or semi-diameters. Hence, the proportion which the arc of any circle bears to the circumference of that circle determines the magnitude of the angle. From this it is evident that the quantity or magnitude of the angles does not depend upon the length of the sides or radii forming them, but wholly upon the number of degrees contained in the arc cut from the circumference of the circle by the opening of these lines. The circumference of every circle is divided by mathematicians into 360 equal parts, called degrees; each degree being again subdivided into 60 equal parts, called minutes, and each minute into 60 equal parts called seconds. Hence, it follows that the arc of a quarter circle or quadrant includes 90 degrees; that is, one-fourth part of 360 degrees.

A **Right Angle** is produced by one straight line standing upon another so as to make the adjacent angles equal, and is an angle of 90 degrees. This is what workmen call "square," and is the most useful figure they employ.

An **Acute Angle** is less than a right angle, or less than 90 degrees.

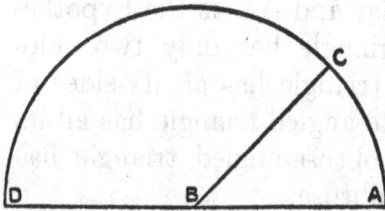
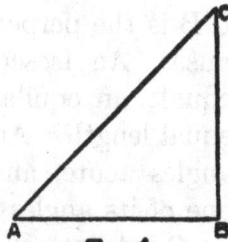

Fig. 3.   Fig. 4.

An **Obtuse Angle** is greater than a right angle, or more than 90 degrees.

The number of degrees by which an angle is less than 90 degrees is called the complement of the angle. For example, in Fig. 2, angle C B D is the complement of the angle A B D because angle A B C, plus angle C B D, is equal to a right angle or one of 90 degrees, or angle A B D.

The **difference between an obtuse angle and a semi=circle,** or 180 degrees, is called the supplement of that angle. For example, in Fig. 3, angle C D B is the supplement of the angle A D B because angle C D B, added to angle A C B, is equal to a semi-circle or an angle of 180 degrees.

Plane Figures are bounded by straight lines, and are named according to the number of sides which they contain. Thus, the space included within three straight lines, forming three angles, is called trilateral figure or triangle.

A **Right=angled Triangle** has one right angle: The sides forming the right angle are called the base and perpendicular; the side opposite the right angle is named hypothenuse. See Fig. 4, A B C is a right-angled triangle—A B is the base, C B is the perpendicular and A C is the hypothenuse. An isosceles triangle has only two sides equal; an equilateral triangle has all its sides of equal length. An acute-angled triangle has all its angles acute, and an obtuse-angled triangle has one of its angles only obtuse.

Quadrilateral Figures are literally four-sided figures; they are also called quadrangles, because they have four angles.

A **Parallelogram** is a figure whose opposite sides are parallel, as A B C D, in Fig. 5.

A **Rectangle** is a parallelogram having four right angles, as A B C D, in Fig. 5.

A **Square** is an equilateral rectangle, having all its sides and angles equal, like Fig. 5.

A **Diagonal** is a straight line drawn between

two opposite angular points of a quadrilateral figure, or between any two angular points of a polygon. Should the figure be a parallelogram,

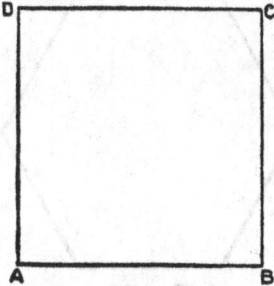

FIG.5.

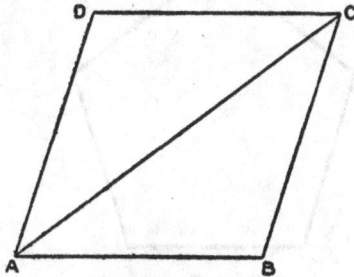

FIG.6.

the diagonal will divide it into two equal triangles, the opposite sides and angles of which will be equal to one another. Let A B C D, Fig. 6, be a parallelogram; join A C, then A C is a diagonal, and the triangles A D C and A B C, into which it divides the parallelogram, are equal.

A **Polygon** is a portion of a plane terminated on all sides by straight lines. A regular polygon has all its sides and angles equal, and an irregular polygon has its sides and angles unequal.

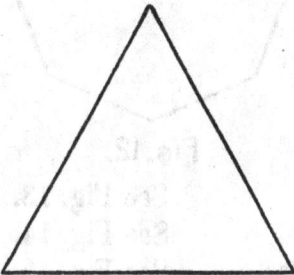

FIG.7.

FIG.8.

Polygons are named according to the number of their sides or angles, as follows:

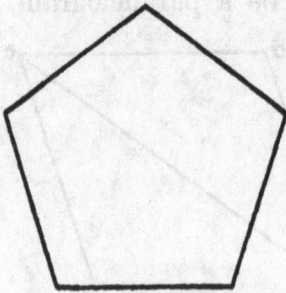

FIG. 9.

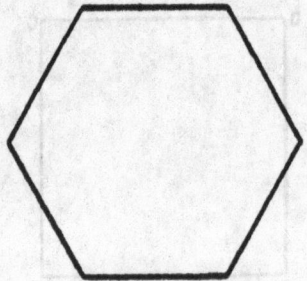

FIG. 10.

A Triangle is a polygon of three sides. See Fig. 7.
A Square is a polygon of four sides. See Fig. 8.
A Pentagon is a polygon of five sides. See Fig. 9.
A Hexagon is a polygon of six sides. See Fig. 10.
A Heptagon has seven sides. See Fig. 11.
An Octagon has eight sides. See Fig. 12.

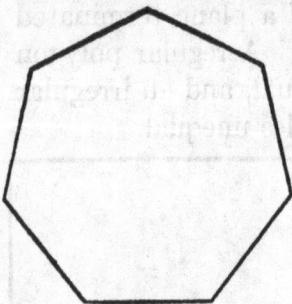

FIG. 11.

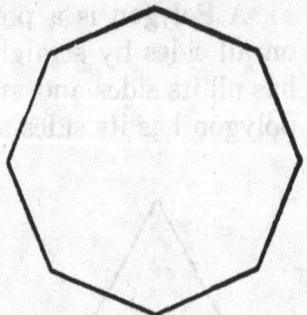

FIG. 12.

A Nonagon has nine sides. See Fig. 13.
A Decagon has ten sides. See Fig. 14.
An Undecagon has eleven sides. See Fig. 15.
A Dodecagon has twelve sides. See Fig. 16.

Figures having more than twelve sides are generally designated polygons, or many-angled figures.

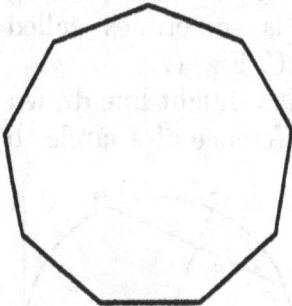

FIG. 13.

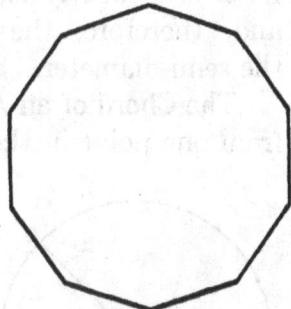

FIG. 14.

A **Circle** is a plane figure bounded by one uniformly curved line, called the circumference, every part of which is equally distant from a point within, called the center, as A in Fig. 17.

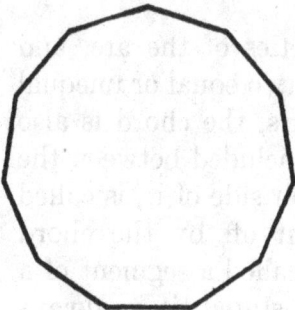

FIG. 15.

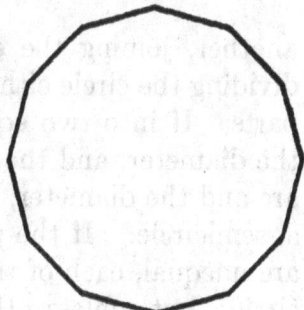

FIG. 16.

The **Radius of a Circle** is a straight line drawn from the center to the circumference; hence, all the radii of a circle are equal, as AB, AD, AC, AE, in Fig. 17.

The **Diameter of a Circle** is a straight line drawn through the center and terminated on each side by the circumference; consequently the circumference is exactly twice the length of the radius, and, therefore, the radius is sometimes called the semi-diameter. See B A C, Fig. 17.

The **Chord of an Arc** is any straight line drawn from one point in the circumference of a circle to

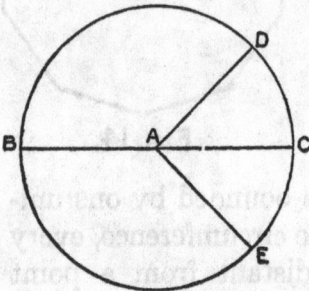

FIG. 17.

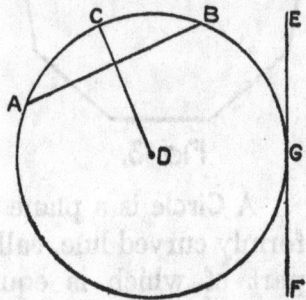

FIG. 18.

another, joining the extremeties of the arc, and dividing the circle either into two equal or unequal parts. If into two equal parts, the chord is also the diameter, and the space included between the arc and the diameter, on either side of it, is called a semicircle. If the parts cut off by the chord are unequal, each of them is called a segment of a circle; but, unless otherwise stated, it is always understood that the smaller arc or segment is spoken of as in Fig. 18 A B is the chord of the arc A C B.

If a straight line be drawn from the center of a circle to meet the chord of an arc perpendicularly, as D C, in Fig. 18, it will divide the chord into two equal parts, and if the straight line be produced to meet the arc, it will also divide it into two equal parts, as A C, C B.

Each half of the chord is called the sine of the half arc to which it is opposite; and the line drawn from the center to meet the chord perpendicularly is called the co-sine of the half-arc. Consequently, the radius, the sine, the co-sine of an arc form a right angle.

A **Tangent** is any straight line which touches the circumference of a circle in one point, which is called the point of contact, as in the tangent E G F in Fig. 18.

An arc is any portion of the circumference of a circle, as A C B in Fig. 18.

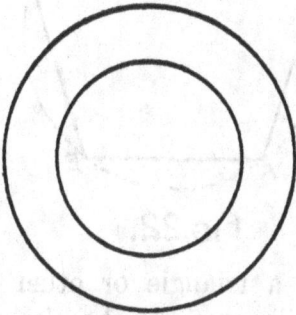

Fig. 19.

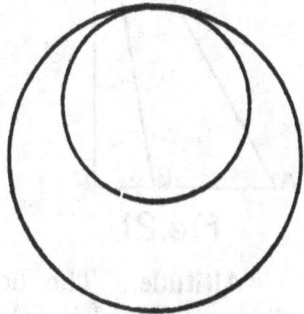

Fig. 20.

It may not be improper to remark here that the terms circle and circumference are frequently misapplied. Thus we say, describe a circle from a given point, instead of saying describe the circumference of a circle—the circumference being the curved line thus described, everywhere equally distant from a point within it, called the center; whereas the circle is properly the superficial space included within that circumference.

Concentric Circles are circles within circles, described from the same center; consequently their circumferences are parallel to one another, as in Fig. 19.

Eccentric Circles are those which are not described from the same center; eccentric circles may also be tangent circles; that is, such as come in contact in one point, only, as Fig. 20.

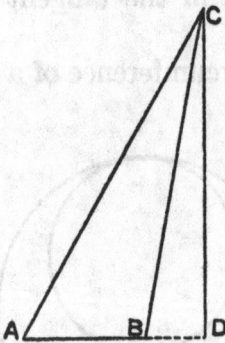

FIG. 21

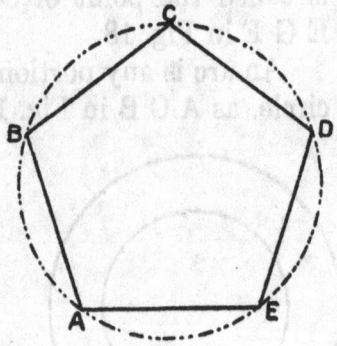

FIG. 22.

Altitude. The height of a triangle or other figure is called its altitude. To measure the altitude, let fall a straight line from the vertex or highest point in the figure, perpendicular to the base or opposite side; or, to the base continued, as at B D, Fig. 21, should the form of the figure require its extension. Thus C D is the altitude of the triangle A B C.

An **Inscribed Polygon** is one which, like A B C D E in Fig. 22, has all its angles in the circumference. The circle is then said to circumscribe such a figure.

We have now described all the figures we shall require for the purpose of thoroughly understanding all that will follow in this book; but we would like to say right here that the student who has time should not stop at this point in the study of geometry, for the time spent in obtaining a thorough knowledge of this useful science will bring in better returns than if expended for any other purpose.

Problems in Geometry.

We will now proceed to explain how the figures we have described can be constructed. There are several ways of constructing nearly every figure we produce; but we have chosen those methods

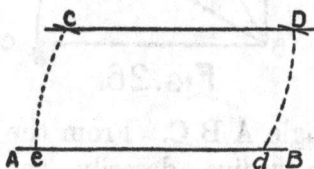

FIG. 23.

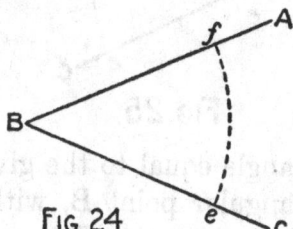
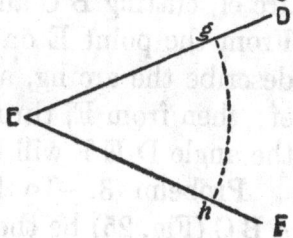

FIG. 24.

that seem to us the best, but to save space have given only those which are most essential in the study of practical carpentry.

Problem 1.—Through a given point C (Fig. 23), to **draw a straight line parallel to a given straight line A B**: In A B (Fig. 23) take any point d, and from d as a center with the radius dC, describe an arc Ce, cutting AB in e, and from C as a center, with the same radius, describe the arc dD, make dD equal to Ce, join C D, and it will be parallel to A B.

Problem 2.—To **make an angle equal to a given rectilineal angle**: From a given point E (Fig. 24), upon the straight line E F, to make an

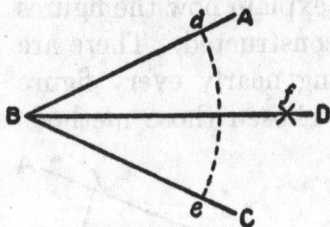

FIG.25.

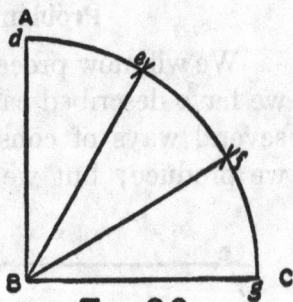

FIG.26.

angle equal to the given angle A B C. From the angular point B, with any radius, describe the arc ef, cutting B C and B A in the points e and f. From the point E on E F, with the same radius, describe the arc hg, and make it equal to the arc ef; then from E, through g, draw the line E D: the angle D E F will be equal to the angle A B C.

Probelm 3.—To **bisect a given angle:** Let A B C (Fig. 25) be the given angle. From the an-

gular point B, with any radius, describe an arc
cutting B A and B C in the points d and e; also,
from the points d and e as centers, with any
radius greater than half the distance between
them, describe arcs cutting each other in f; through
the points of intersection f, draw B f D: the angle
A B C is bisected by the straight line B D; that is,
it is divided into two equal angles, A B D and
C B D.

Problem 4.—To **trisect or divide a right angle
into three equal angles:** Let A B C (Fig. 26) be
the given right angle. From the angular point B,
with any radius, describe an arc cutting B A and

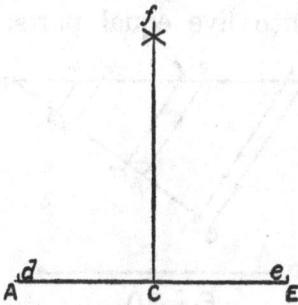
FIG. 27.

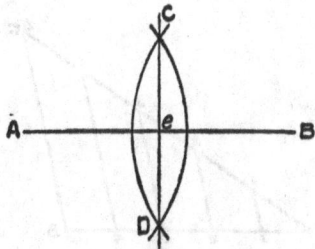
FIG. 28.

B C in the points d and g; from the points d and
g, with the radius Bd or Bg, describe the arcs
cutting the arc dg in e and f; join Be and Bf:
these lines will trisect the angle A B C, or divide
it into three equal angles.

Problem 5.—From a given point C, in a given
straight line A E, **to erect a perpendicular:** From
the point C (Fig. 27), with any radius less than

CA or CE, describe arcs cutting the given line
A E in d and e; from these points as centers,
with a radius greater than Cd or Ce, describe arcs
intersecting each other in f: join Cf, and this
line will be the perpendicular required.

Problem 6.—To **bisect a given straight line:**
Let A B (Fig. 28) be the given straight line.
From the extreme points A and B as centers,
with any equal radii greater than half the length
of AB, describe arcs cutting each other in C and D:
a straight line drawn through the points of intersection C and D, will bisect the line AB in e.

Problem 7.—To **divide a given line into any
number of equal parts:** Let AB (Fig. 29) be the
given line to be divided into five equal parts.

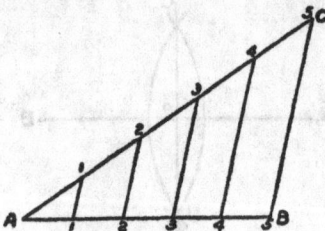
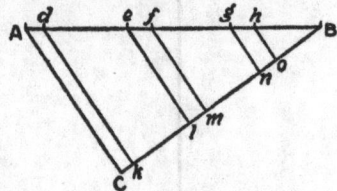

FIG. 29.   FIG. 30.

From the point A draw the straight line AC,
forming any angle with AB. On the line AC,
with any convenient opening of the compasses,
set off five equal parts toward C; join the extreme
points CB; through the remaining points 1, 2, 3
and 4, draw lines parallel to BC, cutting AB in
the corresponding points 1, 2, 3 and 4: AB will
be divided into five equal parts as required.

There are several other methods by which

lines may be divided into equal parts; they are
not necessary, however, for our purpose, so we
will content ourselves with showing how this
problem may be used for changing the scales of
drawings whenever such change is desired. Let
AB (Fig. 30) represent the length of one scale of
drawing, divided into the given parts Ad, de, ef,
fg, gh and hB; and DE the length of another
scale or drawing required to be divided into similar parts. From the point B draw a line BC-DE,
and forming any angle with AB; join AC, and
through the points d, e, f, g and h, draw dk, el,

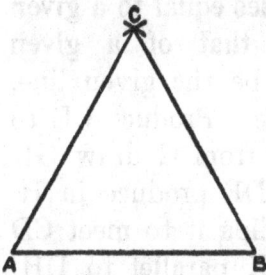

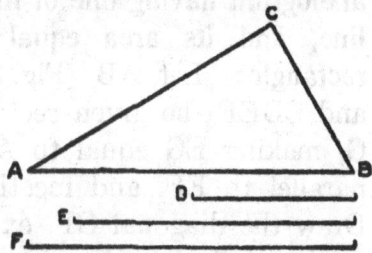

FIG. 31.   FIG. 32.

fm, gn, ho, parallel to AC; and the parts Ck, kl,
lm, etc., will be to each other, or to the whole
line BC, as the lines Ad, de, ef, etc., are to each
other, or to the given line or scale AB. By this
method, as will be evident from the figure, similar
divisions can be obtained in lines of any given
length.

Problem 8.—To describe an equilateral triangle upon a given straight line: Let AB (Fig. 31)
be the given straight line. From the points A
and B, with a radius equal to AB, describe arcs

intersecting each other in the point C. Join CA and CB, and ABC will be the equilateral triangle required.

Problem 9.—**To construct a triangle whose sides shall be equal to three given lines, F, E, D:** Draw AB (Fig. 32) equal to the given line F. From A as a center, with a radius equal to the line E, describe an arc; then from B as a center, with a radius equal to the line D, describe another arc intersecting the former in C; join CA and CB, and ABC will be the triangle required.

Problem 10.—**To describe a rectangle or parallelogram having one of its sides equal to a given line, and its area equal to that of a given rectangle:** Let AB (Fig. 33) be the given line, and CDEF the given rectangle. Produce CE to G, making EG equal to AB; from G draw GK parallel to EF, and meeting DF produce in H. Draw the diagonal GF, extending it to meet CD produced in L; also draw LK parallel to DH, and produce EF till it meets LK in M; then FMKH is the rectangle required.

Equal and similar parallelograms of any dimensions may be drawn after the same manner, seeing the complements of the parallelograms which are described on or about the diagonal of any parallelogram, are always equal to each other; while the parallelograms themselves are always similar to each other, and to the original parallelogram about the diagonal of which they are constructed. Thus, in the parallelogram CGKL the complements CEFD and FMHK are always equal, while the parallelograms EFHG and DFML about the diagonal GL, are always similar to each other, and to the whole parallelogram CGKL.

Problem 11.—**To describe a square equal to two given squares:** Let A and B (Fig. 34) be the given squares. Place them so that a side of each

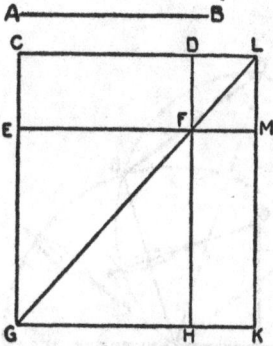

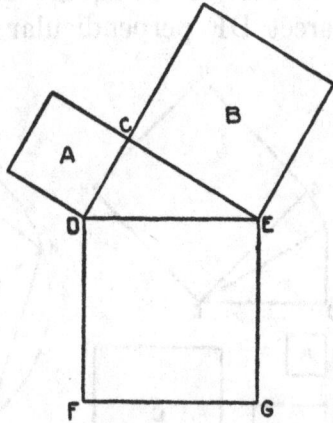

Fig.33. Fig. 34.

may form the right angle DCE; join DE, and upon this hypothenuse describe the square DEGF, and it will be equal to the sum of the squares A and B, which are constructed upon the legs of the right-angled triangle DCE. In the same manner, any other rectilineal figure, or even circle, may be found equal to the sum of other two similar figures or circles. Suppose the lines CD and CE to be the diameters of two circles, then DE will be the diameter of a third, equal in area to the other two circles. Or suppose CD and CE to be the like sides of any two similar figures, then DE will be the corresponding side of another similar figure equal to both of the former.

Problem 12.—**To describe a square equal to any number of given squares:** Let it be required to construct a square equal to three given squares A, B and C (Fig. 35). Take the line DE, equal to the side of the square C. From the extremity D erect DF perpendicular to DE, and equal to the

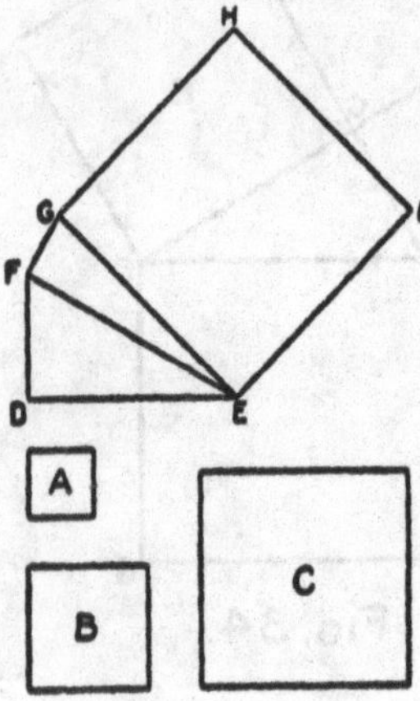

Fig. 35.

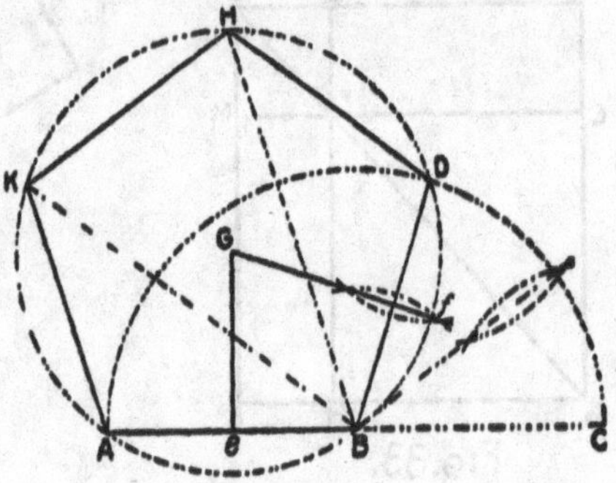

Fig. 36.

side of the square B: join EF; then a square described upon this line will be equal to the sum of the two given squares C and B. Again, upon the straight line EF erect the perpendicular FG, equal to the side of the third square A; and join GE, which will be the side of the square GEHK, equal in area to A, B and C. Proceed in the same manner for any given number of squares.

Problem 13.—**Upon a given straight line to describe a regular polygon:** To produce a regular

pentagon draw AB to C (Fig. 36), so that BC may
be equal to AB; from B as a center, with the
radius BA or BC describe the semicircle ADC;
divide the semi-circumference ADC into as many
equal parts as there are parts in the required polygon, which, in the present case, will be five;
through the second division from C draw the
straight line ED, which will form another side of
the figure. Bisect AB at e, and BD at f, and
draw eG and fG perpendicular to AB and BD;
then G, the point of intersection, is the center of
a circle, of which AB and D are points in the circumference. From G, with a radius equal to its
distance from any of these points, describe the
circumference ABDHK; then producing the dotted lines from the center B, through the remaining divisions of the semicircle ADC, so as to meet
the circumference of which G is the center, in
H and K, these points will divide the circle
ABDHK into the number of parts required, each
part being equal to the given side of the pentagon.

From the preceding example it is evident
that polygons of any number of sides may be
constructed upon the same principles, because
the circumference of all circles, when divided into
the same number of parts, produce equal angles;
and, consequently, by dividing the semi-circumference of any circle into the number of parts
required, two of these parts will form an angle
which will be subtended by its corresponding
parts of the whole circumference. And as all

regular polygons can be inscribed in a circle, it must necessarily follow, that if a circle be described through three given angles of the polygon, it will contain the number of sides or angles required.

The above is a general rule by which all regular polygons may be described upon a given straight line; but there other methods by which many of them may be more expeditiously constructed, as shown in the following examples:

Problem 14.—**Upon a given straight line to describe a regular pentagon:** Let AB (Fig. 37)

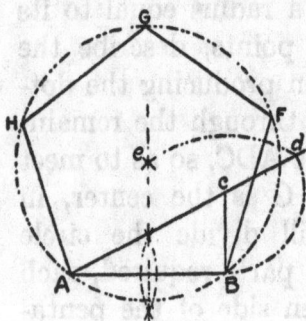

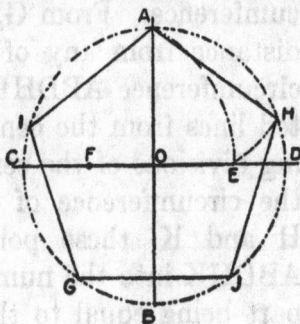

FIG. 37.   FIG. 38.

be the given straight line; from its extremity B erect Bc perpendicular to AB, and equal to its half. Join Ac, and produce it till cd be equal to Bc, or half of the given line AB. From A and B as centers, with a radius equal to Bd, describe arcs intersecting each other in e, which will be the center of the circumscribing circle ABFGH. The side AB, applied successively to this circumference, will give the angular points of the penta-

gon; and these being connected by straight lines
will complete the figure.

Problem 15.—**Another method of inscribing
a pentagon in a circle:** In Fig. 38, draw two diameters AB and CD, at right angles to each other,
as at O. From O as center describe the desired
circumference. Bisect the radius C O at F; with
F as a center and FA as a radius, describe the
arc cutting O D at E. With A as a center and AE
as a radius, describe the arc HE cutting the cir-

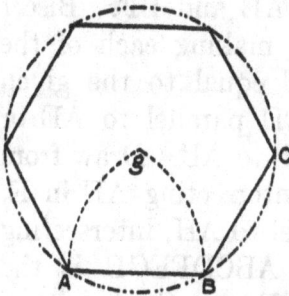

FIG. 39.

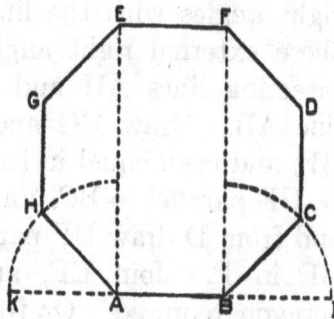

FIG. 40.

cumference of the circle at H, with the same
radius describe an arc cutting the circumference
of the circle at I. With these points as centers
and using the same radius AE describe arcs cutting the circumference of the circle at J and G.
Join HAIG and J and we have the inscribed
regular pentagon.

Problem 16.—**Upon a given straight line
describe a regular hexagon:** Let AB (Fig. 39) be
the given straight line. From the extremities A

and B as centers, with the radius AB, describe
arcs cutting each other in g. Again, from g, the
point of intersection, with the same radius,
describe the circle ABC, which will contain the
given side AB six times when applied to its circumference, and will be the hexagon required.

Problem 17.—To describe a regular octagon
upon a given straight line: Let AB (Fig. 40) be
the given line. From the extremities A and B
erect the perpendiculars AE and BF; extend the
given line both ways to k and l, forming external
right angles with the lines AE and BF. Bisect
these external right angles, making each of the
bisecting lines AH and BC equal to the given
line AB. Draw HG and CD parallel to AE or
BF, and each equal in length to AB. Draw from
G, GE parallel to BC, and intersecting AE in E,
and from D draw DF parallel to AH, intersecting
BF in F. Join EF, and ABCDFEGH is the
octagon required. Or from D and G as centers,
with the given line AB as radius, describe arcs
cutting the perpendiculars AE and BF in E and F,
and join GE, EF, FD, to complete the octagon.

Otherwise, thus:—Let AB (Fig. 41) be the
given straight line upon which the octagon is to
be described. Bisect it in a, and draw the perpendicular ab equal to Aa or Ba. Join Ab, and
produce ab to c, making bc equal to Ab; join also
Ac to Bc, extending them so as to make cE and
cF each equal to Ac or Bc. Through c draw
Ccg, at right angles to AE. Again, through the
same point c, draw DH at right angles to BF,

making each of the lines cC, cD, cG, and cH
equal to Ac and Bc, and, consequently, equal to
each other. Lastly, join BC, CD, DE, EF, FG,
GH, HA; ABCDEFGH will be the regular octagon described upon AB, as required.

Problem 18.—**In a given square to inscribe a
given octagon:** Let ABCD (Fig. 42) be the given
square. Draw the diagonals AC and BD, inter-

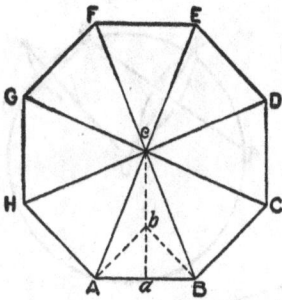
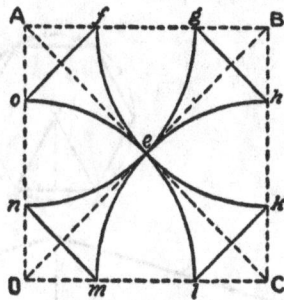

FIG. 41.   FIG. 42.

secting each other in e; then from the angular
points ABC and D as centers, with a radius equal
to half the diagonal, viz.: Ae or Ce, describe arcs
cutting the sides of the square in the points f, g,
h, k, l, m, n, o, and the straight lines of gh, kl,
and mn, joining these points will complete the
octagon, and be inscribed in the square ABCD,
as required.

Problem 19.—**To find the area of any regular
polygon:** Let the given figure be a hexagon; it
is required to find its area. Bisect any two adjacent angles, as those at A and B (Fig. 43), by the
straight lines AC and BC, intersecting in C, which

will be the center of the polygon. Mark the altitude of this elementary triangle by the dotted
line drawn from C perpendicular to the base AB;
then multiply together the base and altitude thus
found, and this product by the number of sides:
one-half of the product gives the area of the whole
figure.

Or otherwise thus: Draw the straight line
DE, equal to six times; that is, as many times

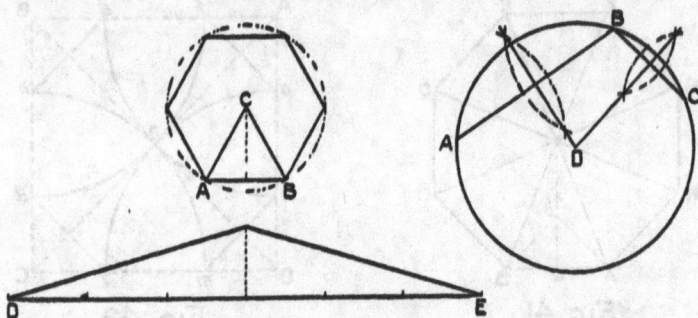

FIG. 43.   FIG. 44.

AB, the base of the elementary triangle, as there
are sides in the given polygon. Upon DE describe
an isosceles triangle, having the same altitude as
ABC, the elementary triangle of the given polygon; the triangle thus constructed is equal in
area to the given hexagon; consequently, by
multiplying the base and the altitude of this triangle together half the product will be the area
required. The rule may be expressed in other
words as follows: The area of a regular polygon
is equal to its perimeter, multiplied by half the
radius of its inscribed circle, to which the sides
of the polygon are tangents.

Problem 20.—To describe the circumference of a circle through three given points: Let A, B, C (Fig. 44) be the given points not in a straight line. Join AB and BC; bisect each of the straight lines AB and BC by perpendiculars meeting in D; then A, B and C are all equi-distant from D; therefore a circle described from D, with the radius DA, DB, or DC, will pass through all the three points, as required.

Problem 21.—To divide a given circle into any number of equal or proportional parts by concentric divisions: Let ABC (Fig. 45) be the given

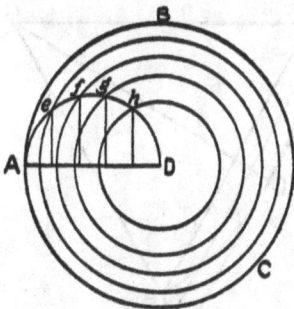
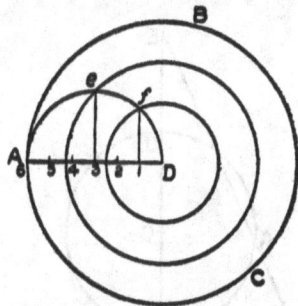

FIG. 45.   FIG. 46.

circle, to be divided into five equal parts. Draw the radius AD, and divide it into the same number of parts as those required in the circle; and upon the radius thus divided, describe a semicircle; then from each point of division on AD, erect perpendiculars to meet the semi-circumference in e, f, g and h. From D, the center of the given circle, with radii extending to each of the different points of intersection on the semicircle, describe successive circles, and they will divide the given circle into five parts of equal area, as required; the center part being also a circle, while the other four will be in the form of rings.

Problem 22.—**To divide a circle into three concentric parts, bearing to each other the proportion of one, two, three, from the center:** Draw the radius AD (Fig. 46), and divide it into six equal parts. Upon the radius thus divided, describe a semicircle: from the first and third points of division, draw perpendiculars to meet the semicircumference in e and f. From D, the center of

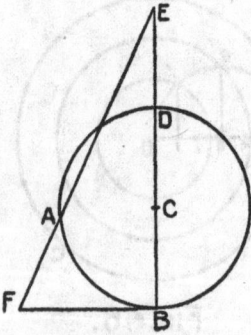

Fig 47.

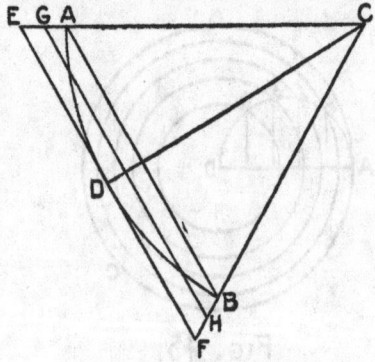

Fig. 48.

the given circle, with radii extending to e and f, describe circles which will divide the given circle into three parts, bearing to each other the same proportion as the divisions on AD, which are as 1, 2 and 3. In like manner circles may be divided in any given ratio by concentric divisions.

Problem 23.—**To draw a straight line equal to any given arc of a circle:** Let AB (Fig. 47)

be the given arc. Find C the center of the arc,
and complete the circle ADB. Draw the diameter BD, and produce it to E, until DE be equal
to CD. Join AE, and extend it so as to meet a
tangent drawn from B in the point F; then BF
will be nearly equal to the arc AB.

The following method of finding the length of
an arc is equally simple and practical, and not
less accurate than the one just given: Let AB
(Fig. 48) be the given arc. Find the center C,
and join AB, BC and CA. Bisect the arc AB in
D, and join also CD; then through the point D
draw the straight line EDF, at right angles to
CD, and meeting CA and CB produced in E and
F. Again, bisect the lines AE and BF in the
points G and H. A straight line GH, joining
these points, will be a very near approach to the
length of the arc AB.

Seeing that, in very small arcs, the ratio of
the chord to the double tangent or, which is the
same thing, that of a side of the inscribed to a
side of the circumscribing polygon, approaches to
a ratio of equality, an arc may be made so small,
that its length shall differ from either of these
sides by less than any assignable quantity;
therefore, the arithmetical mean between the two
must differ from the length of the arc itself by
less than any that can be assigned. Consequently
the smaller the given arc, the more nearly will
the line found by the last method approximate
to the exact length of the arc. If the given arc
is above 60 degrees, or two-thirds of a quadrant,

it ought to be bisected, and the length of the semi-arc thus found being double, will give the length of the whole arc.

These problems are very useful in obtaining the lengths of veneers or other materials required for bending round soffits of doors and window-heads.

Problem 24.—To describe the segment of a circle by means of two laths, the chord and versed sine being given: Take two rods, EB, BF (Fig. 49), each of which must be at least equal in

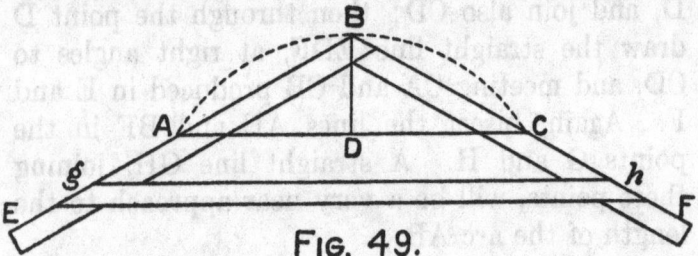

FIG. 49.

length to the chord of the proposed segment AC; join them together at B, and expand them, so that their edges shall pass through the extremities of the chord, and the angle where they join shall be on the extremity B of the versed sine DB, or height of the segment. Fix the rods in that position by the cross piece gh, then by guiding the edges against pins in the extremities of the chord line AC, the curve ABC will be described by the point B.

Problem 25.—Having the chord and versed sine of the segment of a circle of large radius given, to find any number of points in the curve

by means of intersecting lines: Let AC be the chord and DB the versed sine. Through B (Fig. 50) draw EF indefinitely and parallel to AC; join AB, and draw AE at right angles to AB. Draw also AG at right angles to AC, and divide AD and EB into the same number of equal parts, and number the divisions from A and E,

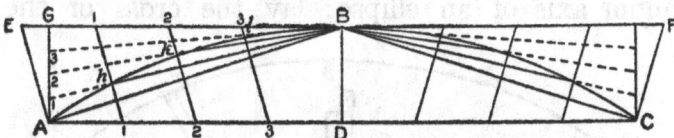

FIG. 50.

respectively, and join the corresponding numbers by the lines, 11, 22, 33. Divide also AG into the same number of equal parts as AD or EB, numbering the divisions from A upwards, 1, 2, 3, etc.; and from the points 1, 2 and 3 draw lines to B and the points of interesection of these with the other lines at h, k, l, will be points in the curve required. Same with BC.

Problem 26.—**To draw an ellipse with the trammel:** The trammel is an instrument consisting of two principal parts; the fixed part in the form of a cross EFGH (Fig. 51), and the movable piece or tracer klm. The fixed piece is made of two rectangular bars or pieces of wood, of equal thickness, joined together so as to be in the same plane. On one side of the frame so formed, a groove is made, forming a right-angled cross. In the groove two studs, k and l, are fitted to slide freely and carry attached to them the tracer klm. The tracer is generally made to slide through the socket fixed to each stud, and provided with a screw or wedge, by which the distance apart of the studs may be regulated. The tracer has another slider m, also adjustable, which carries a pencil or point. The instrument is used as follows: Let AC be the major, and HB the minor axis of an ellipse: lay the cross of the

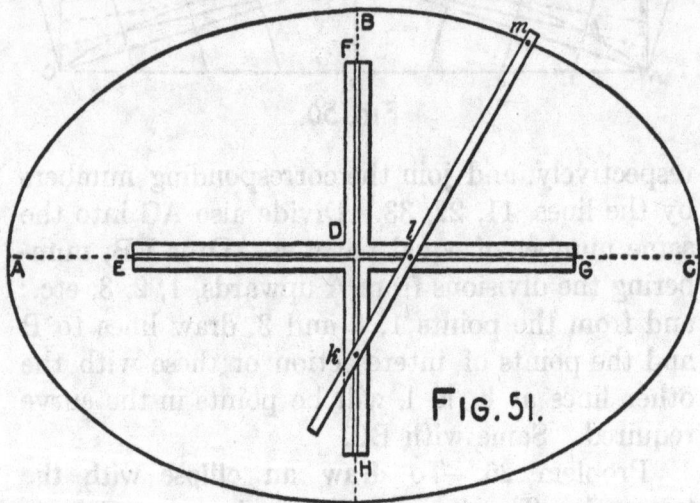

FIG. 51.

trammel on these lines, so that the center lines of it may coincide with them; then adjust the sliders of the tracer, so that the distance between k and m may be equal to half the major axis, and the distance between l and m equal to half the minor axis, then by moving the bar around, the pencil in the slider will describe the ellipse.

Problem 27.—An ellipse may also be described by means of a string: Let AB (Fig. 52) be the major axis, and DC the minor axis of the ellipse, and FG its two foci. Take a string EFG and pass it over the pins and tie the ends together, so that when doubled it may be equal to the distance from the focus F to the end of the axis, B; then putting a pencil in the bight or doubling of the string at H and carrying it around, the curve may

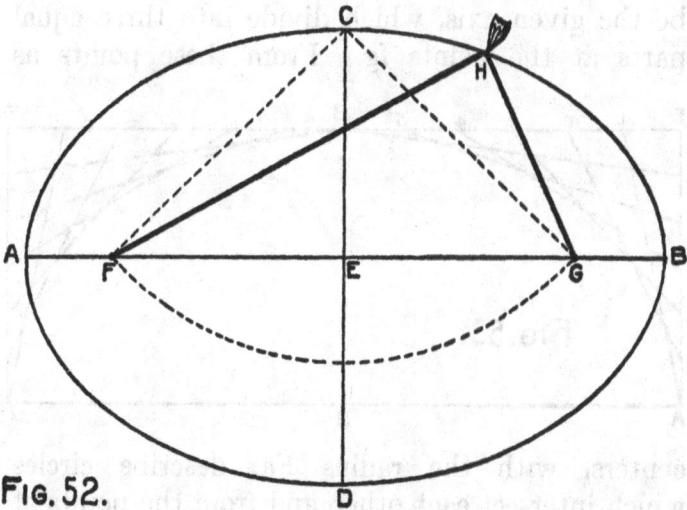

FIG. 52.

be traced. This is based on the well-known property of the ellipse, that the sum of any two lines drawn from the foci to any points in the circumference is the same.

Problem 28.—The axes of an ellipse being given, to draw the curve by intersections: Let AC (Fig. 53) be the major axis, and DB half the minor axis. On the major axis construct the parallelogram AEFC, and make its height equal to DB. Divide AE and EB each into the same number of equal parts, and number the divisions from A and E, respectively; then join A1, 12, 23, etc., and their intersections will give points through which the curve may be drawn.

Problem 29.—To describe with a compass a figure resembling the ellipse: Let AB (Fig. 54) be the given axis, which divide into three equal parts at the points fg. From these points as

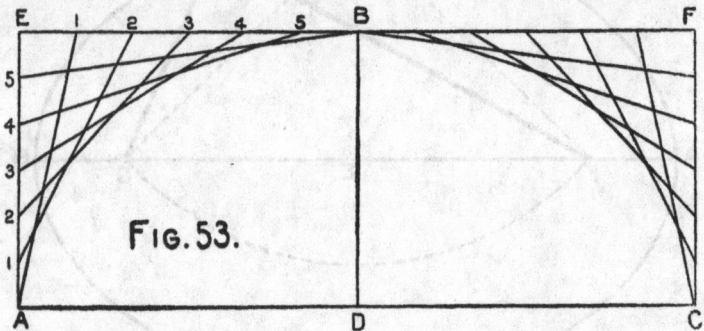

FIG. 53.

centers, with the radius Fa, describe circles which intersect each other, and from the points of intersection through f and g, draw the diameters CgE, CfD. From C as a center, with the radius CD, describe the arc DF, which completes the semi-ellipse. The other half of the ellipse may be completed in the same manner, as shown by the dotted lines.

Problem 30.—Another method of describing a figure approaching the ellipse with a compass: The proportions of the ellipse may be varied by altering the ratio of the divisions of the diameter,

as thus: Divide the major axes of the ellipse
AB (Fig. 55), into four equal parts, in the points
fgh. On fh construct an equilateral triangle fCh,
and produce the sides of the triangle Cf, Ch indefinitely, as to D and E. Then from the centers
f and h, with the radius Af, describe the circles

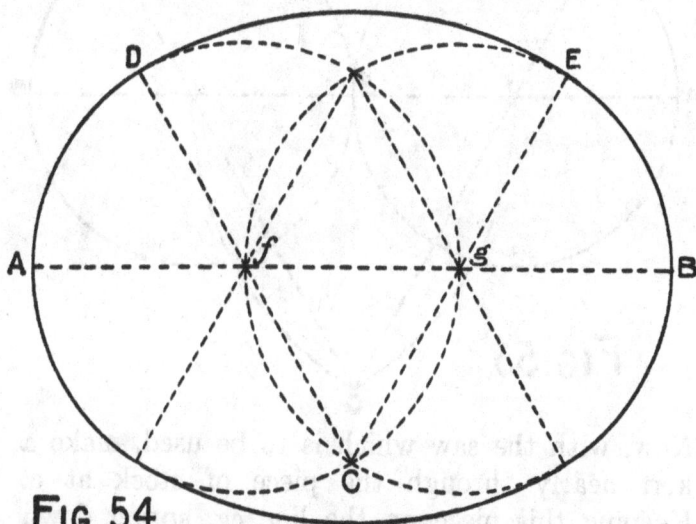

Fig. 54.

ADg, BEg; and from the center C, with the
radius CD, describe the arc DE to complete the
semi-ellipse. The other half may be completed
in the same manner. By this method of construction the minor axis is to the major axis as 14 is
to 22.

Problem 31.—To find how far apart to saw
kerfs to spring a board or moulding: Let ab
(Fig. 56) be the curve, around which it is desired
to spring a piece of stock. Take a piece of stock

dg of the thickness which is to be used. Lay it down so that the edge shall pass through the center c, and mark from c to g, and also at e.

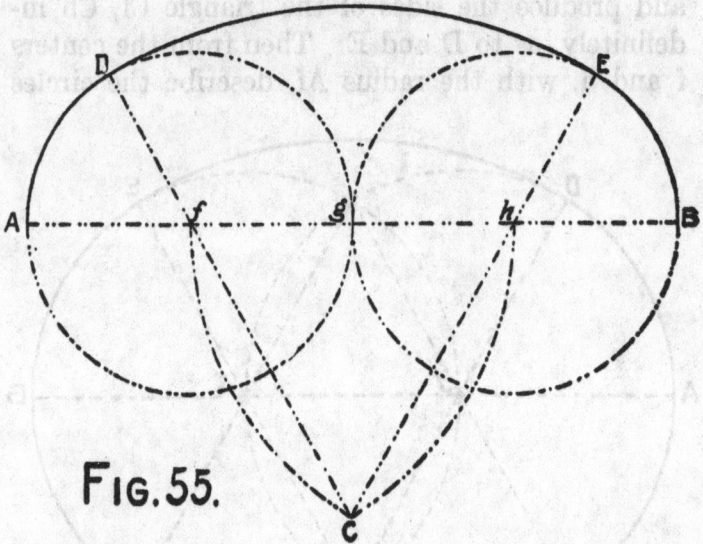

FIG. 55.

Now, with the saw which is to be used, make a kerf nearly through the piece of stock at c. Keeping this piece on the line cg, spring down

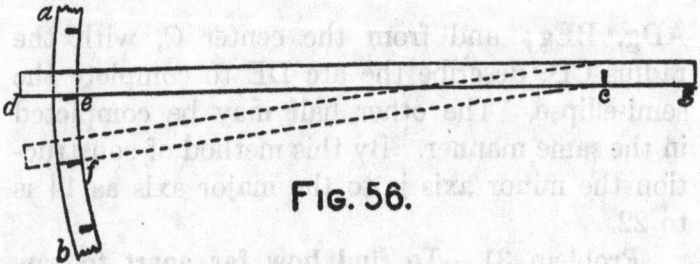

FIG. 56.

the end d until the kerf is closed, then mark the point f: ef will be the distance apart to saw kerfs.

Part II

As it is also the business of the carpenter to prepare patterns for stone cutters, by which they are to cut stones to fit arches, centers for windows and door heads, it is necessary he should have a clear conception of what an arch really is. For if a positive conclusion has not been arrived at, and if the arch principle is not fairly understood, he cannot be expected to design an arch, or to construct it with accuracy or intelligence, even if designed by another. Let us then state once for all, that every curved covering to an opening is not necessarily an arch. Thus, the stone which rests on the piers shown in Fig. 57 is not an arch, being merely a stone hewn out in an arch-like shape; for at its top, the very point A at which strength is required, it is the weakest, and would fracture the moment any great weight was placed upon it.

It is not the province of this work to enter

into a scientific discussion on the arch, but some of its properties must be known to the mechanic before he will be able to construct centers understandingly; and the general principles here laid down will help the workman materially to form correct ideas concerning the work in hand.

The Arch is an arrangement for spanning large openings so arranged that they may, by mutual pressure, support not only each other,

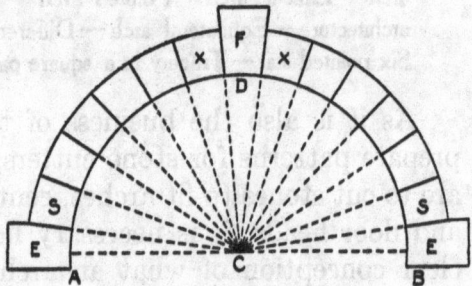

Fig.57.   Fig.58.

but any weight that may be placed upon them.

The leading principles in the construction of an arch are:

1.—That all the stones of which it is formed shall be of the form of wedges; that is, narrower at the inner than the outer end.

2.—That all the joints formed by the meeting of the slanting sides of the wedges should be radii of the circle, circles, or ellipse, forming the inner curve of the arch; and will, therefore, converge to the center or centers from which these are struck. As a rule, the arch answers the same purpose as the beam, but it is widely different in its action and in the effect that it has upon the appearance of a building.

A beam exerts merely a vertical force upon its supports, but the arch exerts both a vertical load and an outward thrust.

Before taking up the construction of the arch, we will define the terms relating to it.

Fig. 58 shows the different parts of an arch. The distance AB is called the span of the arch; the under surface, ADB, is called the intrados, and the outer the extrados; the distance, DC, is the rise; F is the keystone, the blocks X, X, of which the arch itself is composed are called voussoirs, and the lowest ones S, S, the springers. In arches whose intrados are not complete semicircles the springers rest upon two stones E, E, which have their upper surface cut to receive them; these stones are called skewbacks. The highest point in the intrados is called the vertex or crown, and the spaces between the vertex and the springing line AB are called the haunches.

Fig. 59 is the **semi-circle arch,** and was that principally used by the Romans, who employed it largely in their aqueducts and triumphal arches. The semi-circular and segmental arches are the best as regards stability, and are the simplest to construct.

Fig. 60 is a **segmental arch,** and is extensively used over window heads. A true segmental arch is one-sixth of the circumference of a circle, as shown in the figure.

Fig. 61 is an illustration of a segmental arch shown in connection with a window opening.

The Horse=shoe Arch.—This is almost restricted to the Arabian or Moorish style of architecture. In this form of arch the curve is carried below the line of center or centers, for in some cases the arch is struck from one center, and in others from two, as in Fig. 62.

Now it must be supposed that the real bearing of the arch is at the impost A A; for if this were

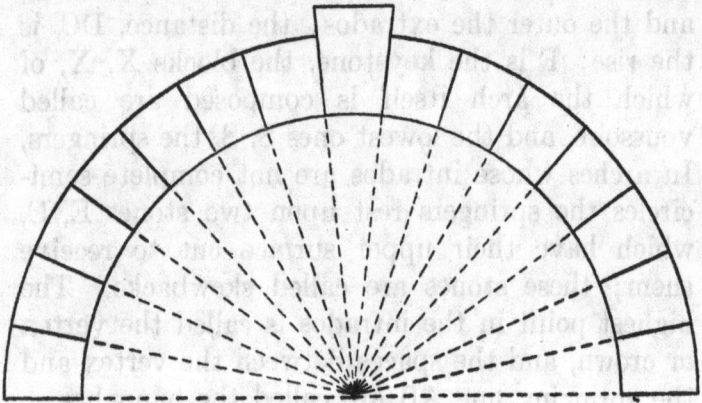

FIG. 59.

really so, it must be seen that any weight or pressure on the crown of the arch would cause it to break at B; but the fact is simply that the real bearings of the arch are at B B, and the prolongation of the arch beyond these points is merely a matter of form and has no structural significancy. The Horse-shoe arch belongs especially to the Mo-

hammedan architecture, from its having originated with that faith and from its having been
used exclusively by its followers.

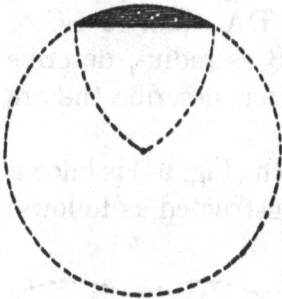

FIG. 60.

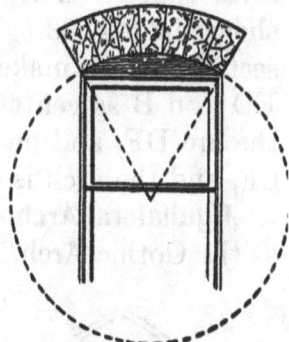

FIG. 61.

Next in point of time, but by far the most
graceful in form, is the **pointed arch,** which is
essentially the middle age style, and is capable of
almost endless variety. Just where it originated

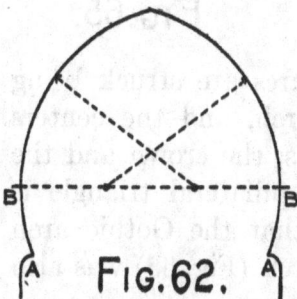

FIG. 62.

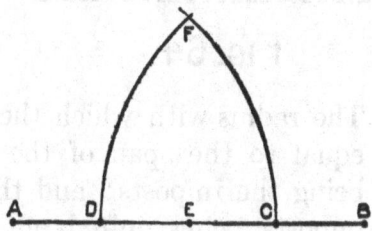

FIG. 63.

is hard to tell, as recent discoveries have shown
that it was used many centuries ago in Assyria.
The greater or less acuteness of the pointed

arch depends on the position of the centers from which the flanks are struck.

The Lancet Arch.—This arch indicates the style called "Early English," and is drawn as shown in Fig. 63. D C is the given span; bisect DC in E, make CB and DA equal to EC or ED; on B as center with DB as radius, describe the arc DF, and on A as center, describe the arc CF, and the arch is complete.

Equilateral Arch.—This arch (Fig. 64) is known as the Gothic Arch, and is constructed as follows:

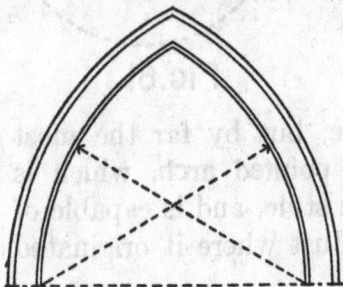

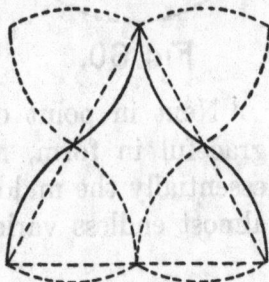

FIG. 64.   FIG. 65.

The radius with which the arcs are struck being equal to the span of the arch, and the centers being the imposts; and thus, the crown and the imposts being united, an equilateral triangle is formed. During the time that the Gothic arch was mostly used the Ogee arch (Fig. 65) was also occasionally used.

A little later on we find the **Tudor arches,** or four-centered arches (Fig. 66), in which two of the centers are on the springing line and two below it. The arches at the later period of this style became flatter and flatter, and this forms one of the features of Debased Gothic, when the beautiful and graceful forms of that style gradually decayed and for a time were lost. Happily in the present century there has been a gradual revival of the Gothic style, and works are now being produced which bid fair to rival in beauty of form and in principle of construction the marvelous buildings of the middle ages.

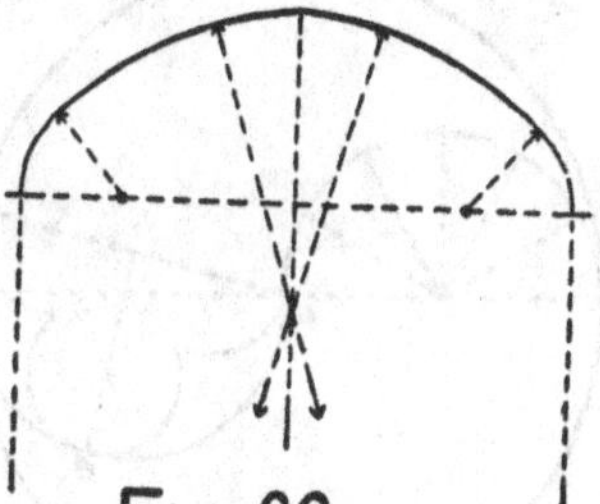

From the examples given the workman will be able to lay out any of the usual arches required in building.

FIG. 66.

There are combinations, however, of these curves which the carpenter may be called upon to construct, such as the ones given herewith.

Fig. 67 is the elementary study upon which the subsequent figure is based. Having drawn the circle, describe on the diameter two opposite semi-circles, meeting at the center, a.

Divide one of these into six equal parts, and set off one of these sixths from i to n. Draw an, and divide it into four equal parts. From the middle point of a n draw a line passing through the center of the semi-circle, and cutting it in c. From c set off on this line the length of one of the fourths of a n. This point and the two in a n will be the centers for the interior curves.

Fig. 68 is the further working out of this elementary figure. It is desirable that a larger circle should be drawn. Then, when the figure has been carried up to the stage shown in the last, all the rest of the curves will be drawn from the same centers.

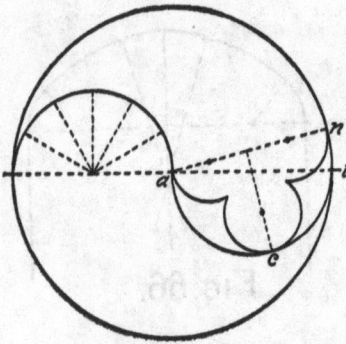

FIG. 67.

FIG. 68.

Fig. 69 and 70 are the elementary forms of the tracery shown in Fig. 71.

We show the method of obtaining these curves in Fig. 69: At any point, as at A, draw a tangent, and A G at right angles to it. From A, with radius O A, cut the circle in B and C, and the tangent in the point F, using B as a center. Bisect the angle O at F, and produce the bisecting line until it cuts A G in H. From O, with radius OH, cut the lines DC and EB in I and J. From H, I and J, with radius HA, draw the three required circles, each of which should touch the other two and the outer circle.

After having constructed Fig. 69 we go one step further and form Fig. 70; having inscribed

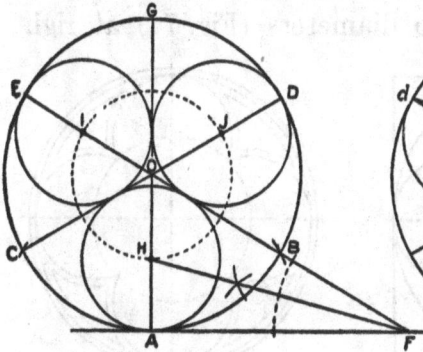

Fig. 69.

Fig. 70.

three equal circles in a circle, join their centers, thus forming an equilateral triangle. From the center of the surrounding circle draw radii passing through the angles of the triangle and cutting the circle in points; as d and two others. Draw ed and bisect it by cg; then the centers for the curves which are in the semi-circle will be on the three lines, dc, cg and ce.

These curves, in Gothic architecture, are called **foliations,** or featherings, and the points at which

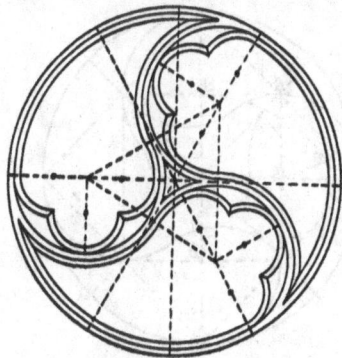

Fig. 71

they meet are called cusps. The completion of this study is shown in Fig. 71.

Fig. 72 is the elementary form of the tracery shown in Fig. 73.

Fig. 74 shows the elementary construction of Fig. 75. Draw two diameters (Fig. 74) at right

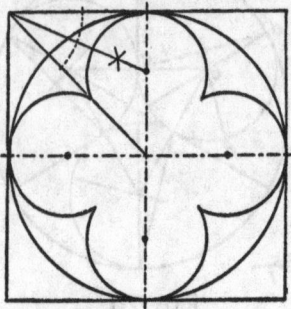

FIG. 72.

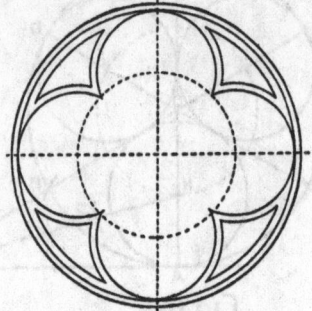

FIG. 73.

angles to each other, and joint their extremities, thus inscribing a square in the circle. Bisect the quadrants by two diameters cutting the sides of the square in the points as g; join these points,

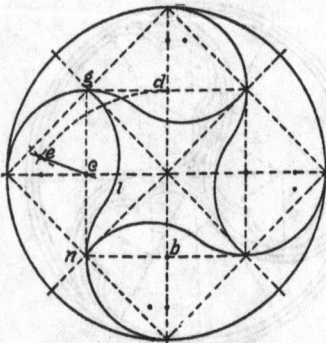

FIG. 74.

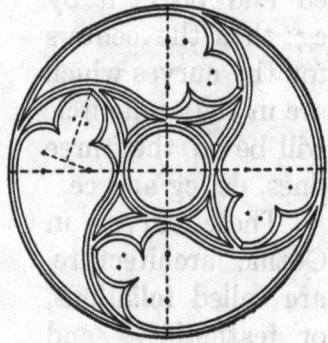

FIG. 75.

and a second square will be inscribed within the first.

The middle points of the sides of this inner square, as bcd, are the centers of the arcs which start from the extremities of the diameters.

From b, with radius bd, describe an arc, and from g, with radius gc, describe another cutting the former one in e. Then e is the center for the

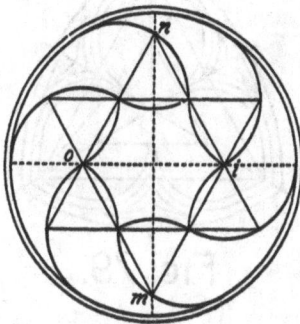
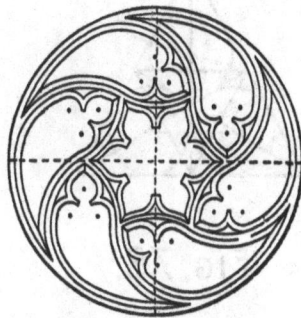

FIG. 76.   FIG. 77.

arc ig, which will meet the arc struck from b, in i. Of course, this process is to be carried on in each of the four lobes.

Fig. 75 is the completed figure.

Fig. 76 shows the skeleton lines of Fig. 77. Divide the diameter into four equal parts, and on the middle two, as a common base, construct the two equilateral triangles oin and oim.

Draw lines through the middle points of the sides of the triangles which, intersecting, will complete a six-pointed star in the circle, the angles of which will be the centers for the main lines of the tracery.

Fig. 77 is the completed figure.

The figures 78 and 79 will be understood without further instruction than is afforded by the examples.

Fig. 80 shows the construction for the tracery in a square panel. From each of the angles of the

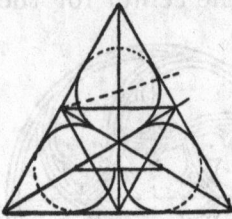

FIG. 78.

FIG. 79.

square (the inner one in this figure), with a radius equal to the length of the side of the square, describe arcs; these intersecting will give a four-

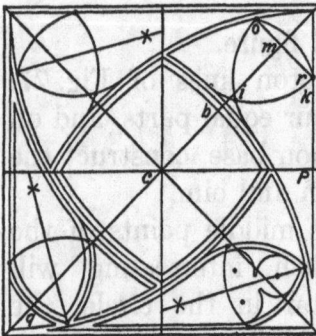

FIG. 80.

FIG. 81.

sided curvilinear figure in the center. Draw diagonals in the square.

From the point where the diagonal intersects the curve b (the middle line of the three here shown) set off on the diagonal the length cb, viz., bm. From q, with radius mq, describe an arc cutting the original arc in o. Make mr equal to m o.

From o and r, with radius o r, describe arcs intersecting each other in i: produce these until they meet the curve p in n.

The foliation and completion, as shown in Fig. 81, will be found simple.

Part III

THE STEEL SQUARE IN CARPENTRY. Brace Rule—Octagon scale—Fence—Square and fence—Length of the common rafter—To find the pitches of roofs—Rafters—Reversed pitches—Laying out rafters—Hip-roof framing—To cut bed moulds for gable to fit under cornice

It is unnecessary to give a full and complete description of the steel square, as every carpenter and joiner is supposed to be the possessor of one of these useful instruments and know how to use it intelligently. The blade of the steel square should be 24 inches long and 2 inches wide, and the tongue 18 inches long and 1½ inches wide. The tongue should be exactly at right angles with the blade.

The next thing to be considered is the use of the figures, lines and scales, as exhibited on the square. It is supposed that the ordinary divisions and sub-divisions of the inch, into halves, quarters, eights and sixteenths are understood by the student; and that he also understands how to use that part of the square that is subdivided into twelfths of an inch.

Perhaps, with the single exception of the common inch divisions on the square, no set of figures on the instrument will be found more useful to the active workman than that known as the board measure, or board rule. If we examine Fig. 82 we will find under the figure 12, on the outer edge of the blade, where the length of the boards, plank or scantling to be measured is given, and the answer in feet and inches is found under the inches in width that the board, etc., measures. For example, take a board 9 feet long and 5 inches wide; then, under the figure 12, on the second line, will be found the figure 9, with is the length of the board; then run along this line to the figure directly under the 5 inches (the width of the board), and we find 3 feet 9 inches, which is the correct answer in "board measure." If the stuff is two inches thick, the sum is doubled; if three inches thick it is trebled, etc.

Brace Rule.—The brace rule is always placed on the tongue of the square, as shown in the central space at X, Fig. 82.

This rule is easily understood; the figures on the left of the line represent the run or the length of two sides of a right angle, while the figures on the right represent the exact length of the third side of a right-angled triangle, in inches, tenths, and hundredths. Or, to explain it in another way, the equal numbers placed one above the other, may be considered as representing the sides of a square, and the third number to the right the length of the diagonal of that square. Thus the

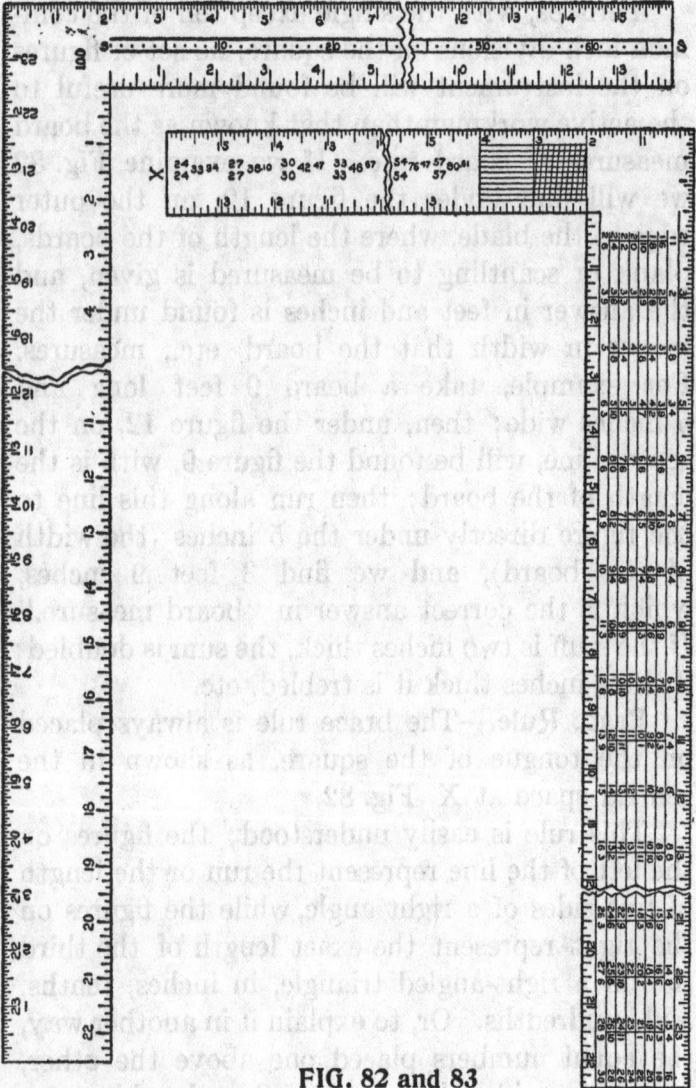

FIG. 82 and 83

exact length of a brace, from point to point, having a run of 33 inches on a post and a run of the same on a girt, is 46.67 inches. The brace rule varies somewhat in the matter of the runs expressed in different squares. Some squares give a few brace lengths of which the runs upon the post and beam are unequal.

Octagon Scale.—The octagonal scale, as shown on the central division of the upper portion of the blade, is on the opposite side of the square to the brace rule, and runs along the center of the tongue as at S S, Fig. 83. Its use is as follows: Suppose a stick of timber ten inches square. Make a center line, which will be five inches from each edge; set a pair of compasses, putting one leg on any of the main divisions shown on the square in this scale, and the other leg on the tenth division. This division, pricked off from the center line on the timber on each side, will give the points for the gauge lines. Gauge from the corners both ways, and the lines for making the timber octagonal in its section are obtained. Always take the same number of spaces on your compasses as the timber is inches square from the center line. The rule always to be observed is as follows: Set off from each side of the center line upon each face as many spaces by the octagon scale as the timber is inches square. For timbers larger in size than the number of divisions in the scale, the measurements by it may be doubled or trebled, as the case may be.

Fence.—A necessary appendage to the steel square in solving problems is the fence, which nearly every carpenter knows the value of. We illustrate in Fig. 84 a wooden fence which was formerly used altogether, and in many places still used although the steel fence can now be purchased in any hardware store.

Fig. 85 shows the square and fence together; the fence can be adjusted so as to slide up and

FIG. 84

down the tongue and blade until it is in the position desired; to show the fence and square in practical use take the square as arranged in Fig. 85 and place it on the prepared board, as shown in Fig. 86. Adjust the square so that the 12-inch lines coincide exactly with the gauge line O, O, O, O. Hold the square firmly in the position now obtained and slide the fence up the tongue and blade until it fits snugly against the

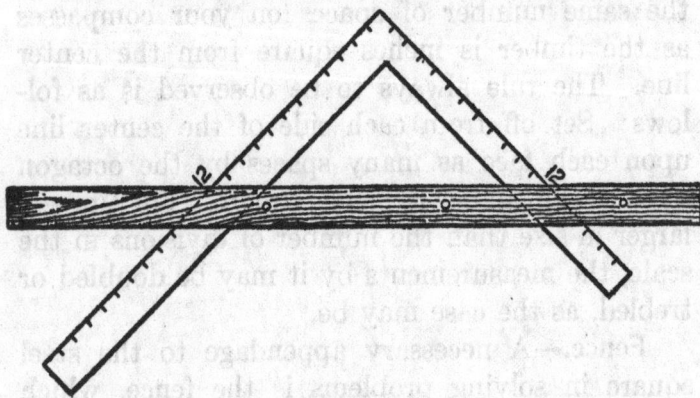

FIG. 85

jointed edge of the board, screw the fence tight on the square, and be sure that the 12-inch marks on both the blade and the tongue are in exact position over the gauge line.

We are now ready to lay out the pattern. Slide the square to the extreme left, as shown on the dotted lines at X, mark with a knife on the outside edges of the square, cutting the gauge

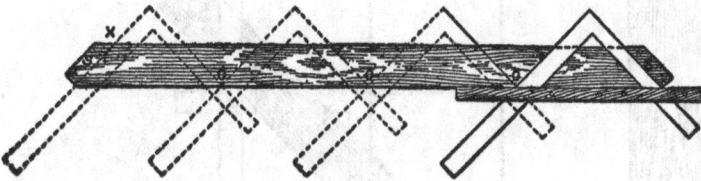

FIG. 86

line. Slide the square to the right until the 12-inch mark on the tongue stands over the knife mark on the gauge line; mark the right-hand side of the square cutting the gauge line as before; repeat the process four times, marking the extreme ends to cut off, and we have the length of the brace and the bevels.

Fig. 87 shows the brace in position, the dotted lines show where the square was placed on the pattern.

We will briefly explain some of the things that can be more readily done with the steel square than by any other method.

One of the best uses to which the steel square can be put is **to find the pitches of roofs,** which is a subject not as well understood as it should be.

The word pitch has reference to the rise given the common rafter in proportion to the span. Therefore, by letting 12 on the tongue represent the

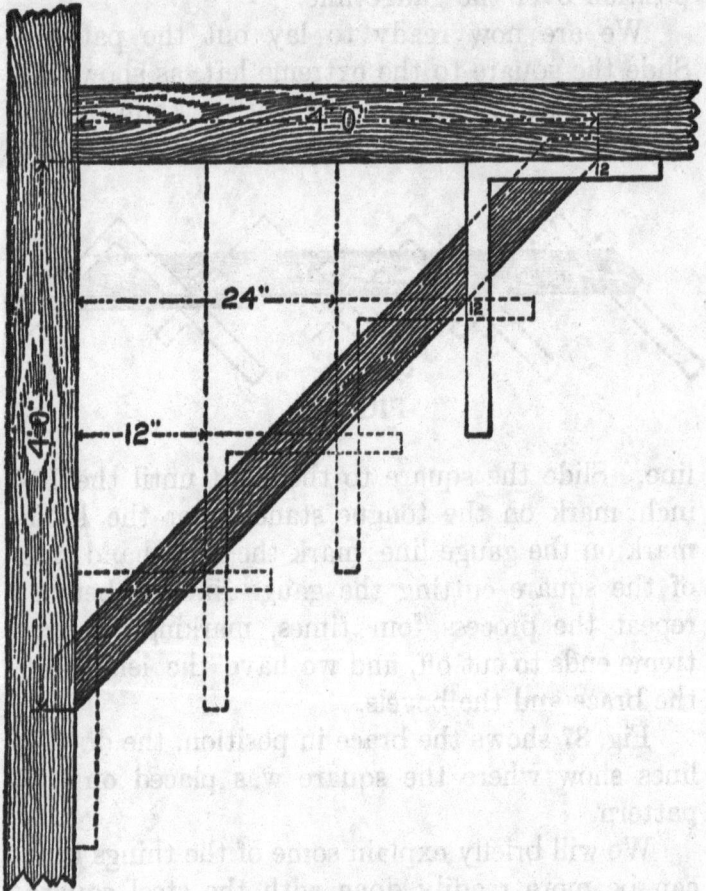

FIG. 87

run of the common rafter, the figures on the blade will then represent the rise in proportion to its length (the blade), as 6 being one-fourth of 24 represents the quarter pitch, 8 represents the one-third pitch, 12 the one-half pitch, etc. See illustration Fig. 88.

For the corresponding hip or valley for the octagon or square-cornered building, substitute

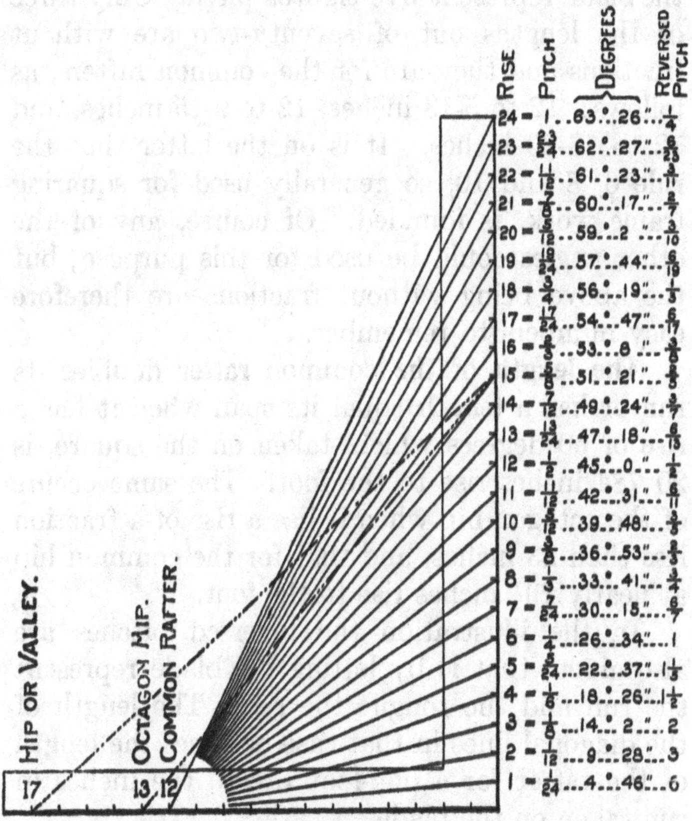
RISE
PITCH
DEGREES
REVERSED PITCH
24 = 1 .. 63° .. 26″ .. 1/4
23 = 23/24 .. 62° .. 27″ .. 6/23
22 = 11/12 .. 61° .. 23″ .. 3/11
21 = 7/8 .. 60° .. 17″ .. 2/7
20 = 5/6 .. 59° .. 2″ .. 3/10
19 = 19/24 .. 57° .. 44″ .. 6/19
18 = 3/4 .. 56° .. 19″ .. 1/3
17 = 17/24 .. 54° .. 47″ .. 6/17
16 = 2/3 .. 53° .. 8″ .. 3/8
15 = 5/8 .. 51° .. 21″ .. 2/5
14 = 7/12 .. 49° .. 24″ .. 3/7
13 = 13/24 .. 47° .. 18″ .. 6/13
12 = 1/2 .. 45° .. 0″ .. 1/2
11 = 11/12 .. 42° .. 31″ .. 6/11
10 = 5/12 .. 39° .. 48″ .. 3/5
9 = 3/8 .. 36° .. 53″ .. 2/3
8 = 1/3 .. 33° .. 41″ .. 3/4
7 = 7/24 .. 30° .. 15″ .. 6/7
6 = 1/4 .. 26° .. 34″ .. 1
5 = 5/24 .. 22° .. 37″ .. 1 1/5
4 = 1/6 .. 18° .. 26″ .. 1 1/2
3 = 1/8 .. 14° .. 2″ .. 2
2 = 1/12 .. 9° .. 28″ .. 3
1 = 1/24 .. 4° .. 46″ .. 6
HIP OR VALLEY.
OCTAGON HIP.
COMMON RAFTER.
17
13
12

FIG. 88

13 and 17 on the tongue, respectively. However, neither is absolutely correct, but near enough for practical purposes.

The lengths taken diagonally from 12, 13 and 17 on the tongue to the figures designating the rise on the blade represent the lengths of the above rafters for a one-foot run. The diagonal lines in the illustration from those figures to 15 on the blade represent five-eighths pitch. Only three of the lengths out of seventy-two are without fractions and they are for the common rafters, as follows: 12 to 5-13 inches, 12 to 9-15 inches, and 12 to 16-20 inches. It is on the latter that the rule 6, 8 and 10, so generally used for squaring frame work, is founded. Of course, any of the other angles could be used for this purpose; but the above being without fractions are therefore easy numbers to remember.

The length of the common rafter doubles its run or has a length equal its span when it has a rise of 60 degrees which, taken on the square, is 20.784 inches rise to the foot. The same occurs of the octagon hip when it has a rise of a fraction less than 23 inches, and that for the common hip at nearly 29½ inches rise to the foot.

In the illustration the reversed pitches are also given, that is by letting the blade represent the run and the tongue the rise. The length of the diagonal lines in that case becomes the length of the rafter for a one-foot rise to the inches in run taken on the blade.

The reader will notice that several of these pitches are transposed and are found in the first column, as follows: The one pitch is the same as the one-quarter when reversed. The three-quarters the same as the one-third. The two-thirds same as three-eighths. The one-half remains the same or unchanged. The low pitches in the first column become very steep when reversed; thus, the one-twenty-fourth pitch becomes six pitches or a rise of 12 feet to a one-foot run. The one-twelfth pitch is equal to a rise of 6 feet to a one-foot run, etc.

Rafters.—Laying out rafters with the aid of the square, we proceed as follows: In Fig. 89 it

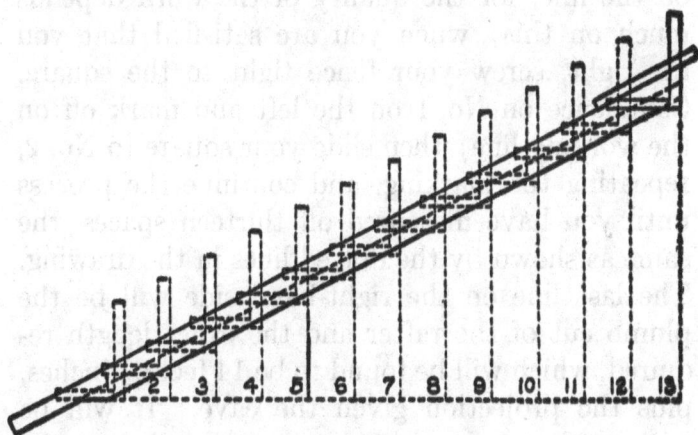

FIG. 89

will be noticed that the rafter is for quarter pitch, and for convenience it is supposed to consist of a piece of stuff 2 inches by 6 inches by 17 feet. That portion of the rafter that projects over the wall of the building and forms the eave, is any length desired. The length of the projecting piece in this case is one foot—it may be more or less to suit the eave, but the line must continue from end to end of the rafter, as shown on the plan, and we will call this line our working line.

We are now ready to lay out this rafter, and will proceed as follows: We adjust the fence to the square the same as for the braces, press the fence firmly against the top edge of the rafter, and place the figure 12 inches on the left-hand side, and the figure 6 inches on the right-hand side directly over the working line, as shown in the plan. Be very exact about getting the figures on the line, for the quality of the work depends much on this; when you are satisfied that you are right, screw your fence tight to the square. Commence on No. 1 on the left and mark off on the working line; then slide your square to No. 2, repeating the marking, and continue the process until you have measured off thirteen spaces, the same as shown by the dotted lines in the drawing. The last line on the right-hand side will be the plumb cut of the rafter and the exact length required, which will be found to be 14 feet 6½ inches, plus the projection given the eave. It will be noticed that the square has been applied to the timber thirteen times.

The reason for this is, that the building is 26 feet wide, the half of which is 13 feet, the distance that one rafter is expected to reach; so, if the building was 30 feet wide, we should be obliged to apply the square fifteen times instead of thirteen. We may take it for granted, then, that in all cases where this method is employed to obtain the lengths and bevels, or cuts of rafters, we must apply the square half as many times as there are feet in the width of the building being covered. If the roof to be covered is one-third pitch, all to be done is to take 12 inches on one side of the square and 8 inches on the other and operate as for quarter pitch.

Whenever a drawing of a roof is to be followed we can soon find out how to apply the square by laying it on the drawing, as shown in Fig. 90. Of course, something depends on the scale to

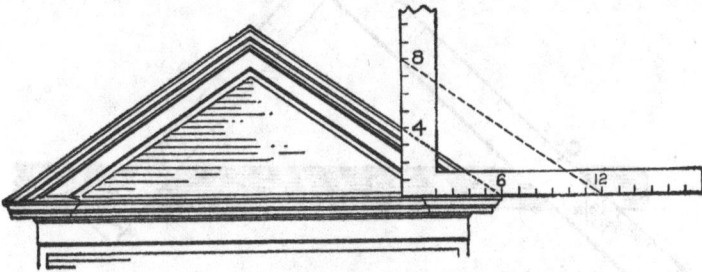

FIG. 90

which the drawing is made. If any of the ordinary fractions of an inch are used, the intelligent workman will have no difficulty in discovering what figures to make use of to get the cuts and lengths desired.

Hip=roof Framing.—In framing a hip-roof we will take the one-third pitch. We first lay off the common rafter, which has been previously explained, and in order to avoid making a plan, we give a formula in figures. The pitch which we have taken is one-third the width of the building to point of rafter from wall plate or base. For example, always 8, which is one-third of 24, on tongue for altitude; 12, one-half the width of 24, on blade for base. This cuts common rafter. Next is the hip rafter. It must be understood that the diagonal of 12 and 12 is 17 in framing, and the hip is the diagonal of a square added to the rise of roof; therefore, we take 8 on tongue and 17 on blade; run the same number of times

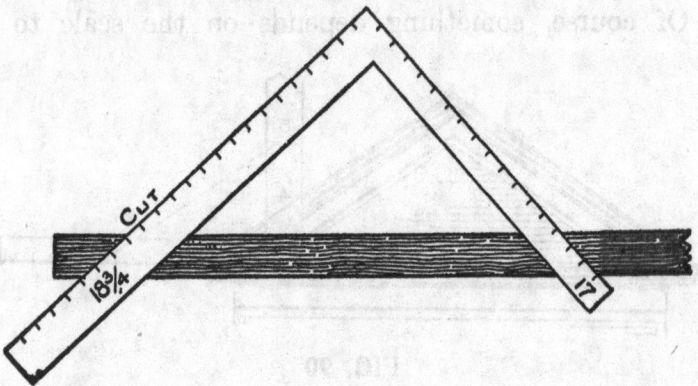

FIG. 91

as common rafter; these figures also give the seat and plumb cut of the hip. To cut jack rafters, divide the number of openings, for common rafter. Suppose we have five jacks, with six openings, our common rafter is 12 feet long, each jack would be 2 feet shorter. First 10 feet, second 8 feet, third 6 feet, etc. The plumb cut is the same as for the cut of common rafter; seat cut also the same as for common rafter. To cut miter to fit hip: Take half the width of building on tongue and length of common rafter on blade; blade gives cut; or, the same result may be found by taking the diagonal of 12 and 8, which is 14 7-16; then take 12 on the tongue, 14 7-16 on blade; blade gives cut. In connection with the hip there must be another cut considered, called the side cut of the hip; though the angle to obtain this cut is across the top or back of rafter. Were there no slope to the roof, this angle where it meets the ridge pole would be an angle of 45 degrees; but when a slope is given this angle becomes

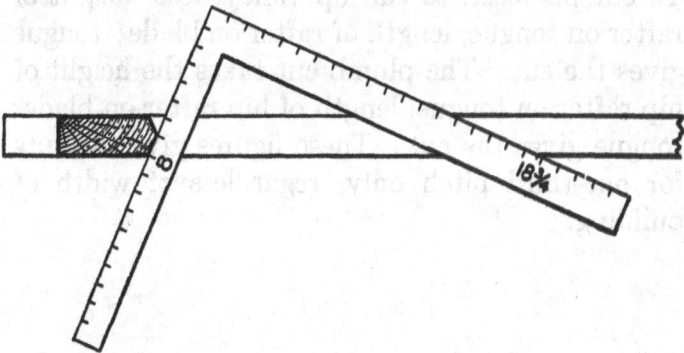

FIG. 92

more acute. The rule is: take the length of the hip on blade and its run on tongue, the blade gives the cut, or take 17 on the tongue and 18¾ on the blade; blade gives the cut, as shown in Fig. 91.

The hip rafter should be beveled so as to be in plane of the common rafter; height of hip on tongue, length of hip on blade; tongue gives bevel, or we take 8 on tongue, 18¾ on blade, tongue gives bevel, as shown in Fig. 92. These figures will cover all cuts in putting on cornice and sheathing.

To cut bed moulds for gable to fit under cornice, take half the width of the building on tongue, length of common rafter on blade; blade gives cut. Machine mouldings will not member, but this gives a solid joint; and to member properly it is necessary to make moulding by hand, the diagonal plumb cut differences. We find many mechanics puzzled to make the cuts for a valley. To cut planceer, to run up valley, take height of rafter on tongue, length of rafter on blade; tongue gives the cut. The plumb cut takes the height of hip rafter on tongue, length of hip rafter on blade; tongue gives the cut. These figures give the cuts for one-third pitch only, regardless of width of building.

Part IV

HOUSE FRAMING.   Framing a house — Good and faulty construction—Constructing the sill—Construction at the bearing of second floor joist — Construction of cornice — Constructing box sills—Framing ordinary studding—Constructing a porch—Cutting window openings — Framing sills — Method of halving sills- -Making a good corner—Framing a joist bearer—Method of setting studding — Constructing a cornice — Cutting openings in frame work—Affixing of joinery work, Etc.

Framing a House.—Some thirty-five or forty years ago when lumber was more plentiful, it was the common practice to build frame houses, great and small, with solid timbers. The sills, plates and corner posts were often hewn from the round timber with broad ax and adz, often taking months in preparing these for the "new house." In fact, it was quite the custom to commence the year before to get out the timbers, preparatory for the day of "house raising." After the timbers were hewn to the desired size, then came the work of laying out the mortise and tenons for joining the different parts together. No nails or spikes being used for this work. The corner posts were usually made out of timbers six or eight inches square with the inner corner hewn out to receive the lath and plaster. Think of doing that kind of work nowadays. This carries us back to the time of the building of our old home, now more than forty years ago; though only a lad we remember the time the trees were being felled in the forest and after a long wait for the timbers to be squared, they were hauled to the building site, and after a time for them to season, the carpenters came, and as though but yesterday, we see them under the old apple trees astride the timbers with auger, chisel and mallet working away from morn till night. It was just so with all of the work connected with the building. The mill work was gotten out by hand, even to the sash and doors. How well they built their works remains as a silent witness; suffice it to say the latter day workmen could gain some good pointers in construction from these old timers. Neither short hours, long hours, strikes nor lockouts worried them. Those were days of toil, days of contentment and peace. How different it is now! When the new house is decided upon, within sixty or ninety days it is ready to move into. The work is divided up into different classes and done by different workmen. The solid timbers are no longer used for the frame work. In its stead the sills and other timbers are built up with joists and studding commonly known as balloon framing, and everything is rushed from start to finish, and in the hurry many things that should be done are overlooked to the detriment of the house. Some of these things may require but little or no extra expense, if attended to at the proper time, but if neglected prove a serious detriment to the building. It is to this phase of the question that we desire most to call attention.

Good and Faulty Construction.—In Fig. 93 are shown two ways of constructing the sill; however, there are several ways, but these will serve our purpose. The one shown at A is marked "faulty," and that at B is marked "good construction." At A the masonry projects a little beyond the base board and the sill is laid without being bedded in mortar. The water follows the wood-work and runs or beats under the sill, the in-

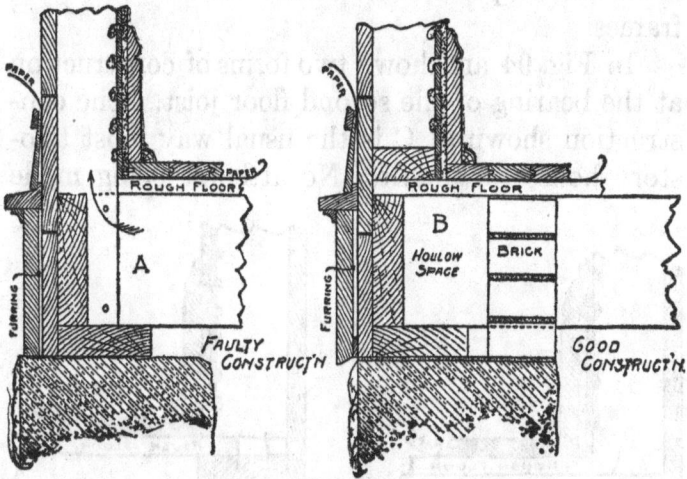

FIG. 93

equalities of the stone holding the water and in a few years the sill is rotted out, to say nothing of the cold that the open crevices will let in. The studding are halved to allow nailing space to the sides of the joists, but in doing this a space between the studding is left open, allowing free circulation, as shown by the course of the arrow. In the construction at B the sheathing is flush with the mason work, and the base sets clear and a little below the top edge of the stone. The back edge of the base is beveled so as to form a drip. The sills are bedded in mortar and the spaces between the joist and studding are cut off. Bricks are used to fill in between the joist but in this allowance should be made for shrinkage of the timbers by leaving the masonry work a little below the top edge of the joist. The building paper extends from sill to plate and under window and door frames.

In Fig. 94 are shown two forms of construction at the bearing of the second floor joist. The construction shown at C is the usual way most two-story houses are built. No attempt being made

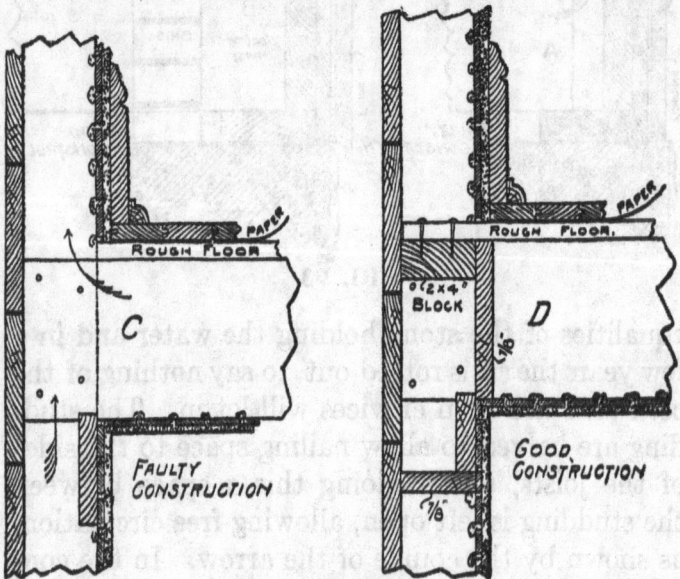

FIG. 94

to cut off the space between either studding or joist. At D is shown what should be done. Two by four-inch pieces are set in between the studding on a level with the top of the joist. The rough flooring should be laid diagonally with the joist and extend over on to these pieces and nailed. This forms a good tie and makes a closed job and by cutting in $\frac{7}{8}$-inch boards between the joist, letting the lower edge lap over the bearing board will cut off the space between the joist. If back plastering is desired a third piece should be cut in between the studding just beneath the bearing board for the back plaster to stop against same.

In Fig. 95 are shown faulty and good construction of cornice and bearing of the ceiling joist. At E is the usual way of construction for cottages

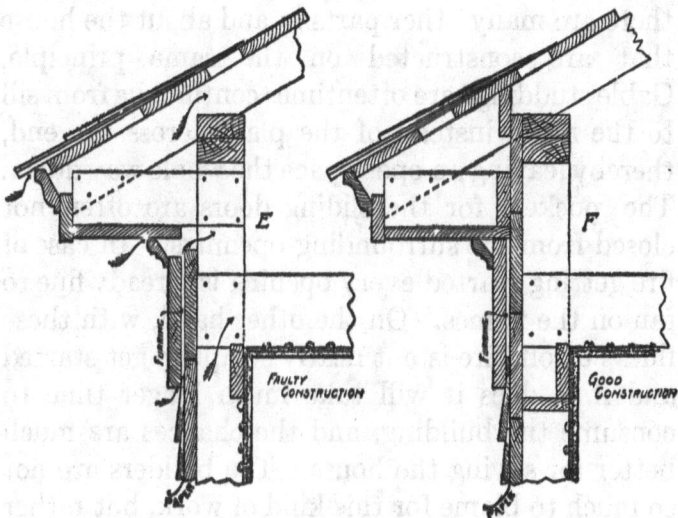

FIG. 95

where the ceiling is lower than the plate. The space between the studding is left open and is otherwise built on a refrigerator plan. Great open cracks are left between the frieze and plancier with the idea of covering with the bed mould. After the natural shrinkage of the different members, small crevices are open up to huge proportion and the frosts and cold winds find their way in to compete with the burning coals for supremacy. As the heat rises and warms the attic, the cold air seeks the lower level, and if by faulty construction, as before mentioned, in the lower parts of the house, it has a perfect current of cold air sweeping through these spaces and the result is a cold house. At F is shown how these defects could have easily been remedied at a very little extra expense. But it does not stop here, for there are many other parts in and about the house that are constructed on the same principle. Gable studdings are oftentimes continuous from sill to the rafter instead of the plate across the end, thereby leaving an open space the whole way down. The pockets for the sliding doors are often not closed from the surrounding openings. In case of fire getting started every opening is a ready flue to fan on the flames. On the other hand, with these flues cut off, fire is not nearly so apt to get started and if it does it will take much longer time to consume the building, and the chances are much better for saving the house. The builders are not so much to blame for this kind of work, but rather the architect who should show these things in the plans and specifications and see that it is done. Otherwise, in these days of close competition, the contractor must figure close, and unless these things are clearly shown and described he cannot reasonably be expected to do it unless all are required to do the same.

Constructing Box Sills.—Fig. 96 shows several ways of constructing box sills, commonly called

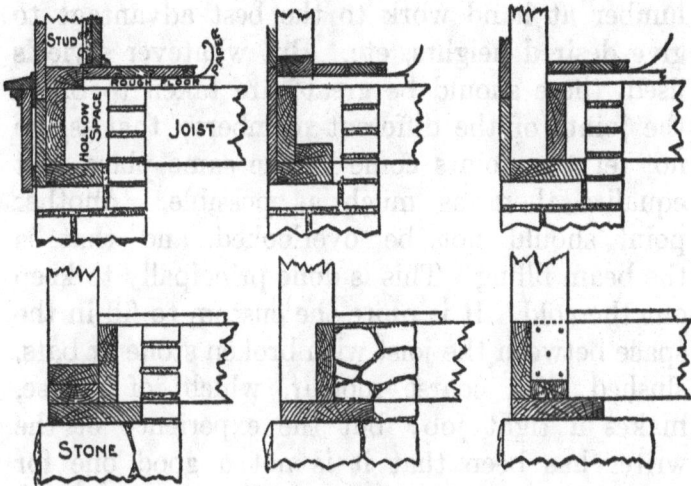

FIG. 96

balloon framing. The first one shown at the upper left-hand corner is the same as that shown in the previous illustration for good construction, and what is said in its favor is not necessary to repeat here. However, the section here shown is that resting at right angles to the joist. The side section should have an extra joist with ⅞-inch blocks set in between for the plate or shoe to rest on, so as to form a projection to receive the ends of the flooring. In long spans there should also be cross-pieces set in between the sill and the adjacent joist, as a further precaution to keep the sill from springing outward. The other types here shown are good forms of construction for general residence work, and are here given to meet the requirements that may arise. It is quite frequently necessary to change, so as to make the lumber at hand work to the best advantage to give desired heights, etc. But whatever style is used, there should be great care taken to break the joints of the different members; that is, do not let two joints come at the same place, but equalize them as much as possible. Another point should not be overlooked, and that is the beam filling. This is done principally to keep out the cold. It is quite the custom to fill in the space between the joist with broken stone or bats, slushed with coarse mortar, which, of course, makes a tight job; but the experience of the writer has been that it is not a good one for several reasons, which are as follows:

First—unless the sill is built water-proof, after forms as previously illustrated, it is only a question of a few years until the sill is rotted out. Then, again, if is is built perfectly tight dry rot is liable to make quite as quick destruction; besides the joist are sure to shrink more or less, so something has to give, and it is usually a bulged floor, as the result. The better method is to fill in with brick, leaving an air chamber back of same and make allowance for the shrinkage by not filling flush with the top edge of the joist at first; then after a time, if a tighter job is wanted, it can be pointed.

Framing Ordinary Studding.—Volumes could be written on roof framing alone and yet not completely cover the subject. We will mention a few simple matters of the most common framing. The best way to frame ordinary studding or joist is to make a good pattern of one, and for two men, one at each end to lay the pattern on and mark off the rest. Of course, in very heavy work it is well to make a pattern out of lighter material. It happens many times that one man is left to frame alone, and it is very inconvenient, especially in long work, to go back and forth to mark both ends. It is very handy to use a pattern on each side and square over and lay off a dozen or so at a time, as illustrated in Fig. 97. Here we show a long studding notched for ribbon to put second floor joist on. This used to be con-

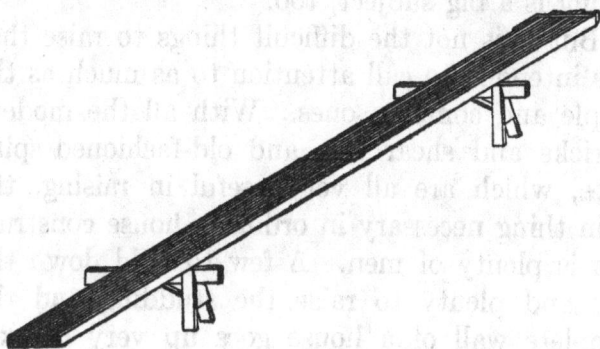

FIG. 97

sidered much better than short studding for each
story, but the majority seem to prefer the short
studding for each story, as it is surely much easier
to handle and, if properly made, is fully as
strong.

Of course, it could be made weak by only
slightly nailing the upper studding and have a
crack in the lining just at the bottom of the upper
studding, etc., but these matters can be easily
avoided and should be. The long studding can
be made weak, and often are by carelessly sawing
them almost in two when sawing for ribbon.
Again, we often see a very weak building of this
kind because many of the joists simply rest on
the ribbon and are not spiked to the studding at
all. Most of these weak points in either style
could and should be avoided. The British
Government has shear legs in their dockyard at
Chatham that will raise a dead weight of 180
tons out of a ship and into the air 60 feet and place
it in another ship, and it makes one realize that
raising is a big subject, too.

But it is not the difficult things to raise that
it is intended to call attention to as much as the
simple and common ones. With all the modern
derricks and shear legs and old-fashioned spike
poles, which are all very useful in raising, the
main thing necessary in ordinary house construction is plenty of men. A few to hold down the
foot and plenty to raise the studding and the
complete wall of a house goes up very quickly
and easily.

While it is very necessary to have plenty of men it sometimes happens that we do not have them. Then by nailing together a section at a time, and having a lining board nailed at the bottom to hold the foot down and a studding at each end with a spike in each to help hold it, it is raised as illustrated in Fig. 98. Where there is

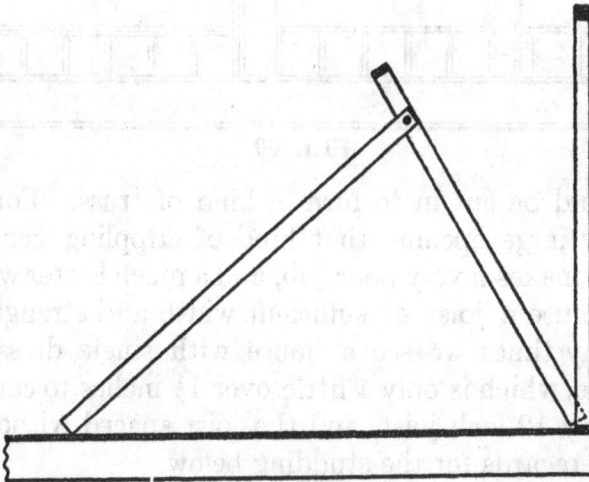

FIG. 98

plenty of help always finish the openings complete before raising, as it is much easier while the studdings are lying down than it is to climb up and cripple them in.

Fig. 99 illustrates a section complete. A common window opening may be made with single studding and headers flat ways; but for larger openings they should be doubled and the header put up on edge and, where necessary, a brace

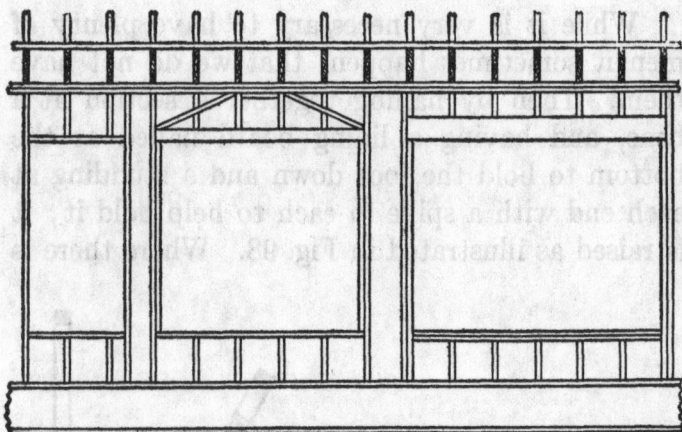

FIG. 99

should be cut in to form a kind of truss. For a very large opening that kind of crippling generally makes a very poor job, and a much better way is to use a joist of sufficient width and strength. Many times we see a house with single dressed plate, which is only a little over $1\frac{1}{2}$ inches to carry heavy 12-inch joist, and the joist spaced without any regards for the studding below.

While it is true that the joist will probably not break through the plate, still how much better it would be to space the joist so they would come over the studding. In fact, it is not only much better, but it is much easier, too, for you to simply lay off the sill, and it is all carried along with that one laying off. Studding and joist are all 16 inches apart all over the building and ready to receive the lath.

There is no time spent at all for laying off the second floor except for laying out for the openings.

Constructing a porch.—Fig. 100 shows the general construction and details in full of the framing and the various parts of an ordinary front porch. A special feature in the framing of this porch is the joists, which in most porches must also answer for the outside sills. In framing porches many carpenters and contractors frame the joists so that there is nothing but the narrow edge of a 2 by 6 or 2 by 8 that rests on the porch piers, and this small bearing, of course, must come all on the outside edge of the piers.

By the method which we refer to in our detail we take a 2 by 6 and a 2 by 8 and spike them together, as shown in the detail at A. This gives a full bearing on the pier, or very nearly so, and it will be found to make a much better job than the old custom of just having the narrow edge of the floor joists resting on the piers.

The facing shown at B should be from one-half to three-quarters of an inch wider than the floor joist, so that it will project a little below the pier. This gives a good place to secure the lattice work from the back side.

The top rail in the porch rail is made of two pieces, as shown at R. It is difficult to get a single piece rail large enough for this kind of a porch. Many of the built up rails consist of from six to eight pieces, which make them more or less expensive. The rail, as shown in the sketch, is designed to work in between the small single rail and the large six and eight piece rails, and will be found a very easy rail to make, an inexpensive

one, and at the same time a good, serviceable rail
and one that will look well. The bottom rail of
this porch can be made in one or two pieces, as

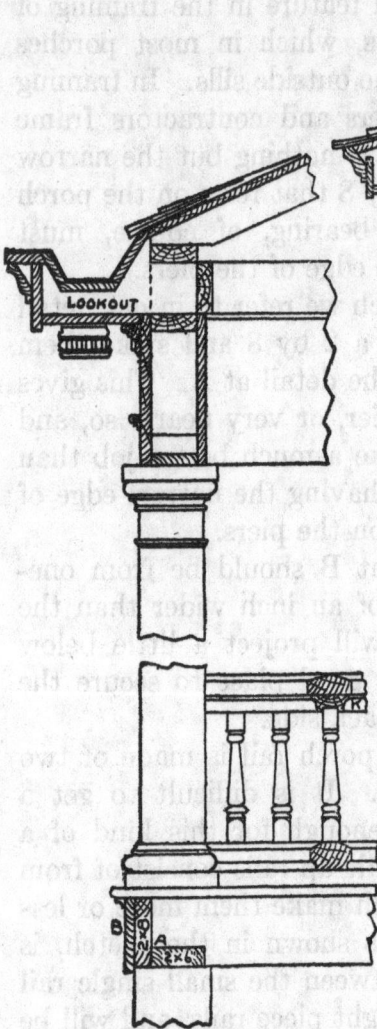

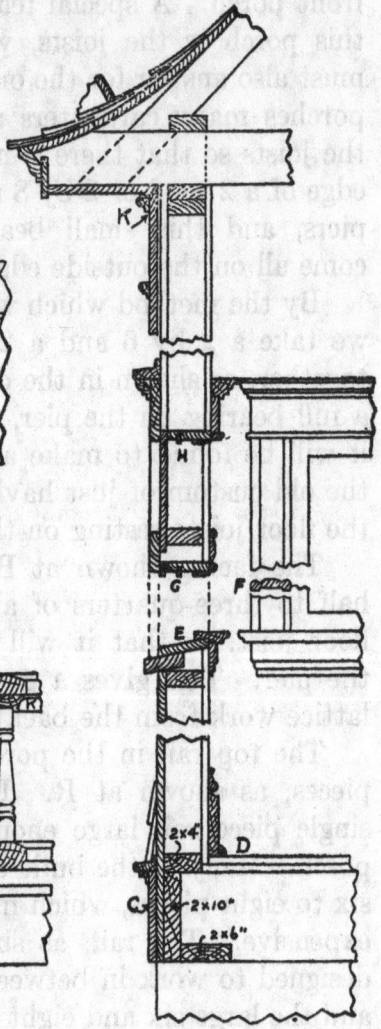

FIG. 100   FIG. 101

desired. If made from a 4 by 6 it could easily be
made in one piece. The porch has a box frieze
and cornice with a sunken gutter in cornice. The
cornice has a wide frieze with a band mold and a
bed and dental mold. A planceer, fascia, crown
mold and cap piece complete the outline of the
cornice.

The gutter is formed with three pieces, a bottom and two side pieces. The side pieces are put
in sloping. No gutter should be formed with
perpendicular sides, making square angles in the
bottom, for they are always causing trouble by
freezing and bursting. Sloping sides will allow
the ice to expand without any danger of injuring
the gutter.

Lookouts are nailed to the ceiling joists and
allowed to project in front as far as required for
the cornice, and a plate spiked on top of the
lookouts supports the rafters. This porch is designed for a shingle roof, but the framing of the
roof is such that any pitch can be used, even to a
very flat pitch for tin roof, or any kind of a flat
roof it is desired to have can be applied to this
construction, for the pitch can be varied to suit
without in any way interfering with any other
part of the design.

Fig. 101 represents a section of house framing
from sill to cornice, showing the ordinary window
construction. The sills in this are framed similar
to the porch, but in addition have a 2 by 4 plate
put on top, which laps over on the joists, as shown.
The advantage of this kind of sill and framing is

that it saves the time required to cut gaines in solid sills. This kind of a sill, made of a 2 by 6, 2 by 10 and a 2 by 4, requires a little more lumber than a solid 6 by 6 sill, and if well put together we consider it fully as good as the solid sill cut full of gaines for the joists. In some large cities there are building laws prohibiting this form of construction. We presume this is principally on account of making the walls more susceptible to the spreading of fire than for lack of strength. We have good reason to believe that sills made in this way, that is, well made, are fully as strong as the solid sill ordinarily put together. If the walls were bricked up inside to level with top of floor joists, then all danger of the spreading of fire is avoided, there is something to hold the buildings solid to the foundation wall, and the result cannot fail to be satisfactory in every particular.

At C in Fig. 101 is shown the outside base water table and mold. At D the double floor, base, base mold and quarter round, the ordinary base finish of the average job. At E is shown a section through the sill of an ordinary window in a frame building, the sill, subsill, stool and apron. At F is shown a section of the casing. At G is shown a section through the side jambs and casings, showing outside casing, blind stop, jamb, stops, inside casing, studding and weight box. At H is shown a section through the head, showing outside casing, cap and mold, jamb and stops, inside casing, cap trim, etc. Above this, at

K, is shown the general construction of the cornice, etc. In this the ceiling joists extend out over the building as far as required to make the cornice. A plate on top of the ceiling joists supports the rafters. The roof is concaved at the bottom by nailing on circular pieces to the bottom ends of the rafters, as shown. A standing gutter is placed on the third course of shingles on the roof. This makes a very nice finish and one much used at the present time. The porch finish and main cornice, as here shown, will work in harmony on the same house, and are well adapted to houses of the hip roof class.

Cutting Window Openings.—There are many carpenters who cannot really figure out the proper size to cut window openings. That is, how wide the space should be between studding and how high between headers. Windows are nearly always listed and sizes given glass measure, as 24 x 32, 2-light, which means that the window has two lights with glass twenty-four inches wide and thirty-two inches high. The width of the opening between studding should be ten inches wider than the width of the glass which, in this case, would be thirty-four inches between studding. The height between headers should be eleven inches more than the length of the two lights, thus two lights of thirty-two inches each will be 64 inches and eleven inches added to this will make 75 inches, or six feet and three inches. In regular two-light windows six inches is allowed for wood or sash in height, two inches top rail,

one inch meeting rail and three inches for bottom
rail. To this add one inch for top jamb, one inch
for subsill, two inches for sill and one inch for
play, which, all told, makes the eleven inches. If
these figures are carefully followed no trouble in
setting frames will ever occur. In large frames,
such as cottage front frames, with glass 40 by 44
or approximately this size, allow eleven inches
more than the width of the glass if the windows
are to be hung with weights, because these windows will require extra large weights and, consequently, will require a little more room; but the
average window will have plenty of room for
weights when ten inches is allowed for the sash,
jambs and weights.

Framing Sills.—Referring to the sketches,
Fig. 102 represents a method of framing sills for
light framing without gaining the joists into the
sills. This method consists of using a 2 by 4 or
2 by 6 flat on the wall, and spiking a 2 by 8 or
2 by 10 on the face edge, as shown by A and B.
This will allow the floor joists to be spiked through
into the ends of the frame, as shown. The floor
joists are to be gained on the bottom to have a
good bearing on the piece A, and should be cut
to fit tight against the face piece B. After these
two pieces are thoroughly spiked together, then
put on the third piece C and thoroughly spike the
same. Allowances should be made in cutting the
timbers so that they will work out right and
bring the finish out flush with the wall, or as intended if otherwise shown. In most cases where

houses are sheathed with seven-eighths sheathing
it is best to keep the frame back one inch from
the wall, so that when the sheathing is put on it
will come flush with the wall, then there is nothing to project over the wall at the bottom but the
outside base, as shown.

This method of framing makes a very strong
frame, and if well spiked it is fully as strong as
solid timbers of the same dimensions. Two
2 by 8's and one 2 by 4 are just a little more than

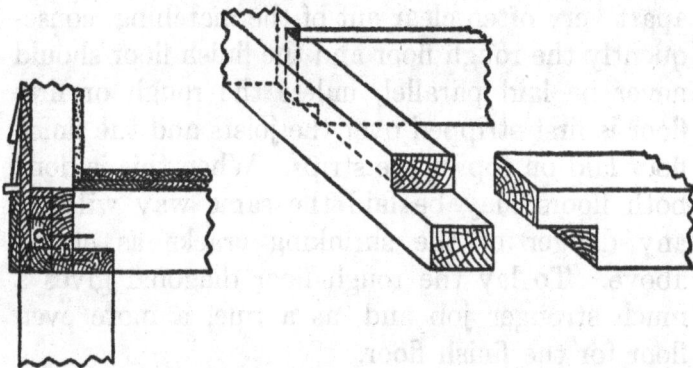

FIG. 102   FIG. 103

equal in cross section to a 6 by 6, and it is safe
to say that they are stronger than a 6 by 6 gained
out to receive the joist. The built-up sill has
much the advantage in making splices. In splicing, the joint can be so broken that when put
together it will be like one continuous sill all
around the building. The cut shows the sheathing, outside base, water table, siding and a double
floor. Where double floors are laid the first rough

floor should be laid diagonal, or stripped, before laying the finish floor. Where both floors are laid the same way the shrink of the lower, or rough floor, is often enough to cause great cracks to appear in the finish floor. This is due to the fact that the rough floor boards are usually 12 inches wide, and the finish flooring boards only about three inches wide, thus each twelve-inch board will take four of the three-inch boards, and the joint in the finish floor that comes nearly over the joint in the twelve-inch board below will draw apart very often clear out of the matching, consequently the rough floor and the finish floor should never be laid parallel, unless the rough or first floor is first stripped over the joists and the finish floor laid on top of the strips. When this is done both floors may be laid the same way without any danger of the shrinking cracks as stated above. To lay the rough floor diagonal gives a much stronger job and, as a rule, a more even floor for the finish floor.

Method of Halving Sills.—Fig. 103 shows the usual method of halving sills together at corners in framing, and of gaining floor joists into the sills as practiced in light framing. There is very little of the old-fashion mortice and tenon framing any more. Principally the timbers are just halved together and spiked into place. When floor joists are gained into sills and girders they are usually gained into the sills or girders so that the joists have a bearing of two or two and one-half inches on the sills or girders, and the depth of the gain is usually two inches less than the width of the joist it is to take, and this two inches is cut out of the joist on the lower edge to let it come flush with the sill. This may be seen at A in Fig. 103. It is often the case that girders can be set below the sills in the center of buildings and thus avoid the necessity of cutting gains in the girders. When this can be done it is better than to cut the gains, for it gives the full strength of the timber used.

Making a Good Corner.—Fig. 104 shows a simple and inexpensive way of making a good corner for ordinary house framing. Take two

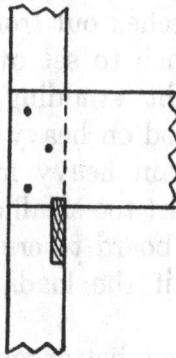

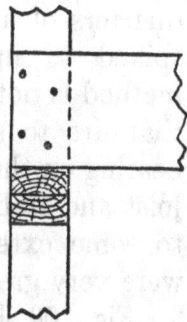

FIG. 104   FIG. 105   FIG. 106

2 by 4's and spike them together, as shown at A and B, and on the inside corner spike on a 2 by 2 on each side to receive the lath. This will make a good solid corner for the lathing and has less lumber in it than the way some builders make corners, and is also better than the way some are

constructed. The pieces C and D are only short
blocks put in to make a nailing for the base in
finishing. This is a matter that should never be
overlooked in building a house. No mechanic
can do a good job of putting down base unless
there is something to nail it to. All the short
blocks about a building can be utilized in a similar
manner, as shown in the sketch, to make nailing
places for base corners and by the sides of doors,
etc.

Framing a Joist Bearer.—Fig. 105 shows the
usual manner of framing a joist bearer for second
floors in ordinary residences. A 1 by 4, or 1 by 6,
is usually gained into the studding, and the joists
are sized and notched out from one-half to three-
quarters of an inch to set over it, and are then
spiked to upright studding, as shown. This
method is not good on heavy work and for floors
that are to sustain heavy loads; it makes the
bearing on the joist too small and would cause the
joist and ribbon board to crush into one another
to some extent if the loads placed upon them
were very great.

Fig. 106 shows a better method. This method
consists of putting a double plate around for the
joists to rest on. The double plate makes a good
bearing for the joists, one that will sustain any
weight. By using the double plate it is much
easier to keep the walls straight than with a
ribbon board, which bends in and out at the
slightest cause and is often hard to get straight,
and harder still to keep straight. Then, again,

the double plate serves as a first stop between stories. The double plate ought always to be used on the houses that are full two stories in height; it is far better than the ribbon board, and it will result in giving a straighter and better job all around.

Fig. 107 shows a method of framing the foot of rafters for an ordinary square, plain cornice. On the average work there is probably more of this put on than any other kind. Double plates should always be used in order to bring the plate straight, even if it is not needed otherwise. This

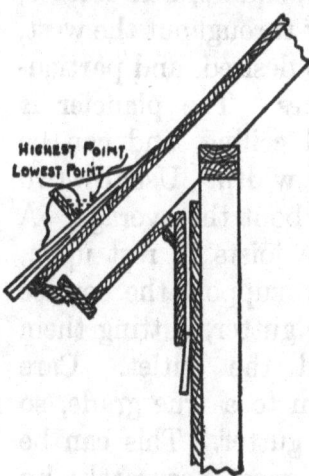
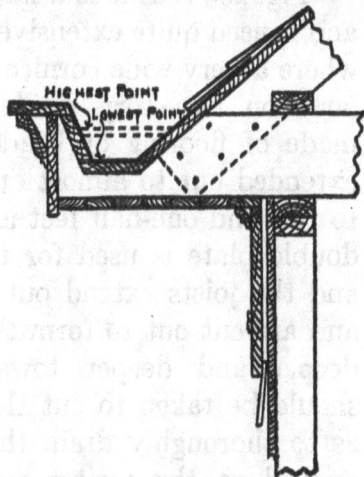

FIG. 107   FIG. 108

cornice provides for a frieze, plancier, fascia, crown and bed mold, as shown. The standing gutter shown on the roof should be placed either on the second or third course of shingle; the second course, unless for some cause it cannot be so placed. This gutter must be lined with tin, and the tin should extend up under the shingles at least five inches. The first course of shingles starting from the gutter must be a double course, the same as starting from the bottom. The pitch in the gutter to run the water to the outlet is usually obtained by dropping the end, where the outlet is to be, down the roof a little. This is all right for short lengths, but for long runs it does not give sufficient fall to the gutters. When the fall or pitch is not enough it can be increased by putting tapering pieces in the bottom, as shown.

Fig. 108 . This is a large, square, box cornice, and is used quite extensively throughout the west, where a very wide cornice is desired, and particularly on two-story residences. The plancier is made of flooring or beaded ceiling, and can be extended out to almost any width. Usually two to two and one-half feet is about the average. A double plate is used for the joists to rest upon, and the joists extend out to support the cornice and are cut out to form the gutter, cutting them deeper and deeper toward the outlet. Care should be taken to cut them to a true grade, so as to thoroughly drain the gutter. This can be arrived at the easiest and most accurately by striking a chalk line on the ends of the joists (before the fascia is put on), from the high point to the low point. This will show the proper depth to cut each joist to have a perfect grade. Care should be taken not to let the line sag, in striking the line.

The single plate above the joists supporting the rafters can be varied a little if necessary. For example, if a narrower gutter is desired, the plate can be thrown out some, say four inches or about, but if this is done it should be remembered in cutting the rafters, for if thrown out it widens the run of the rafters, and would necessitate cutting the rafters to meet the requirements in the case.

At A is shown a band mold broken around the frieze. This is usually placed about four and one-half inches above the frieze from the bottom of the mold, so that when window frames are set with the side casings up against the frieze, the part of the frieze below the mold forms the head casing for the frame and gives it a finished appearance.

Method of Setting Studding.—It is indeed a simple thing to set studding and yet it is surprising how many mechanics, and many of them good ones, set them apparently without any thought. This causes them a good deal of bother and work, as it results in the plaster cracking in the corners.

Fig. 109 shows a corner post made out of four studding, or the center one can be made out of short pieces—almost any carpenter knows how to make one—yet apparently few know how to make one practically as good, in less time and less lumber by nailing them together.

Fig. 110 shows an easy way and would be all right if the thickness of two was the exact width

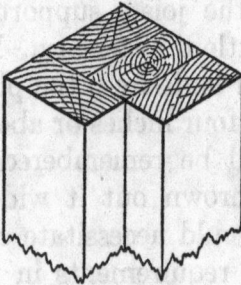

FIG. 109

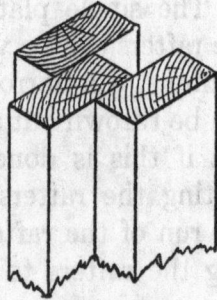

FIG. 110

of one, but many times they are not, and so it is better to nail them together, as in Fig. 111. We have made them this way for years and we consider them a very good corner.

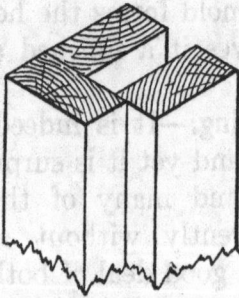

FIG. 111

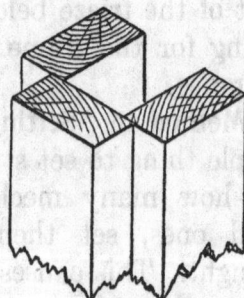

FIG. 112

Fig. 112 makes the best partition corner we know of. It is made much easier than one with blocks nailed in between two studding and then one nailed on the blocks, and my observation has been that it does not crack the plastering.

Fig. 113 is practically the same as Fig. 112. It is made for partitions which are made the flat way of the studding.

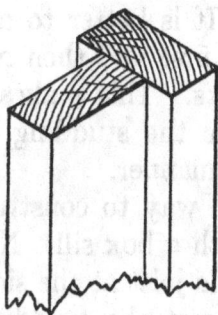 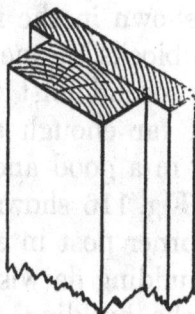

FIG. 113   FIG. 114

Fig. 114. While it is not quite as good as the ones just mentioned, it does very nicely, especially if one is running short of studding. It is simply to nail a sheathing board on the back.

Fig. 115 carries out the same idea in double joist under partition for floors, as Fig. 114 does

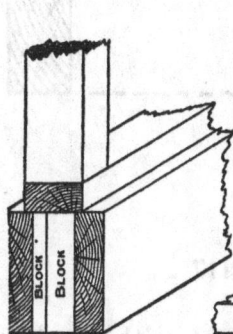 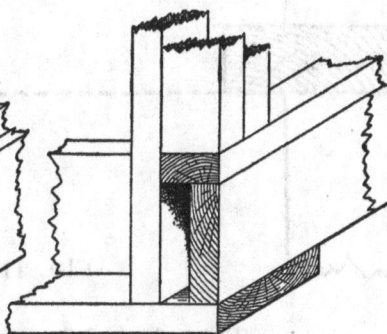

FIG. 115   FIG. 116

for receiving the lath and plaster. This saves the necessity of nailing pieces on to the sides of the joists to receive the ends of the flooring.

If you wish to make the joists rigid this may be done by nailing three-inch blocks in between,

as shown in the illustration. It is better to nail
the blocks to one of the joists first and then nail
the second joist to these blocks. This makes it
just far enough apart to spike the studding on
top in a good and substantial manner.

Fig. 116 shows a very good way to construct
a corner post in connection with a box sill. Nail
a studding flatwise on top of the joist at the sides
of the building on which to rest the studding.
Set the end joist in just enough to receive the end
studding, making a good, strong and tight job.

Fig. 117 shows a very good way to make the
top of the end studding by putting on a 2 by 6

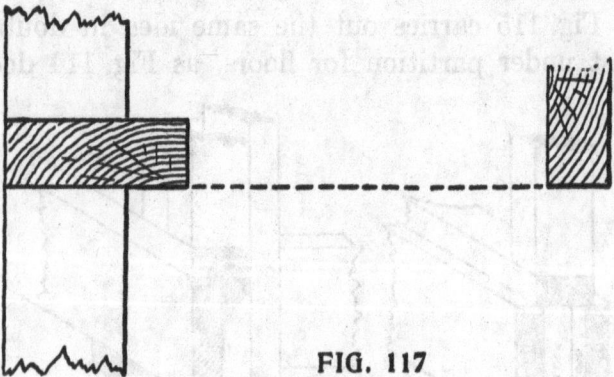

FIG. 117

for a plate. This should be set with its lower
edge even with the bottom of the ceiling joists.
This makes a solid angle and prevents the plaster
from cracking.

Constructing a Cornice.—There is nothing that
adds to the appearance of a building like the
cornice. A building with beautiful walls and a

good roof is indeed the kind every one should have, yet a poorly constructed cornice on that kind of a building would mar the effect of the entire building.

A cornice to be well made and look nice and pleasing to the eye does not necessarily need to always be a massive or expensive one, in fact, for some buildings it should be the reverse.

While the gutter is nearly (so to speak) out of sight, to the human eye, yet where the human eye can see it there is nothing that looks worse or entirely racks the whole human frame more than to see one standing full of water, and that kind, we are sorry to say, is only too common. Not only does the sight of that kind rack the human frame, but in time, and only too short a time, it wrecks the frame of the building as well.

Fig. 118 illustrates a simple cornice, and is, as you see, put on (where the rafter does not project beyond the plate) by simply nailing a board at top edge to the plate and rafters and let the bottom edge project, and at the gable end a board nailed one edge at center of rafter and the other edge gives the projection. Let the bottom cornice project and cut the end cornice at top edge of it and break a fascia board around the whole cornice and it makes a very good, tight and cheap cornice and answers very well for small light work, where the projection does not have to be but a few inches. Of course this kind of cornice on a large house would not do at all, but on cheap out-buildings, such as coal sheds, hen houses, etc., it answers very nicely. This kind is generally used on only those small, cheap buildings and used without gutter, although, of course, they could put a gutter on them.

Fig. 119 illustrates what is called rafter finish, and used to be railroaded through this country on all kinds of buildings, and while it was claimed

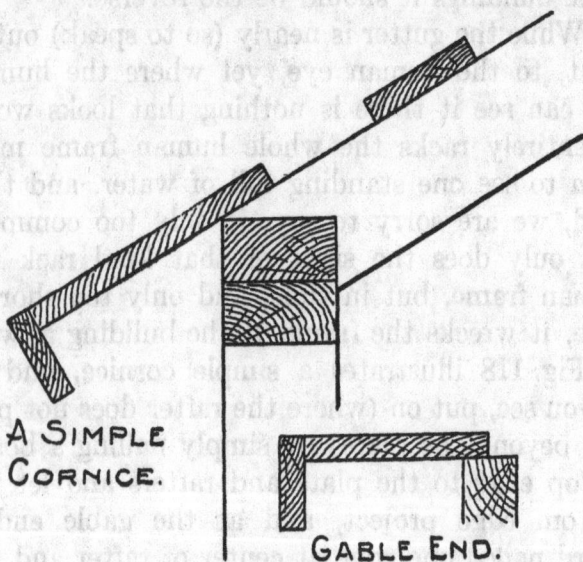

A SIMPLE CORNICE.

GABLE END.

FIG. 118

to be cheap, many very expensive ones were put on with several members, molds, brackets, etc.

We have illustrated the two ways of putting it on; the lower one is generally the best, as the drip drops over the edge of the house, while in the upper one the drip drops on the lower part of the cornice, and in a few years a nice house with expensive cornice, with molds and brackets, is all rotted away.

The main objection to the lower one is that sometimes in a big rain the water comes down the roof so fast it splashes over the gutter, and to overcome that the gutter is set at an angle between the two here shown.

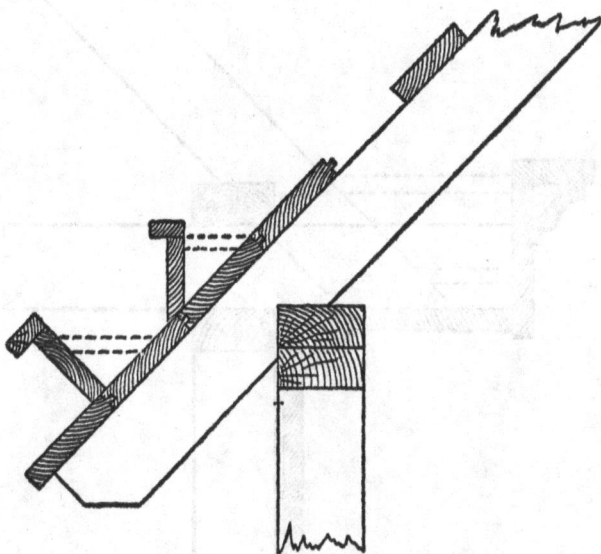

FIG. 119

The fall in these gutters is sometimes made entirely by putting one end up higher on the roof than the other, though to put them level and put a bottom board in (as dotted lines show) near the top of gutter and tapering down to nothing at the down spout makes a better appearing job; yet, for a very long gutter it is well to do both—

put one end higher than the other and put the bottom board in also.

It is much better to construct a gutter and give it all the fall you think necessary, and then

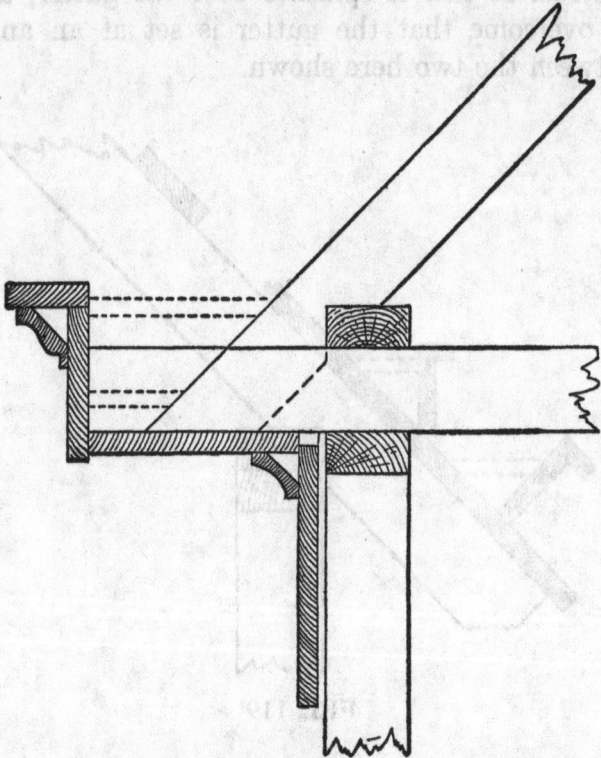

FIG. 120

add an inch more rather than make it an inch less. Give all gutters plenty of fall and you will not only be pleased, but the owner of the building will be also. These gutters are sometimes put on top of the shingles. It makes a more lasting job, but is far from ornamental, or convenient, either, when it comes to re-shingling.

Fig. 120 illustrates the real cornice and can be constructed cheaply, as illustration shows, or can be made very expensive with many members, mouldings, brackets, etc., and it will practically last forever if a good gutter is kept in it, as there is no part of it exposed.

With dotted lines we have shown both ends of the gutter board in bottom.

Cutting Openings in Frame Work.—It sometimes seems to me that there is not a trade or calling that the boy falls into as naturally as the building trade, especially carpentering. We never saw a boy but what knew something about driving nails, and quite a large per cent have at least tried to use a saw. So it comes naturally, and they easily commence to nail on a few boards, which gives them a taste, and they start in to be a builder. Step by step they move along from rough work to siding, cornice and finally to inside finish, and if they have a desire to go ahead with the work they lay it out for themselves as well as others.

They get hold of some good work on the steel square, and if they are natural mechanics they soon become a good framer. And yet as easy as it is, and as natural also, to be a good carpenter and a building foreman, it is indeed surprising to know that there is probably not one carpenter in a hundred who can cut out the openings in the frame work, and get them so that when the frames come they will fit perfectly. We have heard good foremen say the only right way was to wait until the frames came and measure them and then cut out the openings. Some sheath a house all up solid, and when the frames come they cut out, which is perhaps a sure way, but it never seemed a very pleasant, easy or cheap way. Others set the studding all up, and then before they sheath cut out for openings, claiming it is a nice way, as the studding are all evenly spaced all over the building. Some of that theory is good, but we could always cut a studding better when it was lying down level on a pair of trestles than standing plumb nailed in a building.

We always found it a much better way to lay the openings out and frame them complete before they are raised.

For a common 2 feet 8 inches by 8-foot door, get the exact height the bottom of the door should be from top of joist, then measure up 8 feet, which will give the top of door, then allow two or three inches for head jamb and space above the door. Always allow plenty of room; don't get the header so low that the lugs will have to be all cut off the jambs.

Now, as the majority of windows are the same height as doors, the measurement, when once gotten right, is good for many openings. Two inches on each side of door is generally enough, or four inches more than door measurement, is the measurement between studding, but as it is well to cut the double studding in under the

header, the outside or main studding should be
set about eight inches wider than the door.

Figs. 121 and 122 show it perhaps plainer than
words, and also show an opening ready for a
two-light 36 by 36-inch window.

A very good general rule, one easy to remember and one which works nicely on ordinary 5-inch

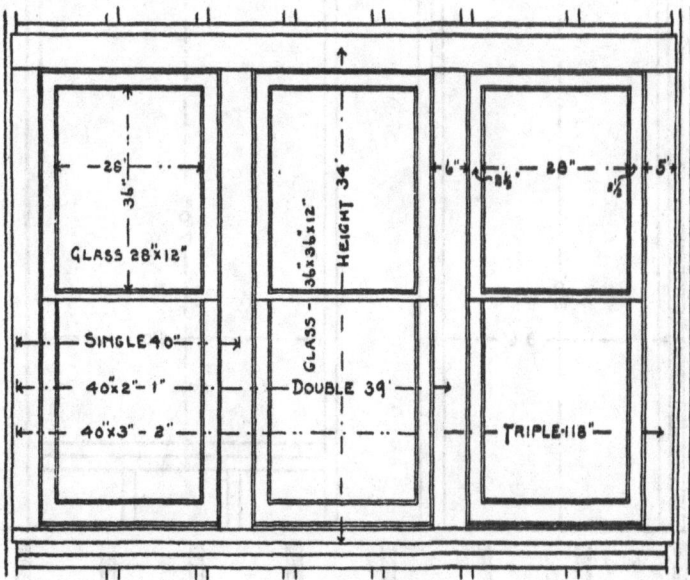

FIG. 121

casings, is to set the studding and headers just
one foot more than glass measurement. For a
house where the siding is put on the studding, in
order to give room to nail and not split the ends
of siding, it is well to make the opening a half or
an inch wider.

The double and triple windows are the ones that seem to give the most bother; and yet it is very simple. A double window (with a seven-inch mullion) is just double what a single one is. But

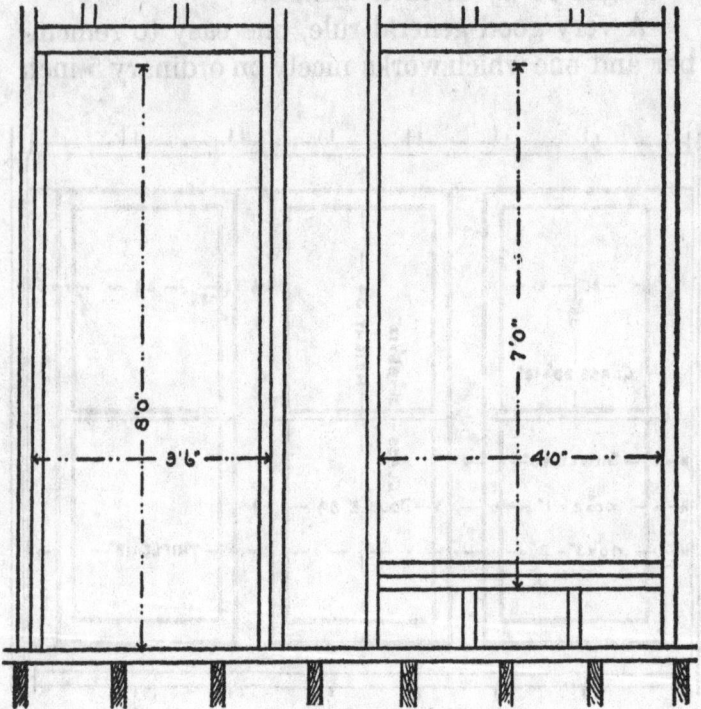

FIG. 122

perhaps the most common is as illustrated, a six-inch mullion on the outside, which makes them all alike, if you wish, on the inside.

We hope that a careful study of these illustrations will make it plainer to many carpenters when they go to lay out their next openings.

Affixing of Joinery Work.—Joinery work is fixed to either plugs, grounds or backings. **Plugging** is done by means of wood wedge-shaped plugs, which are driven into the vertical joints of the bricks, afterwards cut off to a level face and flush with that of the plaster, and to which the finished woodwork is secured by nailing, as shown in Fig. 123.

Grounds are used in better work, being wrought and splayed to form a key for the plaster.

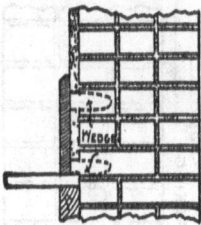 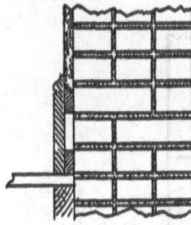 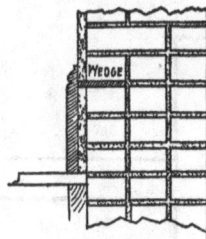

FIG. 123   FIG. 124   FIG. 125

They are nailed to plugs in the bross joints, as Fig. 124, or to wood pads, as shown in Fig. 125, wh ch the brick-layer builds in the wall for the carpenter to secure his work. Metal plugs are now quite generally used in lieu of wood; they are creased and made clamp-shape, and when built into the wall they form a perfect key. The nail is driven directly into the clamp and is held more securely than if driven into wood. The cost is reduced to the minimum and, all things considered, they are really cheaper than wood. In the best class of work the grounds should be mitered at the angles, and those at door and window openings should be beveled or splayed to receive and key the plastering, and being fixed perfectly plumb or perpendicular to the plugs or other nailing, they form an excellent plane to work to.

Backings are pieces of wood framed in between the studding or other work to form a solid backing for shelving, plumbing or other fixtures. Sometimes these pieces are framed in lieu of nailing, as shown in Fig. 126; but for all practical purposes this is unnecessary work, because if

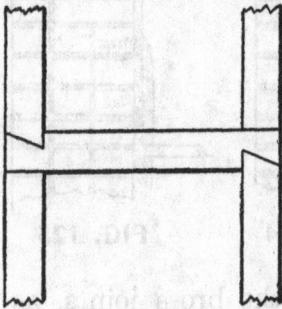 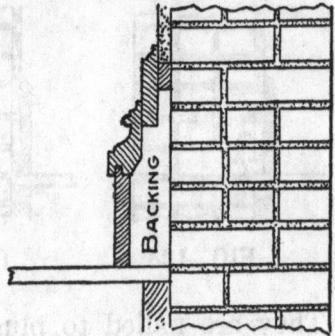

FIG. 126   **FIG. 127**

properly framed and nailed will be sufficient for most any kind of work.

Shaped backings are those which are cut and notched to receive irregular framing or different members, as skirtings, etc., used in the best class of work (see Fig. 127).

All first class work should be fixed to grounds in lieu of plugs alone, which are only fit for interior work; and when the grounds are fixed perfectly true, vertically and horizontally, there is no difficulty in affixing the joinery without fitting, cutting and scribing; and, moreover, the plasterer finishes his work with greater accuracy, as he has perfectly true and substantial grounds to work to.

Attaching Woodwork to Stone Walls.—The cement building block industry is fast becoming a great and growing business. Almost every issue of a trade journal brings to light some new machine for the manufacture of cement building blocks or some new use for the product.

Portland cement is one of the very best building materials that has ever been produced. It is fire-proof, water-proof, strong and durable beyond all doubt when it is properly mixed and applied, and with the cheapness of it and the high price of lumber it is sure to be used in place of wood in many cases where it is practical to apply it.

For the outside walls of buildings it is a first class article when manufactured into hollow concrete building blocks. These blocks are now being made to resemble natural stone. Rock face, smooth face, tooled face or any kind of face desired can be had at a very small expense. The usefulness to which this material can be put is almost unlimited. Artistic designs of residences, stores, school houses, churches and all kinds of buildings can be executed in fine shape, giving the buildings the appearance of magnificent stone structures.

Is it durable? This is a question that will be

aksed over and over again by the doubting ones. There is no question as to its durability when the article is properly mixed and applied. No better proof need be asked for than a comparison of a good cement sidewalk with one of stone, and finding the cement walk wearing and lasting better than natural stone. Sidewalks get more wear and rougher usage than stone put into a building, and a cement block that would stand in a walk would stand in ordinary building construction. Of course there might be such a thing as putting too great a weight on a building block, but this would not be likely to happen in ordinary building, and in cases where there is great weight to support it can be distributed so that the weight will not all be concentrated in a few places about the building. The crushing strength of the cement block is fully the equal of ordinary brick work and no fear of crushing need be entertained. Blocks that crumble easily indicate that the proper quantity of cement has not been used in their manufacture. For good blocks not less than one part of cement to five parts of sand and broken stone should be used. The mixture to be what is known as a 5 to 1 mixture and should be thoroughly mixed.

In some parts of the country a cement stone block building can be put up quite as cheaply as a frame building. First there is the saving in paint, for the outside wall requires no painting. Second, the blocks are hollow, producing a dry wall without furring on the inside, and plaster can be applied directly to the blocks on the inside, thus saving the furring and the lathing. Third, the blocks being uniform in size, only a very thin coat of plaster is necessary. Fourth, all the frame work, sheathing and siding is saved. Taking all these items into consideration, the cement block house can be erected nearly or quite as cheaply as the frame house.

The use of cement building blocks, where no furring is used, makes it necessary to provide some means of securing the woodwork or finish of a building, such as the window casings, baseboards, etc. It is not practical to nail the woodwork directly into the blocks, and it is doubtful if such a thing could be done. Woodwork can be nailed more or less to a brick wall, but not very successfully, and nailing finish to a cement stone block wall is something that few would care to try.

In Fig. 128 we show a wall nine inches thick, which is about the least thickness that will admit a window frame in good shape. The jamb stone at A just comes about flush on the inside of the box frame, leaving just a little for the plaster. This makes it all right for nailing the side casings on the frame, leaving just a little of the casing to reach over on the main wall and completely covering the frame box. This makes the nailing for the side casings all right. With the head casing it is different. In this case there is nothing to nail the cap trim of the frame to except at the very lower edge. And in order to get a good job of finishing some means must be provided to nail the top edge.

As the stones are made in a mold it would be very easy to put a 1 by 2-inch strip in the right position in the mold to form a recess in the lintel stone, as shown at B, in which could be driven a

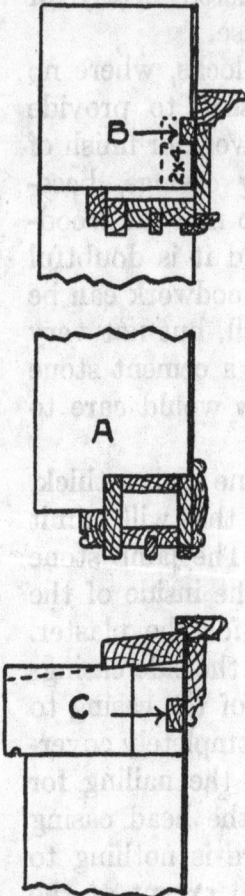

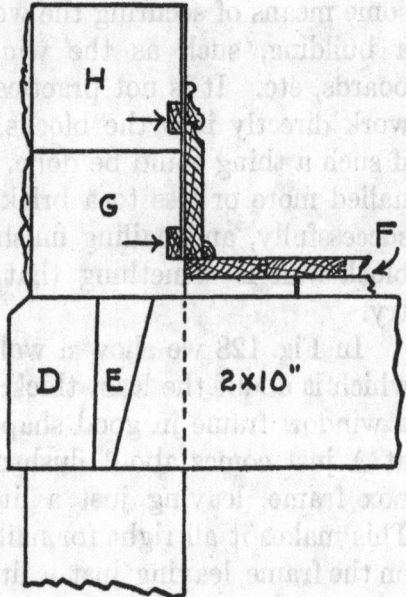

FIG. 128   FIG. 129

1 by 2 furring strip which, if put in tight, would make a sufficient nailing for top of cap. Again, the furring strip could be molded right into the lintel when it was being made, and thus save the time of driving it in after the stone was laid. Again, the lintel stone could be molded with a recess on the lower inside edge just sufficient to take in a piece of 2 by 4, which might, perhaps, be still better, as shown by dotted lines. The window sill shown at C might be backed with a smaller stone molded with a recess to receive a 1 by 2 furring strip to make a nailing place for the bottom of the apron to the window. The stool and top part of the apron could be readily nailed to the wooden sill of the window frame.

In Fig. 129, D represents a plain bevel edge wide base course, the depth of which is the same as the joist used for the floor. E shows the backing stone, which should be made just the right size to fit in between the joists. The floor, which is double, is shown at F. Above are shown two courses of stone, G and H. These two courses are recessed at the proper height to receive furring strips to make nailing places for the base, top and bottom, as shown.

With furring strips driven into these recesses in the stone blocks, a continuous nailing strip is provided for the base, and one that ought to hold the woodwork securely in place. Woodwork cannot be attached to stone walls unless there is some means provided for securing it, and it would seem that this is a practical and inexpensive way should be jointed rounding, while a rip saw should be jointed perfectly straight, although many of them are jointed rounding, and some joint them hollowing.

Fig. 5 illustrates a good way to file a cut-off saw that has gotten into very bad sbape. After

FIG. 5.

the teeth are all made even and the same pitch (or rake), then give it proper bevel.

Fig. 6 shows a saw filed with about the right pitch and bevel for ordinary hardwood.

Fig. 7 shows the proper pitch, bevel and fleam, which is the bevel on the back of the tooth, for ordinary soft wood. It also shows the file, which as you will note should point towards the point of

FIG. 6.

the saw. Not only is that my opinion, but all the best authorities I have ever read on the subject give it the same way; still I am free to admit many good mechanics file just the other way.

Put it in flat way directly under the rafter and flush with the outside of rafters, as shown by A and B in Fig. 130. By doing this all those short pieces which are required in the old way are put

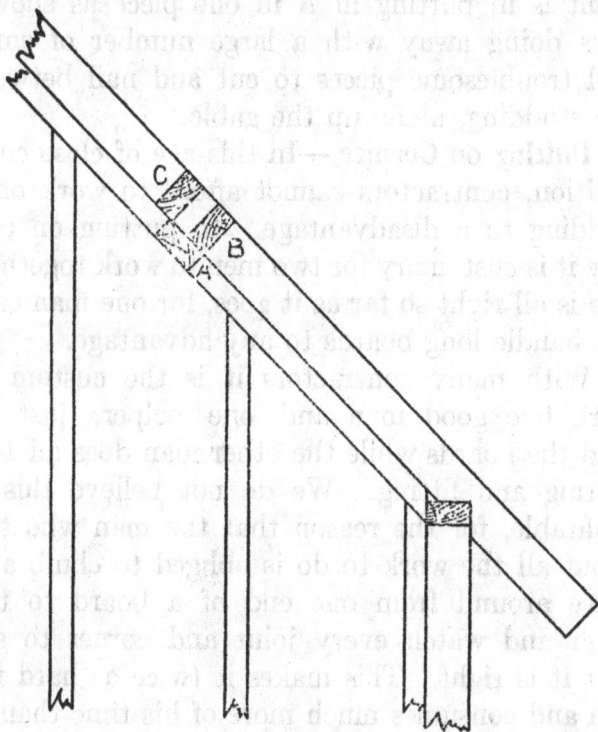

FIG. 130

in at once, with only one cut to make, whereas the old way may require six to eight cuts.

After the 2 by 4, A, is put in under the rafter, then cut in the gable studding in the usual manner. At B is shown a short rafter which need extend only so far as the plastering goes, and which should be set as shown in the sketch to receive the lath that are nailed to the studding. This is not an extra piece, for by the old method it should be there just the same. The improvement is in putting in A in one piece as shown, thus doing away with a large number of small and troublesome pieces to cut and nail between the studding, along up the gable.

Putting on Cornice.—In this age of close competition, contractors cannot afford to work on a building to a disadvantage. In putting on cornice it is customary for two men to work together; this is all right so far as it goes, for one man cannot handle long boards to any advantage.

With many contractors it is the custom to work one good man and one helper, just to hold the boards while the other man does all the cutting and fitting. We do not believe this is profitable, for the reason that the man who has about all the work to do is obliged to climb and chase around from one end of a board to the other and watch every joint and corner to see that it is right. This makes it twice as hard for him and consumes much more of his time than it would if he had a man to help him who was equally as good as himself. Either man could then cut and fit a joint whenever necessary. In this way one would not be obliged to wait for the other so much. In putting on cornice one has to wait for the other more or less, but this is unavoidable.

On the average job we believe that three men working together will accomplish as much as four

men working in pairs. The way to work three
men is to have the man who understands the cutting best, work on the ground, making all the cuts
and passing the boards up to the two men on the
scaffold to nail on. The boards can be passed up
and down and cut on the ground much quicker
and better than they can on the scaffold. The
men on the scaffold should, of course, have a saw
and square and occasionally cut a board for themselves when they can just as well as not; but
mainly let the man on the ground do almost all
of the cutting. If he understands his business he
can make nearly all the cuts right the first cut,
and keep the two men on the scaffold constantly
at work, and there need be but very little loss of
time on account of one waiting for the other. In
our opinion this way of working is far more satisfactory than the way men usually work at putting on cornice.

Part V

ROOFS AND ROOF CONSTRUCTION.
Lean-to or shed
roof — Saddle roof — Simple form of truss — Scissors truss —
Hammer beam — Principles of roofs — King-post roofs — Flat-
pitched roofs — Queen post roofs — Pressure on roofs — To
find dimensions of tie-beam — To find dimensions of king-post
— To find dimensions of struts — To find dimensions of the
queen-post — To find dimensions of a straining beam — To
find dimensions of purlins — To find dimensions of common
rafters — Hip-roofs — Principles to be determined in hip-roofs
— To find backing of a hip-rafter — How to find the shoulder
purlin — To pierce a circular roof, Etc.

It is scarcely necessary to say that the roof of
a building is that covering which is to protect
the inhabitants and their property from the
effects of the weather and that, in addition to this,
it should be so constructed that it may shelter
the walls, foundation and fabric generally from
snow and rain.

Roofs are of various forms and pitches; the
high pitched roofs are more generally found
through the north, as they discharge the rain with
greater facility and the snow lies on their surface
for a much shorter time. When constructed on
sound principles, the roof is one of the principal
ties of a building, as it binds the exterior walls to
the interior and to the partitions; while a badly
designed roof will have the tendency to give way
or to force the walls out of the perpendicular.

The most simple form of roof is that known as the **lean-to** or **shed roof.** This is illustrated in Fig. 131, and it derives its name from the fact that it is the roof usually used on a small annex or shed built against or leaning against the main building.

The roof most in use and also very simple in its construction is the **saddle roof** or **gable roof,** as it is often called.

This is illustrated in Fig. 132, and shows that the roof has a double slope, and the highest point

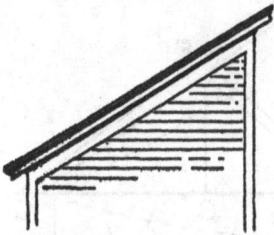

FIG. 131

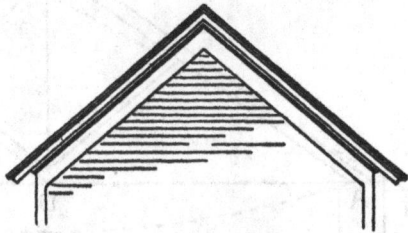

FIG. 132

where they meet is called the ridge of the roof.

Before going into the detailed construction of roofs, it will not be out of place to explain some of the principles involved in roof construction.

In Fig. 133, if AB, CB be two rafters, placed on walls A and C, and meeting in a ridge B, even by their own weight, and much more when loaded, these rafters would have a tendency to spread outwards at A and C, and to sink at B. If this tendency be constrained by a tie established betwixt A and C, and if AB, BC be perfectly rigid and the tie AC incapable of extension, B will

become a fixed point. This, then, is the ordinary
couple roof, in which the tie AC is a third piece of
timber, and which may be used for spans of
limited extent; but when the span is so great that
the tie AC tends to bend downward or sag, by
reason of its length, then the conditions of stability obviously become impaired. Now, if from
the point B a string or tie be let down and attached to the middle D, of AC, it will evidently

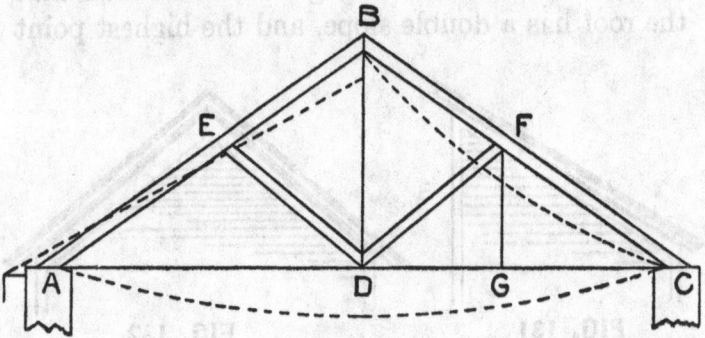

FIG. 133

be impossible for AC to bend downwards so long
as AB, BC remain of the same length: D, therefore, like B, will become a fixed point, if the tie
BD be incapable of extension. But the span may
be increased or the size of the rafters AB, CB be
diminished, until the latter also have a tendency
to sag; and to prevent this, pieces DE, DF remain
unaltered in length. Adopting the ordinary
meaning of the verb "to truss," as expressing to
tie up, we truss or tie up the point D, and the
frame ABC is a trussed frame. In like manner,

F being established as a fixed point, G is trussed
to it.

In every trussed frame there must obviously
be one series of component parts in a state of
compression and the other in a state of extension.
The functions of the former can only be filled by
pieces which are rigid, while the place of the latter
may be supplied by strings. In the diagram the

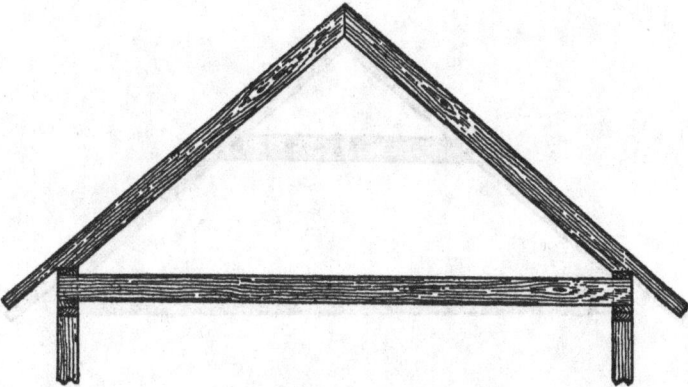

FIG. 134

pieces AB, BC are compressed, and AC, DB are
extended; yet in general the tie DB is called a
king post, a term which conveys an altogether
wrong idea of its duties. Thus we see how the
two principal rafters, by their being incapable of
compression, and the tie beam by its being incapable of extension, serve, through the means of
the king post, to establish a fixed point in the
center of the void spanned by the roof, which prevents the rafters from bending, and serve in the
establishing of other fixed points; and a combi-

nation of these pieces is called a king post roof.

The most simple form of truss is that shown in Fig. 134, and is called the common rafter—so named, we presume, because it is used in all classes of building. When it becomes necessary to add to its strength, the first thing done is to nail on a cross piece, as shown in Fig. 135, commonly called a tie or collar beam. This piece also

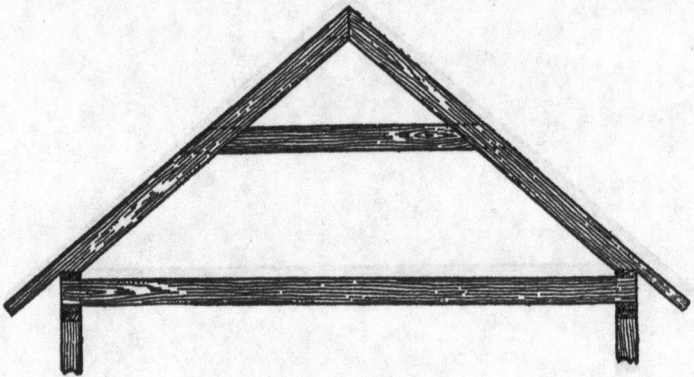

FIG. 135

serves as the ceiling joist where it is desired to finish a room in the attic. Some times a vertical piece is added at the center, as shown in Fig. 136. This, of course, stiffens the truss, but it does not add as much to its strength as is generally supposed. This is a common form used for one and a half story houses. The cross piece has a double purpose here; that it, to keep the side walls from spreading outwards and also forms the ceiling. It can be greatly strengthened by the addition of two extra pieces set brace shape from the center of the collar beam, as shown in Fig. 137. The lower the collar beam is placed the stronger will be the truss, and should not in most cases be placed above one-third the length of the common rafter.

Another form of truss that is generally used for small church buildings is that shown in Fig.

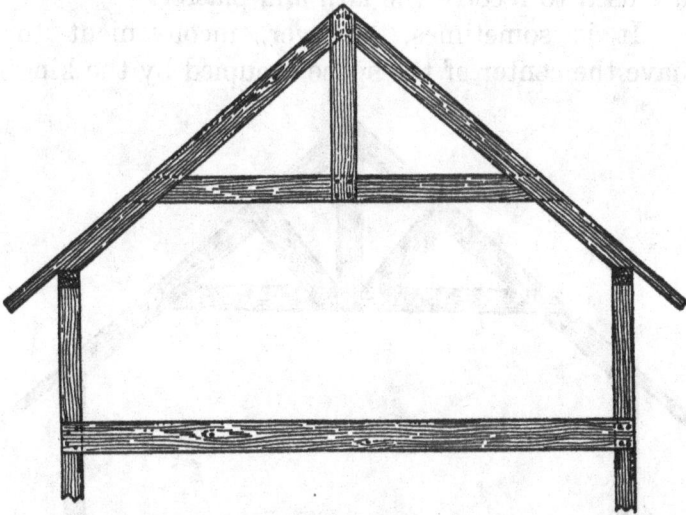

FIG. 136

138, commonly called scissors truss. This is suitable for a building 34 feet wide, shingle roof, and rafters set on 24-inch centers. The timbers required will be 26 feet in length for the common tie rafters and 24 feet for the collar beam. At the seat of the rafters is a 2 by 6-inch piece circled out to form a cove, as shown. This piece should be thoroughly spiked to the studding and to the side of the tie rafter and to the under edge of the common rafter. However, only one-third of these pieces will catch the rafters, owing to their being spaced on 24-inch centers, thus requiring the other pieces to be framed in between the rafters. Other timbers in the truss are 1 by 6-inch fencing plank. All parts should be thoroughly spiked together. Cross pieces of 2 by 2-inch stuff are used to receive the lath and plaster.

It is sometimes, however, inconvenient to have the center of the space occupied by the king

FIG. 137

post, especially where it is necessary to have apartments in the roof. In such a case recourse is had to a different manner of trussing. Two suspending posts are used, and a fourth element is introduced; namely, the straining beam a b (Fig. 139), extending between the posts. The

principle of trussing is the same. The rafters are
compressed, and the tie beam and posts, the latter
now called queen posts, are in a state of tension.

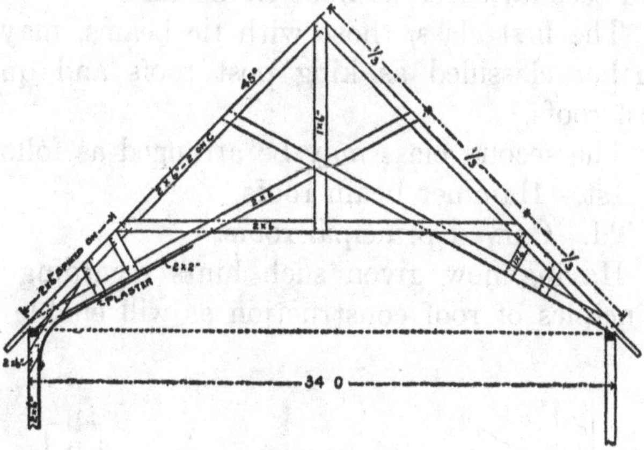

FIG. 138

In some roofs, for the sake of effect, the tie
beam does not stretch across between the feet of
the principals, but is interrupted. In point of

FIG. 139

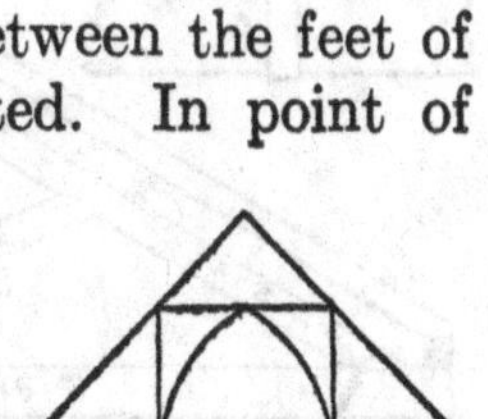

FIG. 140

fact, although occupying the place of, it does not
fill the office of a tie beam, but acts merely as a
bracket attached to the wall (Fig. 140). It is
then called a hammer beam.

The **Principles of Roofs** may, therefore, in respect to their construction, be divided broadly into two classes: First, those with tie beams; and second, those without tie beams.

The first class, those with tie beams, may be further classified as king post roofs and queen post roofs.

The second class may be arranged as follows:

1st. Hammer beam roofs.

2d. Curved principal roofs.

Having now given such hints regarding the principles of roof construction as will enable the

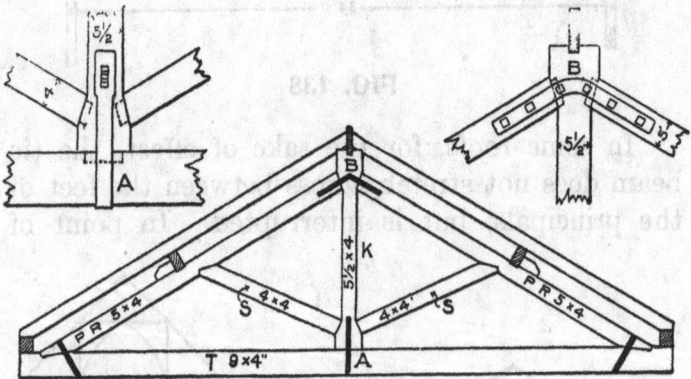

FIG. 141

workman to build any ordinary roof intelligently, we proceed to describe the methods of construction.

King=Post Roofs.—This form of roof is practically the beginning of all trusses, which are complete framings in themselves, spanning from wall to wall, and doing duty for the cross walls, in

that they support, in their turn, the ridge and
purlins which require a bearing every eight or ten
feet.   Trusses should be no more than eight or nine
feet apart and have a nine-inch bearing on each wall.

Fig. 141 represents a king-post roof truss. P R
is the principal rafter, 5 inches deep and 4 inches
thick; T the tie beams 9 by 4 inches; S the struts, 4
by 4 inches; and the king-post K, is 7 by 4 inches; the
cuts to give a bearing for the struts are also shown.

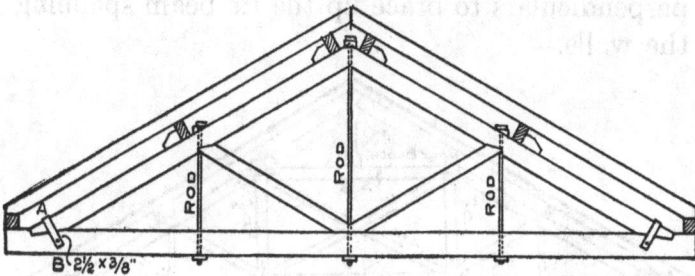

FIG. 142

Flat=Pitched Roofs are not so strong as those
that are pitched higher.   The nearer to the perpendicular that wood is fixed the stronger it is.
This is shown by the fact that the horizontal
thrust of a pair of rafters is proportionate to the
length of the oblique line drawn, at right angles
from the foot of the rafters, to the perpendicular
dropped from the apex.

The joints of a king-post truss, in fact, all
consist of mortises and tenons entering but a short
distance into the timbers; and they have all
beveled shoulders which ought, wherever possible,
to be at right angles to the incline of the roof.

Fig. 142 is the joint between king-post and principal rafters at the apex supporting the ridge, a pair of 2½ by ⅜-inch wrought iron straps, bolted from side to side, completing the joint; or, in practice, a through-bolt from A to B will answer the same purpose though not so good. King-post trusses are suitable for spans up to thirty feet.

Queen=Post Roofs.—Queen-post trusses are used for spans over thirty feet, and contain two perpendiculars to brace up the tie beam spanning the walls.

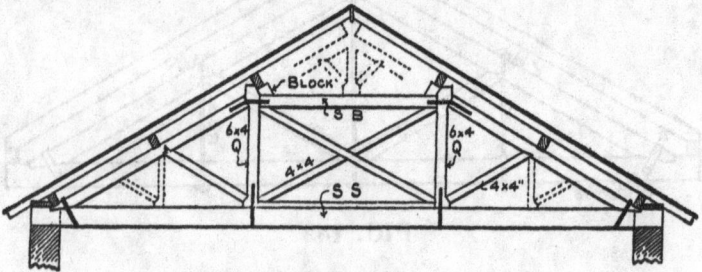

FIG. 143

Fig. 143 is a queen-post truss for a thirty-two feet span. The same form is suitable up to about forty-two feet span, and beyond that size princesses or intermediate posts and struts have to be inserted between the queens and the heels of the roof, as shown by the dotted lines; and it is sometimes necessary to frame a small king-post truss (also shown on the figure by dotted lines) above the straining beam SB, to support the ridge. SS is the straining sill and Q the queens. The other members are known by the same names as in other trusses.

A good rule to ascertain the thickness of queen post trusses is as follows: Divide the span by 8, and the quotient will be the required thickness (in inches), making up for odd parts by adding another inch for heavy-tiled roofs and omitting such fractions for slates.

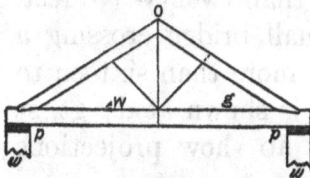

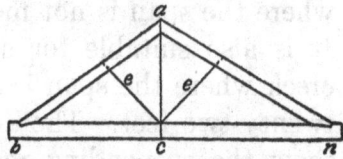

FIG. 144   FIG. 145

Taking the tie-beam for thirty-two feet span at 11 inches deep, and principal rafters at 6 inches, by adding one inch for every five feet additional span we can arrive at their depth for the different roofs.

In Fig. 144 we show a roof that is at once

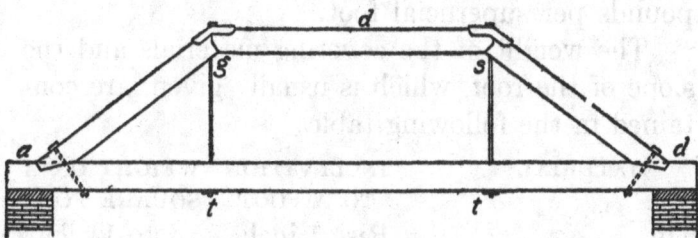

FIG. 146

strong and cheap for spans from twenty to thirty feet. pp shows the wall plates, w the wall, o the ridge and head of suspending rod; W and g show where the suspending rods may be placed if the span exceeds twenty-five feet.

Fig. 145 shows a roof with unequal sides; ac shows the suspending rod; ee may be braces of wood or rods of iron; b and n are resting points. This is suitable for a span from twenty to thirty feet.

Fig. 146 is suitable for a roof with a deck and where the span is not more than twenty-five feet. It is also suitable for a small bridge crossing a creek where the span is not more than sixteen to twenty-two feet. The deck is shown at d; gt, st show the suspending rods; ab show projections for gutters and ease-offs.

In estimating the pressure on the roof, for the purpose of apportioning the proper sizes of timber to be used, not only the weight of the timber and the slate or other covering must be taken, but also the weight of snow which, in severe climates, may be on its surface, and also the force of the wind, which we may calculate at forty pounds per superficial foot.

The weight of the covering materials and the slope of the roof, which is usually given, are contained in the following table:

MATERIAL	INCLINATION TO A FOOT	WEIGHT ON A SQUARE FOOT
Tin.................	Rise 1 inch	$\frac{5}{8}$ to $1\frac{1}{4}$ lbs.
Copper..............	" 1 "	1 to $1\frac{1}{2}$ lbs.
Lead...............	" 2 "	4 to 7 lbs.
Zinc...............	" 3 "	$1\frac{1}{4}$ to 2 lbs.
Short pine shingles....	" 5 "	$1\frac{1}{2}$ to $2\frac{1}{2}$ lbs.
Long cypress shingles.	" 6 "	4 to 5 lbs.
Slate...............	" 6 "	5 to 9 lbs.

With the aid of this table, and taking into account the pressure of the wind and the weight of snow, the strength of the different parts may be calculated from the following rules, which were deduced by Mr. Tredgold, from experience; they are easy of application and useful for simple cases. Mr. Tredgold assumes 66½ lbs. as the weight on each square foot. It is customary to make the rafters, tie-beams, posts and struts all the same thickness.

To find the dimensions of the principal rafters in a king=post roof of pine timber:

Rule:—Multiply the square of the length in feet by the span in feet, and divide the product by the cube of the thickness in inches; then multiply the quotient by .96 to obtain the depth in inches.

Mr. Tredgold gives also the following rule for the rafters as more general and reliable:

Multiply the square of the span in feet by the distance between the principals in feet, and divide the product by 60 times the rise in feet; the quotient will be the area of the section of the rafter in inches.

If the rise is one-fourth of the span, multiply the span by the distance between the principals, and divide by 15 for the area of section.

When the distance between the principals is 10 feet, the area of section is two-thirds of the span.

To find the dimensions of the tie=beam, when it has to support a ceiling only.

Rule.—Divide the length of the longest unsupported part by the cube root of the breadth, and the quotient multiplied by 1.47 will give the depth in inches.

To find the dimensions of the king-post:

Rule.—Multiply the length of the post in feet by the span in feet; multiply the product by .12. which will give the area of the section of the post in inches. Divide this by the breadth for the thickness, or by the thickness for the breadth.

To find the dimensions of struts:

Rule.—Multiply the square root of the length supported, in feet, by the length of the strut in feet, and the square root of the product multiplied by .8 will give the depth; which multiplied by .6 will give the thickness.

In a Queen-Post Roof.

To find the dimensions of the principal rafters:

Rule.—Multiply the square of the length in feet by the span in feet, and divide the product by the cube of the thickness in inches; the quotient multiplied by .155 will give the depth.

To find the dimensions of the tie-beam:

Rule.—Divide the length of the longest unsupported part by the cube root of the breadth, and the quotient multiplied by 1.47 will give the depth.

To find the dimensions of the queen-posts:

Rule.—Multiply the length in feet of the post by the length in feet of that part of the tie-beam it supports: the product multiplied by .27 will give the area of the post in inches; and the breadth and thickness can be found as in the king-post. The dimensions of the struts are found as before.

To find the dimensions of a straining beam:

Rule.—Multiply the square root of the span in feet by the length of the straining beam in feet, and extract the square root of the product; multiply the result by .9, which will give the depth in inches. The beam, to have the greatest strength, should have its depth to its breadth in the ratio of 10 to 7; therefore, to find the breadth, multiply the depth by .7.

To find the dimensions of purlins:

Rule.—Multiply the cube of the length of the purlin in feet by the distance the purlins are apart in feet, and the fourth root of the product will be the depth in inches, and the depth multiplied by .6 will give the thickness.

To find the dimensions of the common rafters when they are placed 12 inches apart:

Rule.—Divide the length of bearing in feet by the cube root of the breadth in inches, and the quotient multiplied by .72 will give the depth in inches.

Beams acting as struts should not be cut into or mortised on one side, so as to cause lateral yielding.

Purlins should never be framed into the principal rafters, but should be notched. When notched they will carry nearly twice as much as when they are framed.

Purlins should be in as long pieces as possible.

Rafters laid horizontally are very good in construction and cost less than purlins and common rafters.

The ends of tie-beams should be kept with a free space around them to prevent decay.

It is an injudicious practice to give an excessive camber to the tie-beam; it should only be drawn up when deflected, as the parts come to their bearings.

The struts should always be immediately underneath that part of the rafter wheron the purlin lies.

The diagonal joints of struts should be left a little open at the inner part to allow for the shrinkage of the heads and feet of the king and queen posts.

It should be specially observed that all cranks or bends in iron ties are avoided.

And, as an important final maxim, every construction should be a little stronger than strong enough.

Hip=roofs.

The principles to be determined in a hip roof are eight, namely:

- 1st.　Span or width of building to be roofed.
- 2d.　Run of the building, which is one-half the span.
- 3d.　The rise given the common rafter.
- 4th.　The angle that the common rafter makes with the level of the plate; that is, the pitch of the roof.
- 5th.　The length of the common rafter.

6th. The angle that the hip-rafter makes with
       the adjoining sides of the roof.

7th. The length of the hip-rafters.

8th. The distance from the corner of the build-
       ing to the center line of the first jack;
       that is, the common difference.

The 1st, 2d and 3d being given the others may
be found, as will be shown in the following illustrations:

Let ABCD, Fig. 147, be the plan of the roof.
Draw GH parallel to the sides, AD, BC, and in

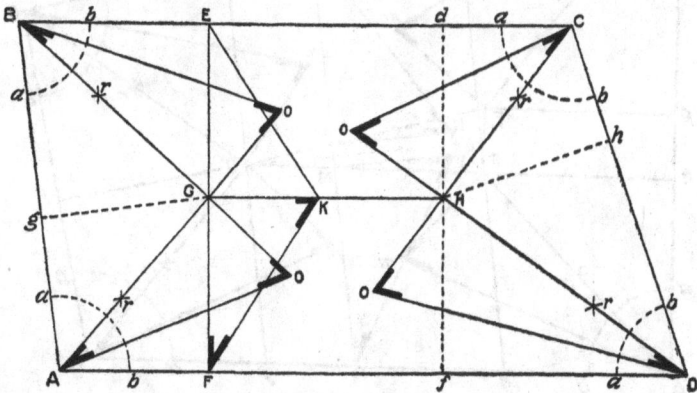

FIG. 147

the middle of the distance between them. From
the points A, B, C, D, with any radius, describe
the curves ab, ab, cutting the sides of the plane
at a, b. From these points, with any radius,
bisect the four angles of the plan at r, r, r, r, and
from A, B, C, D, through the points, r, r, r, r, draw
the lines of the hip rafters, AG, BG, CH, DH,
cutting the ridge line, GH, in G and H, and pro-

duce them indefinitely. The cross lines, EF, df, are the seats of the last entire common rafters. Through point in the ridge-line, make GK equal to the height or rise of roof, and join EK, FK; then EK is the length of the common rafter. Make Go, Ho, equal to GK, the rise of the roof, and join Ao, Bo, Co, Do, for the length of the hip

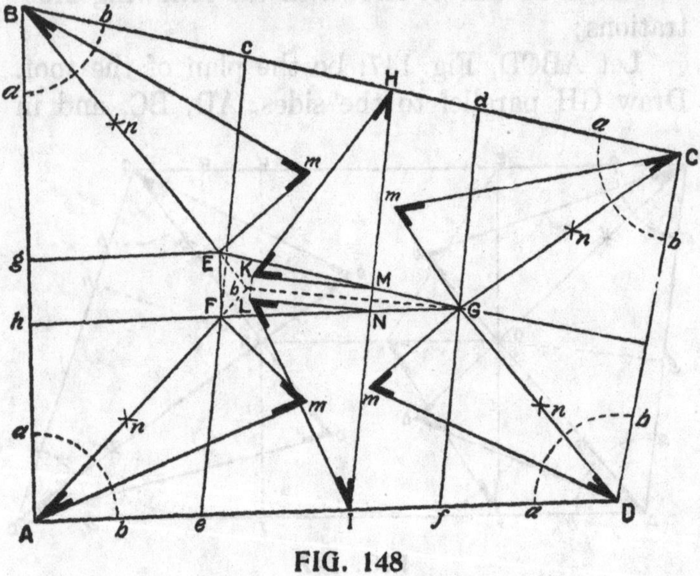

FIG. 148

rafters. If the triangles, AoG, Bog, be turned round their seats, AG, BG, until their perpendiculars are perpendicular to the plane of the plan, the points, OO, and the lines Go, Go, will coincide, and the rafters, Ao, Bo, be in their true positions.

If the roof is irregular, and it is required to keep the ridge level, we proceed as shown in Fig. 148.

Bisect the angles of the two ends by the lines Ab, Bb, CG, DG, in the same manner as in Fig. 147; and through G draw the lines GE, GF, parallel to the sides, CB, DA, respectively, cutting Ab, Bb, in E and F; join EF; then the triangle EGF, is a flat, and the remaining triangle and

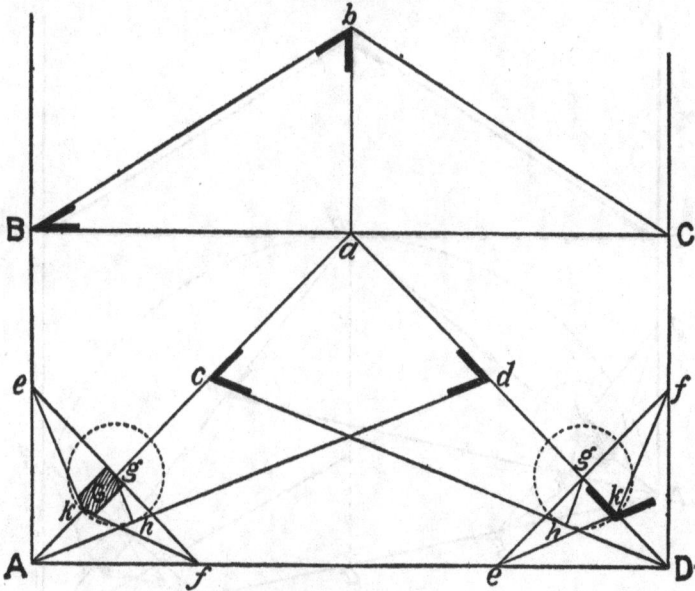

FIG. 149

trapeziums are the inclined sides.   Join Gb, and draw HI perpendicular to it; at the points M and N, where HI cuts the lines GE, GF, draw MK, NL perpendicular to HI, and make them equal to the rise; then draw HK, IL for the lengths of the common rafters. At E, set up Em perpendicular to BE; make it equal to MK or NL, and join Bm, for the length of the hip rafter, and proceed in the same manner to obtain Am, Cm, Dm.

To find the backing of a hip rafter, when the plan is right-angled, we proceed as shown in Fig. 149. Let Bb, bC be the common rafters, AD the width of the roof, and AB equal to one-half the

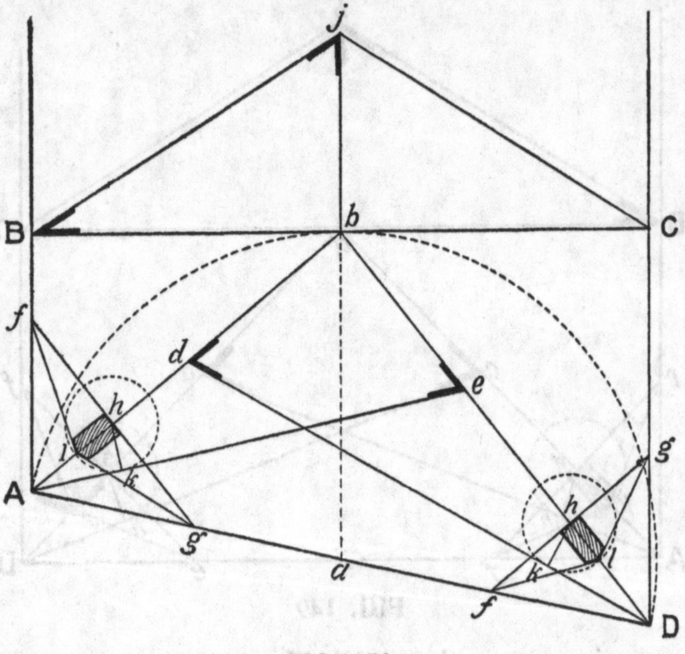

FIG. 150

width. Bisect BC in a, and join Aa, Da. From a set off ac, ad equal to the height of the roof ab, and join Ad, Dc; then Ad, Dc are the hip rafters. To find the backing from any point h in Ad, draw the perpendicular hg, cutting Aa in g; and through g draw perpendicular to Aa the line ef, cutting

AB, AD in e and f. Make gk equal to gh, and join ke, kf; and the angle ekf is the angle of the backing of the hip rafter C.

Fig. 150 shows the method of **obtaining the backing of the hip** where the plan is not right-angled.

Bisect AD in a, and from a describe the semi-circle, AbD; draw ab parallel to the sides AB,

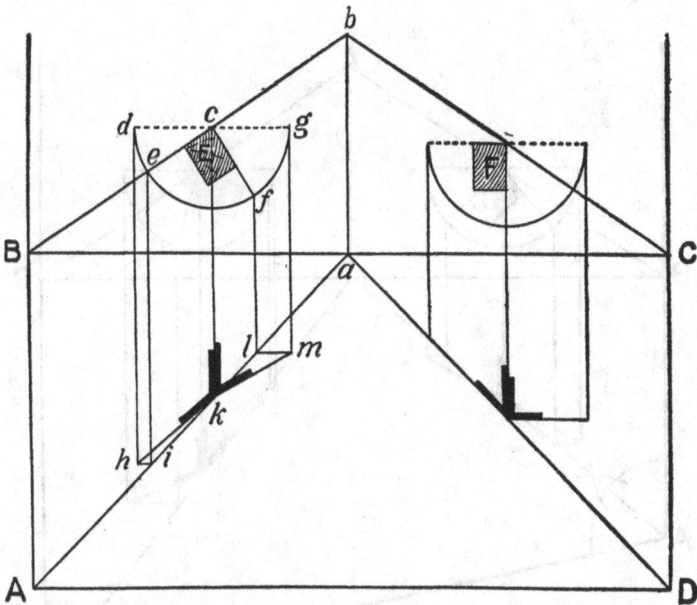

FIG. 151

DC, and join Ab, Db, for the seat of the hip rafters. From b set off on bA, bD, the lengths bd, be, equal to the height of the roof bc, and join Ae, Dd, for the lengths of the hip rafters. To find the backing of the rafter: In Ae, take any

point k, and draw kh perpendicular to Ae.
Through h draw fhg perpendicular to Ab, meeting AB, AD, in f and g. Make hl equal to hk,
and join fl gl; then fl, gl is the backing of the
hip.

Fig. 151 shows **how to find the shoulder purlins:**

First, where the purlin has one of its faces in
the plane of the roof, as at E.   From c as a center,

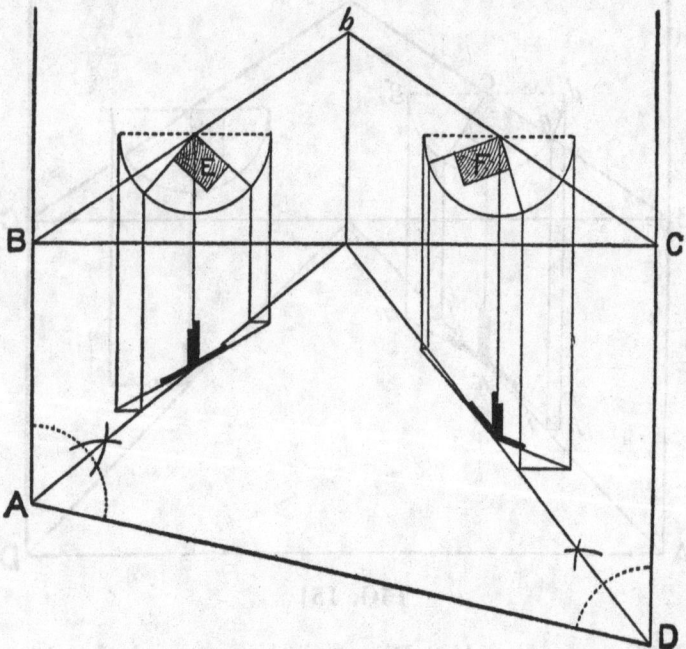

FIG. 152

with any radius, describe the arc dg; and from
the opposite extremities of the diameter draw dh,
gm, perpendicular to BC. From e and f, where

the upper adjacent sides of the purlin produced
cut the curve, draw ei, fl parallel to dh, gm; also
draw ck parallel to dh. From l and i draw lm
and ih parallel to BC, and join kh, km. Then
ckm is the down bevel of the purlin and ckh is
its side bevel.

When the purlin has two of its sides parallel
to the horizon, it is worked out as shown at F.
It requires no further explanation.

When the sides of the purlin make various
angles with the horizon, Fig. 152 shows the application of the method.

It sometimes happens, particularly in rail-
road buildings, that the carpenter is called upon
**to pierce a circular or conical roof with a saddle
roof,** and to accomplish this economically is often
the result of much labor and perplexity if a correct method is not at hand.

The following method, shown in Fig. 153, is an
excellent one and will, no doubt, be found useful
in cases such as mentioned:

Let DH, FM be the common rafters of the
conical roof, and KL, IL the common rafters of
the smaller roof—both of the same pitch. On
GH set up Ge equal to ML, the height of the
lesser roof, and draw ed parallel to DF, and from
d draw cd perpendicular to DF. The triangle
Ddc will then, by construction, be equal to the
triangle KLM, and will give the seat and the
length and pitch of the common rafter of the
smaller roof B. Divide the lines of the seats in
both figures, Dc, KM, into the same number of

equal parts; and through the points of division in E, from G as center, describe the curves ca, 2g, 1f, and through those in B draw the lines 3f, 4g, Ma, parallel to the sides of the roof and inter-

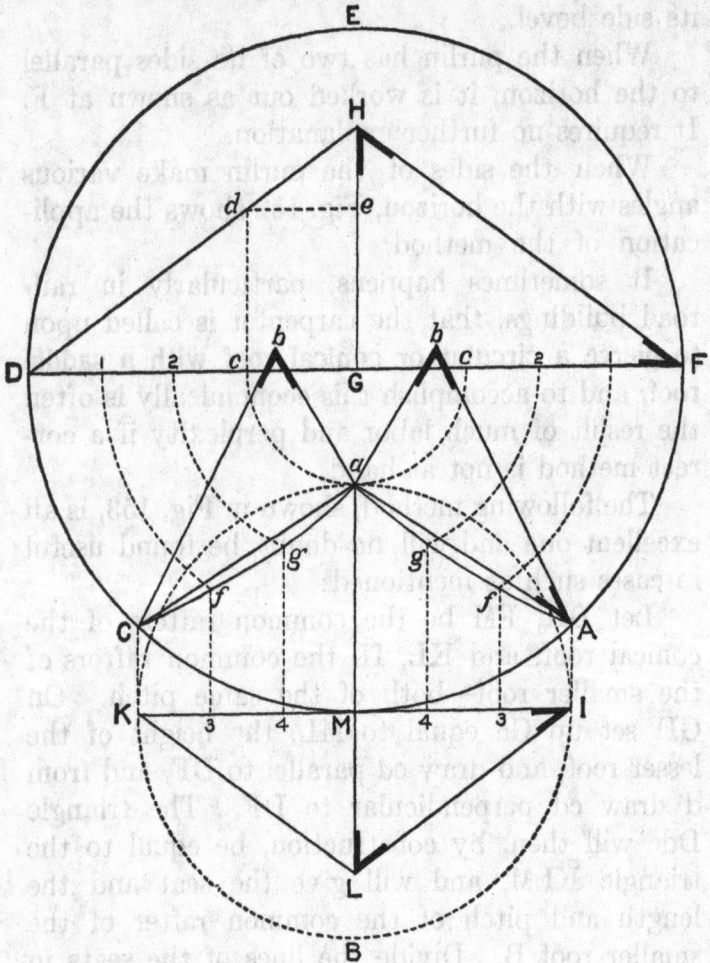

FIG. 153

secting the curves in fga. Through these points trace the curves Cfga, Afga, which give the lines of interesction of the two roofs. Then to find the valley rafters, join Ca, Aa; and on a erect the lines ab, ab perpendicular to Ca and Aa, and make them respectively equal to ML; then Cb, Ab is the length of the valley rafter.

Fig. 154 shows **a section of a mansard roof with concave sides,** and the manner of framing the same when it is to be erected on a brick or stone building. Pc is the wall; c the wall plate; AB the floor joist; hi is the side rafter; aie the ceiling joist; ao the top rafter; Bbd the bracket to nail cornice to; b the gutter, and ri the studding, which will be required if it is desirable to finish the roof story for sleeping rooms.

The wall plate is made of two thicknesses of two-inch plank nailed together and lap-jointed at the ends. The joints should receive the longitudinal piece h, and the ends of each should be sawed off square at or near the dotted line k. They should then be put into place, nailed to the wall plate, and the piece h should be firmly nailed to each. The lower end of the side rafters is cut out at the toe to rest on the piece h. The upper ends are also cut to receive the piece i, to which they should be firmly nailed.

If it is required to lath and plaster on the ceiling joists, they should be notched to rest on the piece i; but if the room is to remain rough, it will be as well to nail beveled pieces on each, as shown by the dotted line at s. The end of each ceiling joist should be sawed in shape to receive the moulding a, with which it is usual to finish the upper part of the roof. The top rafters may rest either on a longitudinal piece laid on the ceiling joists or on the piece i—the latter being the better method.

The curved portions of the side rafters are made separate from the straight part and are

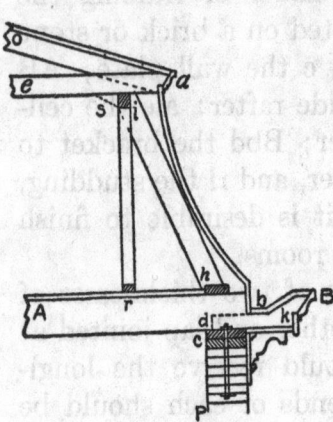

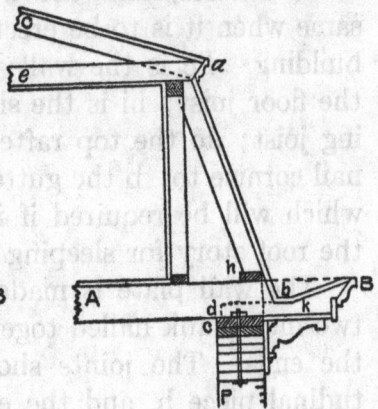

FIG. 154   **FIG. 155**

most generally formed of two thicknesses of inch stuff, first sawed the right shape and nailed together, and then spiked to the straight part of the rafter. When so much of the roof has been put up, it will be as well to mark on the end of the floor joists the proper depth for the gutter. This will be best done by holding a straight-edge on the ends of the joists, with incline sufficiently to allow water to run off, and marking on each joist the depth it will require to be cut down.

The vertical part of the gutter is cut down in a line with the lower ends of the side rafters. The cornice brackets, which are cut of a shape suitable for receiving the different parts of the cornice, are made of inch stuff and are nailed to the floor joists, as shown by the dotted lines and nailmarks at dk. The best method to pursue in putting them up is to first nail one on to the joist at either extremity of the roof, then stretch a line tight between the same points on each, and nail up the intervening brackets, with the same points touching the line. If the line is tightly stretched and proper care is taken in nailing up the brackets the cornice will be perfectly straight.

In Fig. 155 we have a section of a similar roof with straight sides. The different parts are lighter than those of Fig. 154, and the construction is adapted for a balloon frame building. The letters in Fig. 155 denote the same parts as the same letters in Fig. 154, and the explanation of Fig. 154 will answer for Fig. 155 so far as the same letters are concerned. Pc is the balloon frame studding; c, a longitudinal piece for the floor joists to rest upon. The studs are cut out at the top to receive the piece c, and will thus prevent the frame from spreading.

Since there is no curve on the rafter, the face of it comes flush with the inside of the gutter. Hence the side rafters are cut out at the heel to rest on the piece h, instead of the toe, as in Fig. 154. The piece h is beveled in order that the thrust on the side rafters shall not throw the

lower ends out. The inside of the gutter is also
made inclining so as to give as much substance
as possible between the gutter and the piece h.
The remaining parts are the same as those in

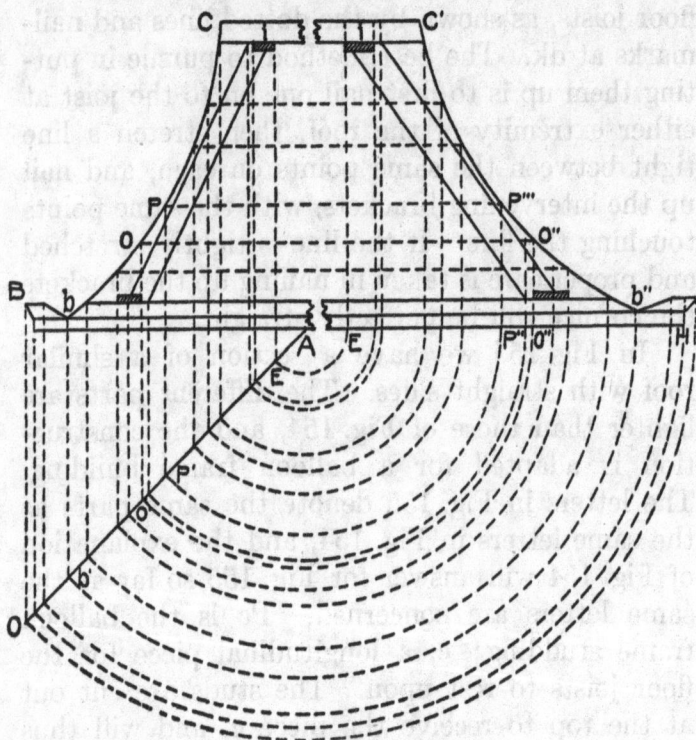

FIG. 156

Fig. 154, and the same description of those parts
will answer for both cuts.

Fig. 156 shows how to find the angle=rafter
and angle=cornice bracket, when the section as
above described has been drawn. Let ABc repre-

sent the given section on the drafting board or
floor. Draw the line AO at an angle of 45 degrees
with AF. Then from any points C, P, O, etc., of
the section as shown, draw lines perpendicular to
AF, and intersecting AO.

In order to transfer the distances AE, AP, etc.,
on AO to AH, it is most convenient, in our small
illustration, to describe arcs with A as a center;
but in practice, since the distance AO will be
several feet, it will be best to lay a straight edge
along the line AO, and mark the points A, E, P,
etc., on it; then change the position of the straight-
edge and lay it along AH, the point before on A
being made to coincide with it again, and transfer
the marks to the floor or board on the line AH
at E', P'', etc. When this has been done, draw
lines from these marks and perpendicular to AH.
Now draw lines from the points C, P, O, etc., on
the section ABC, but parallel to FH, and intersecting the lines which are perpendicular to AH.
Note the interesection of any two of these lines
which are produced from the same point of the
section, and this intersection will be the similar
point of the angle-rafter. Perhaps the subject
will be better understood if we follow the details
of finding a single point of the angle-rafter; such
for instance, as that corresponding to the point P
of the given section. From P draw PP' perpendicular to AF, and intersecting AO at P'. Make
the distance AP'' on AH equal to AP' on AO,
either by describing an arc with A as a center
and AP' as a radius, or by transferring the point

P' to P" on a straight-edge, as before stated.
From P" draw P" P"' perpendicular to AH.
Then from P on the section draw a line PP"'
parallel to FH, until it intersects the line P" P"'
in the point P"'. This point P"' will be the
point of the angle-rafter corresponding to the
point P of the section. After finding all the points
in a similar manner, they must be joined by the

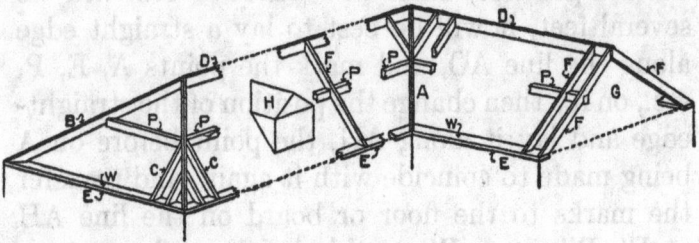

FIG. 157

requisite curved line, and a pattern rafter cut to
fit. It will be apparent from inspection that the
angle bracket is found in the same manner.

Details of Roofs. Hips are the external angles
by the junction of the roof and its return round the
ends, where the walls are not carried up to the
underside of the rake of roof to form gables.

Valleys are the converse of hips, and jack-
rafters are the short rafters which run between
the hips or valleys and the wall-plates.

Dormers are gables on a small scale, or vertical windows placed on the incline plane of the
roof.

Fig. 157 illustrates a returned roof, with hips
and a valley, showing generally the positions of

the parts herein before described, as particularized in the accompanying reference—A, valley; B, hip; C, jack rafters; D, ridge; E, eaves; F, common rafters; G, gable; H, dormer; P, purlins; W, wall plate.

Octagonal Roofs.—Fig. 158 represents an octagonal roof. In its construction the following suggestion on laying out an octagon must be referred to.

To find the side of an octagon when the side of the square is given: Multiply the side of the

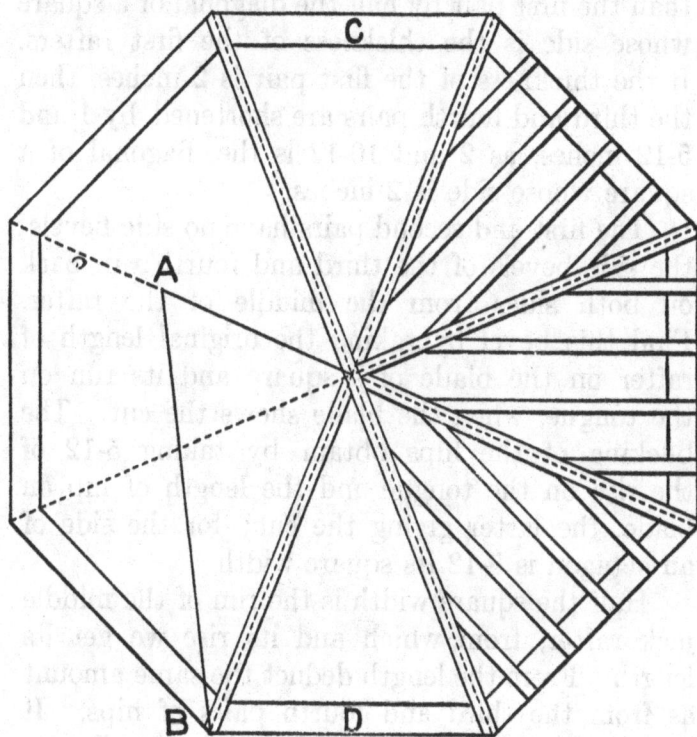

FIG. 158

square by 4.97 and divide by 12. The quotient is the side of the inscribed octagon.

To continue with an octagon roof: the length of hips is found as usual from the rise and run; the run being half the diagonal of the octagon.

Cut the first pair full length to butt against each other; the next pair are to be cut up at right angles to those, and each is to be cut shorter than the first pair by half the thickness of first pair, measured square back from the down bevel. The third and fourth pairs are to be cut shorter than the first pair by half the diagonal of a square whose side is the thickness of the first rafters. If the thickness of the first pair is 2 inches, then the third and fourth pairs are shortened by 1 and 5-12 inches, as 2 and 10-12 is the diagonal of a square whose side is 2 inches.

The first and second pairs have no side bevels; the side bevels of the third and fourth run back on both sides from the middle of the rafter. Find this bevel by taking the original length of rafter on the blade of a square and its run on the tongue, when the blade shows the cut. The backing of the hips obtain by taking 5-12 of the rise on the tongue and the length of hip on blade, the latter giving the cut; for the side of an octagon is 5-12 its square width.

Half the square width is the run of the middle jack rafter, from which and its rise we get its length. From the length deduct the same amount as from the third and fourth pairs of hips. If there are to be two jacks between the middle one and the corner, we divide the length of the side into three parts, also the rise, whence are obtained as before the distance of rafters apart, and the rise of the shortest jack. Divide half the square width of octagon by 3 to find the run of shortest jack. Just as the square is laid on to find the length of a jack, it gives the down and lower end bevels; while the side bevel is obtained by taking the length of the middle jack on blade and half one side of the octagon on the tongue; the blade giving the cut.

The following illustrations and reading matter are from Mr. Woods, and are largely self-explanatory. They show the lengths and bevels of hips and jacks of an octagonal tower

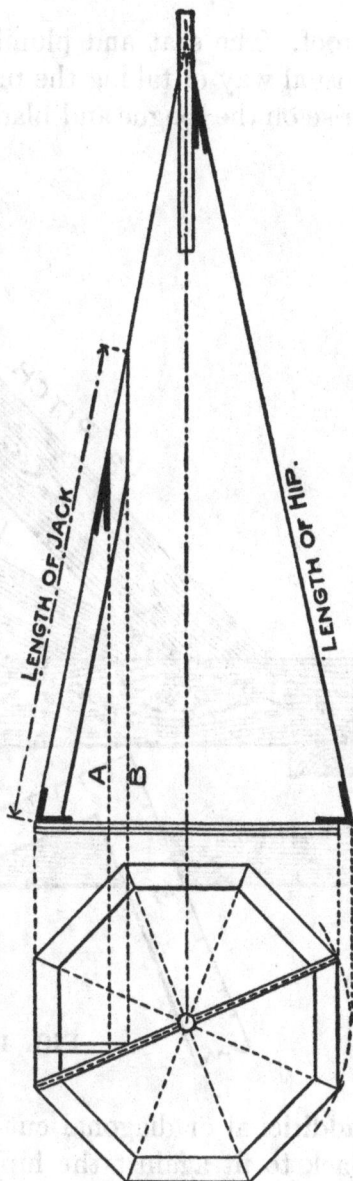

FIG. 159

roof. The seat and plumb cuts are found in the usual way of taking the proportion of the run and rise on the tongue and blade, but there must be an

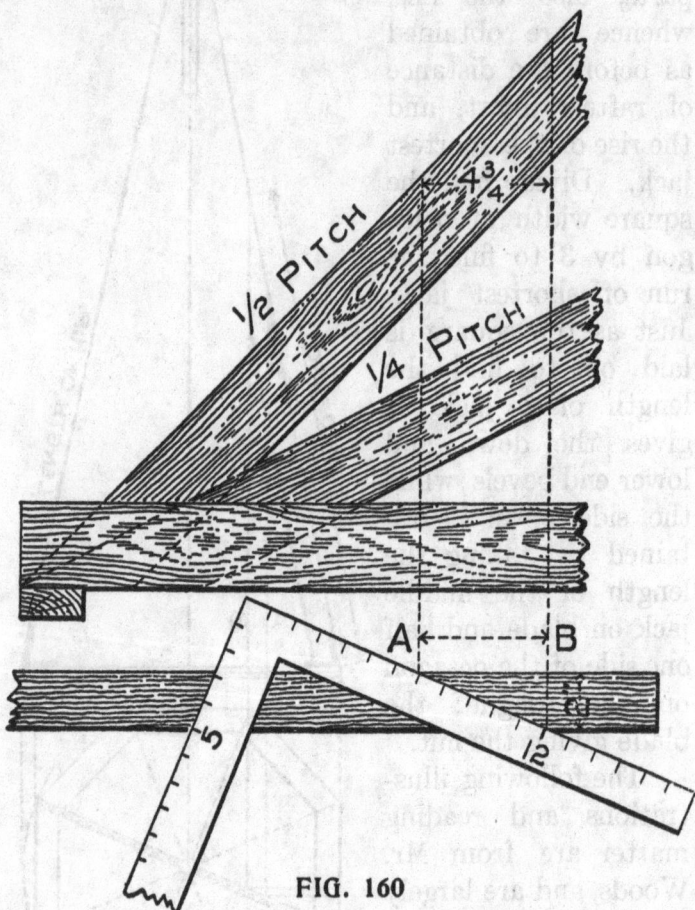

FIG. 160

additional or diagonal cut across the back of the jack to fit against the hip, as shown in Fig. 159, by the dotted lines at A-B. These lines are always vertical and the same distance apart regardless of the pitch given. The diagonal line from A to B across the back of the jack determines the angle. Fig. 160 illustrates this point. If there was no pitch at all then 5 and 12 would give the cut. These figures also give the starting lines A and B which, since the rafters are of the same thickness, will remain at right angles the same distance apart. Thus, if the rafter be 2

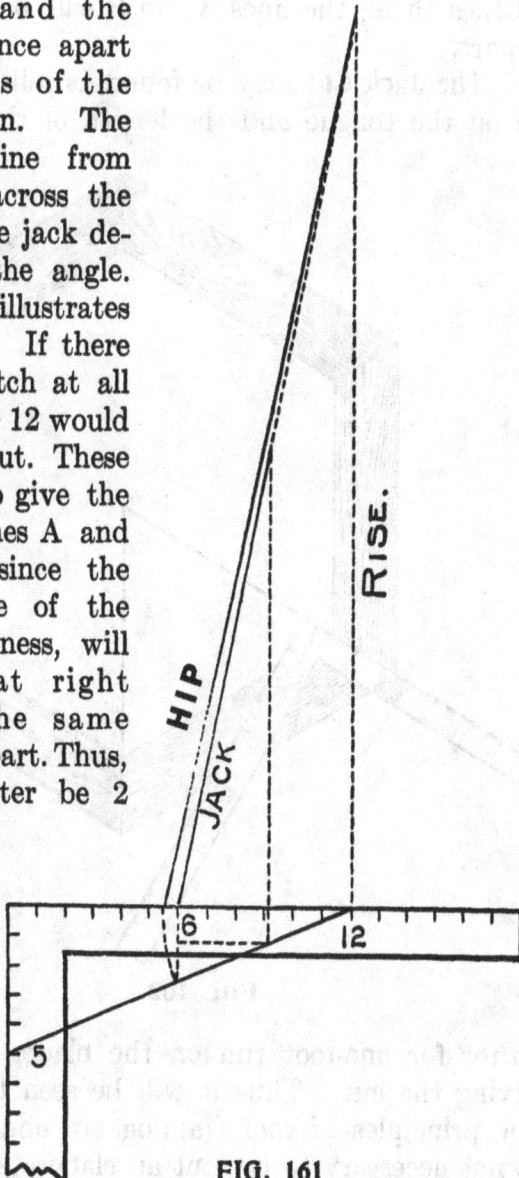

RISE.

HIP

JACK

6

12

5

FIG. 161

inches thick, the lines A and B will be $4\frac{3}{4}$ inches apart.

The Jack Cut may be found as follows: Take 5 on the tongue and the length of the common

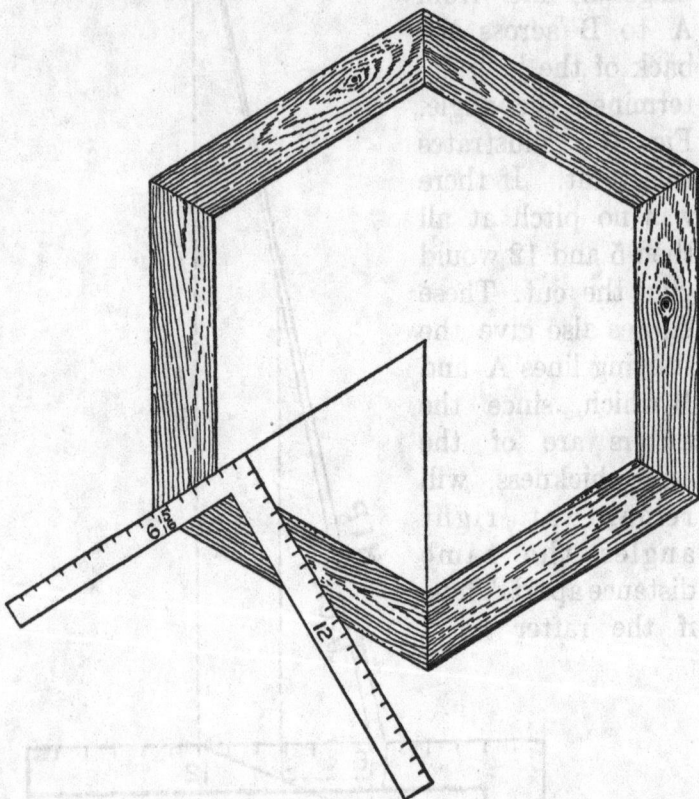

FIG. 162

rafter for one-foot run on the blade, the blade giving the cut. Thus it will be seen that when the principles of roof framing are understood it is not necessary to lay out an elaborate diagram.

A simple line drawing, like that shown in Fig. 161, illustrates all that is contained in Fig. 159.

Hexagonal Roofs.—To make a hexagon, in Fig. 162, take 6 and 15-16 on the tongue and 12 inches on the blade, and apply as shown on a base line which forms a square around the figure.

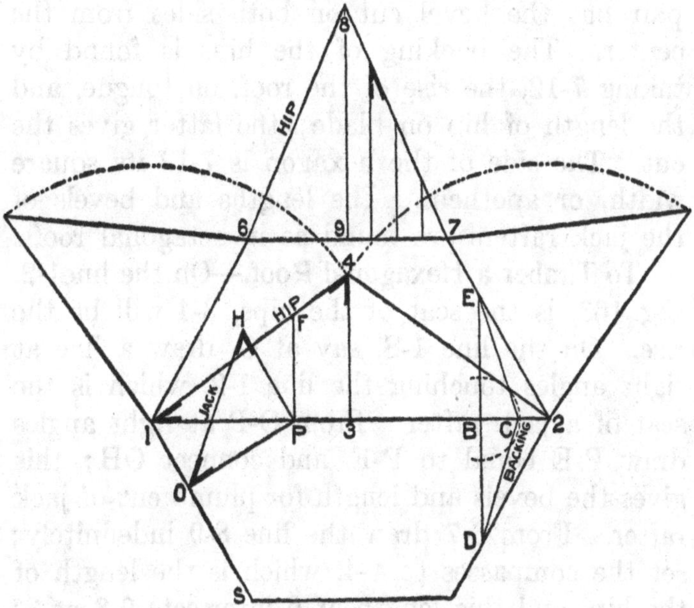

FIG. 163

The side of a hexagon equals the radius of the circumscribing circle. The square width is determined from two parallel sides; a diagonal of the figure is a line from opposite angles.

The first pair of hips are set up as in an octagonal roof. The second and third pairs have a side bevel. To find this, take half the side of the hexagon on the tongue, and half the square width added to the gain of the hip rafter in running that distance, on the blade. The tongue gives the cut. Strike the bevel across the rafter. Now, the second and third pairs are to be measured back shorter than the first pair, on their middle lines, just half the length of this bevel. The third pair has the bevel cut on both sides from the center. The backing of the hips is found by taking 7-12, the rise of the roof, on tongue, and the length of hip on blade; the latter gives the cut. The side of the hexagon is 7-12 its square width, or apothem. The lengths and bevels of the jack rafters are found as in octagonal roofs.

To Timber a Hexagonal Roof.—On the line 1-2, Fig. 163, is the seat of the hips, 3-4 will be the rise. On the line 1-S, say at O, draw a line at right angles touching the line 1-2, which is the seat of a jack rafter. From O-P at right angles draw P-E equal to P-F, and connect OH; this gives the bevels and length for plumb cut of jack rafter. From 6-7 draw the line 8-9 indefinitely; set the compasses to 1-4, which is the length of the hip, and this length at 6 intersects 9-8 at 8; repeat from 7, which is the covering for one side of the roof, the intermediate lines being the lengths of jack rafters and bevels for side cut. On the line 1-2, say at B, take B for a center, touching the line 4-2, for radius, describe the arc BC, through B, at right angles to 1-2, draw the line DE, and from C to E and C to D will be found for backing the hips.

Part VI

A Queen Rafter and How it is Used

Question: Enclosed you will find a sketch of what we call a queen rafter. It is used for barns up to 34 feet wide without putting in purlins.

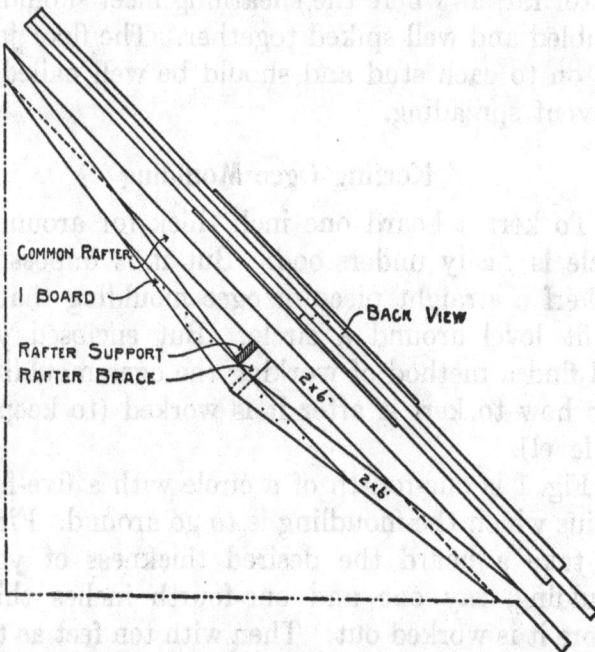

COMMON RAFTER

I BOARD

BACK VIEW

RAFTER SUPPORT
RAFTER BRACE

2x6

2x6

Would like to know which is the best way to support a plain roof on a square pitch barn 30 feet wide, the floor being 6½ feet below the plate.

Answer: The accompanying illustration is prepared from the sketch, and while we believe it possesses considerable merit for stiffening the roof, there is nothing in it to keep the sides of the building from spreading, as it does not form a tie, which is necessary when the loft floor is so far below the plate. We would recommend for a barn of this width (30 feet), to use 2 by 6 rafters set on 24-inch centers, and put on the sheathing diagonally toward the center close and well nailed. The center rafters where the sheathing meet should be doubled and well spiked together. The floor joist lap on to each stud and should be well nailed to prevent spreading.

Kerfing Ogee Moulding

To kerf a board one inch thick for around a circle is easily understood. But it is impossible to kerf a straight piece of ogee moulding that is to fit level around a circle. But enclosed you will find a method of working the ogee moulding, also how to kerf it after it is worked (to keep it up level).

Fig. 1 is one-fourth of a circle with a five-foot radius which the moudling is to go around. First, we take a board the desired thickness of your moulding, say one and one-fourth inches thick before it is worked out. Then with ten feet as the radius describe the arc a-b, Fig. 2, and measure in the width of your moulding, and with same radius describe another arc as c-d. Now as one-eighth of the circumference of a circle with a ten-

foot radius is equal to one-fourth of the circumference of a circle with a five-foot radius, you
will notice in Fig. 2 that only one-eighth of the
circle is taken, allowing a few inches for cutting
away of kerfs. You are now ready to make it
in the shape of ogee, but this will not yet fit

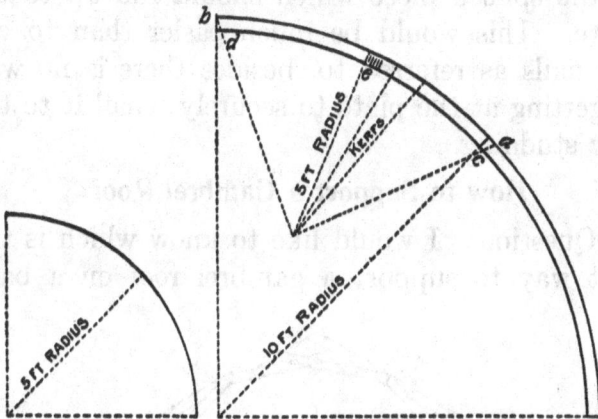

Fig. 1.   Fig. 2.

around the curve; you must kerf it first. Now
with five feet as a radius, as shown in Fig. 2 from
e to f, with f as a center, let all the kerfs radiate
from it, as shown in Fig. 2 in the sketch. All
this being done, it is best to steam it before bending it.

Adding One=Half Story

Question: Which would be the best way to
raise a house from one story to a story and a
half? The roof is a good one and do not care to
take it off. Simply to cut nails under plate and
raise the rafters with plate to the required height,

then put studding under the required length. The house is built like a T.

Answer: It would be better to cut the studding off about one foot below the plate, then raise the roof and splice the studding. The piece of studding left hanging to the plate can be nailed to the spliced piece which should run up to the plate. This would be much easier than to cut the nails as referred to, besides there is no way of getting at the plate to securely renail it to the new studding.

How to Support a Gambrel Roof

Question: I would like to know which is the best way to support a gambrel roof on a barn

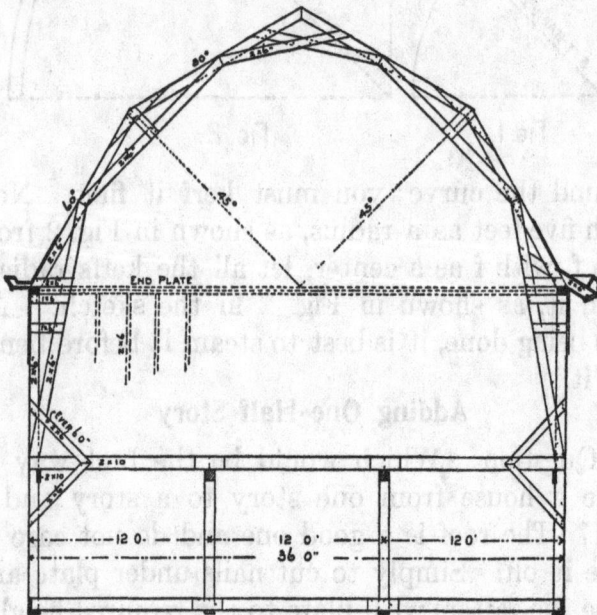

36 feet wide and posts 20 feet high. The floor
being 11 feet below the plate. Please give sketch
of same.

Answer: The space from floor to plate being
11 feet, will require extra strong bracing. Would
recommend using 6-inch studding set on 24-inch
centers and brace as shown in the illustration.
The rafters to set directly over the studding and
braced to same. The floor joist should be tied
to each other, either by letting them lap, or by
nailing a board on the side. All parts should be
framed accurately and well nailed.

How to Lay Out a Gothic Ceiling

Question: Enclosed find a sketch of a Gothic
ceiling over a pulpit in a church 12 feet wide and
8 feet in depth. There will be two hips, seven
main arch rafters and many cripples.

Will you give a rule for finding the shape of
the hip, also the shape and length of the cripples?

Answer: In Fig. 1 is shown the plan. This
simply shows the number of rafters contained in
the roof and would show the same for ogee or
any other shaped rafters or for any pitch given
the rafter.

In Fig. 2 is shown an elevation of the Gothic.
This should be laid out full size on a level surface
or floor—though it is only necessary to draw
one-half of this diagram, as that part enclosed by
A-B-C. Then A-B will be the shape of the main
or common rafters. Now the cripples, or jacks,
are simply a part of the common rafter and their

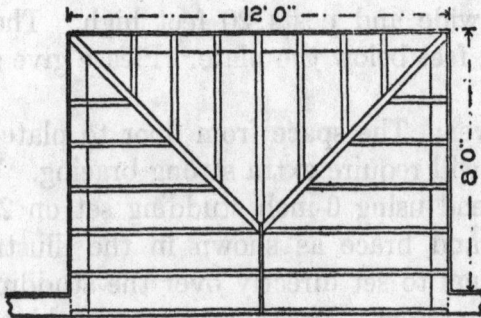

FIG. 1.   PLAN.

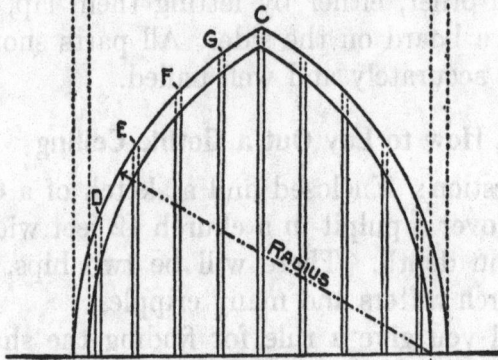

FIG. 2.   ELEVATION.

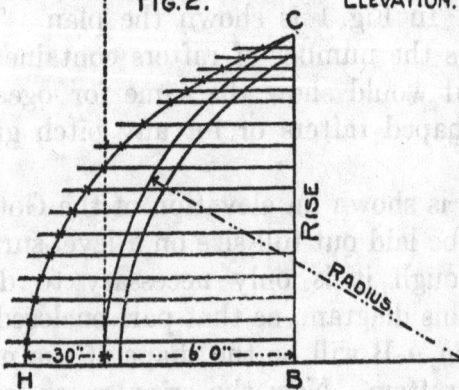

FIG. 3.   DEVELOPING THE HIP.

lengths are as from A to D for the first jack, A to E for the second, and continue on to the line C-B, which is a common rafter and consequently is the same as A-C. The dotted lines across the side of the common rafter represent the distance apart the plumb lines will be for the side cut of the jack, which in the case of a square corner is the width of the jack. In Fig. 3 is shown how to find the corresponding shape for the hip. In this the common rafter is shown the same as in Fig. 2. Lay off any number of parallel lines extending beyond the curve of the rafter as shown. Now measure the length of these lines from the rise to the curve and add 5-12 to that length. In other words, the first or bottom line is 6 feet long, to this add 5 inches for each foot, or 30 inches to the line beyond the curve and check. Proceed in like manner for all of the lines, adding 5 inches for each foot and 5-12 of an inch for each inch. After all these lines have been thus measured, draw an off-hand curve through the checks and this will be the shape of the hip.

Constructing a Circular Porch

The method shown herewith, I have used for several years. It looks better than the mitered seam, is perhaps as good and is cheaper.

In Fig. 1 A is a 4 by 6-inch timber, which gives a good bearing for the ends of the flooring boards. B shows the method of finishing the floor where steps come on the circle.

Opening C in Fig. 2 shows how the heels of

the ceiling joists are put in to give support for the heels of the rafters.

F in Fig. 2 shows the method I use for putting a ceiling planceer on a quarter circle porch.

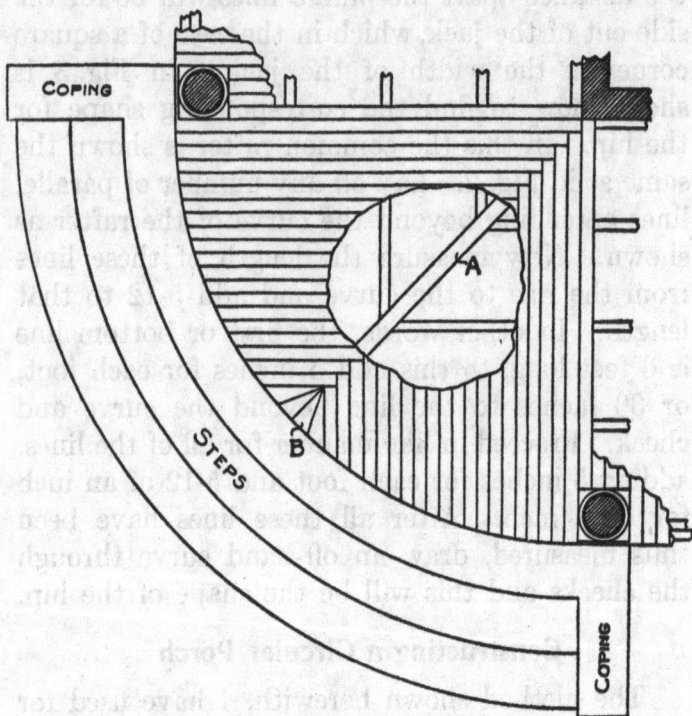

FIG. 1

D shows cripples fitted in between the joists to which the ends of the ceiling boards are nailed. I put the planceer on and saw to the circle afterwards as shown at G.

The ends of the ceiling boards are pared with

a gouge to match the boards running the other
way. E shows a neat way to make a ceiling.

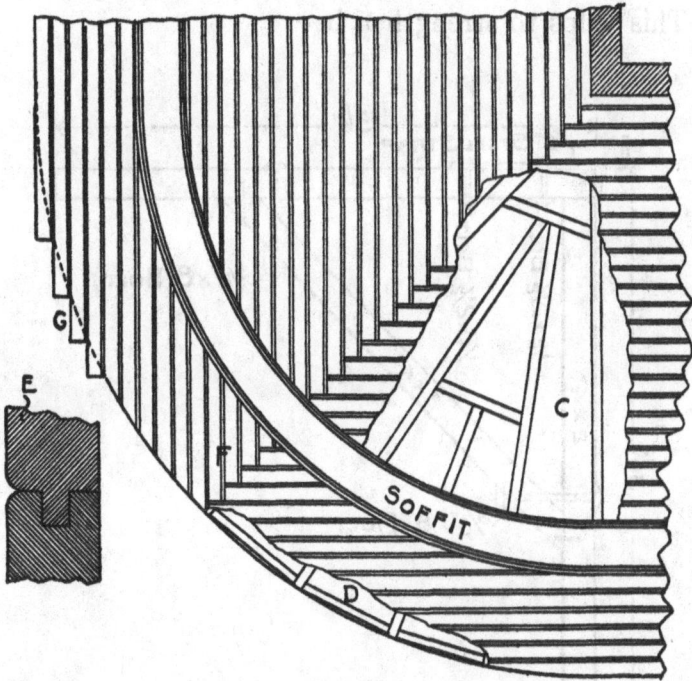

FIG. 2

Making Scaffold Brackets

We show herewith a sketch of a scaffold
bracket. The arms and brace are made of 2 by 4-
inch surfaced white pine. The brace is nailed to the
arms and is also fastened to same by a bolt $\frac{3}{8}$ of
an inch by 8 inches. There is a piece of a wagon
tire one foot long bolted with two bolts $\frac{1}{4}$-inch by
$2\frac{1}{2}$ inches to the 2 by 4-inch with a rounded hook
on the end which extends $\frac{7}{8}$ of an inch beyond the

corner and 2 inches up. There is a piece of iron
⅛ by ⅞ of an inch and 4 inches long nailed on the
sides of each arm where they come together.
This helps to strengthen it.

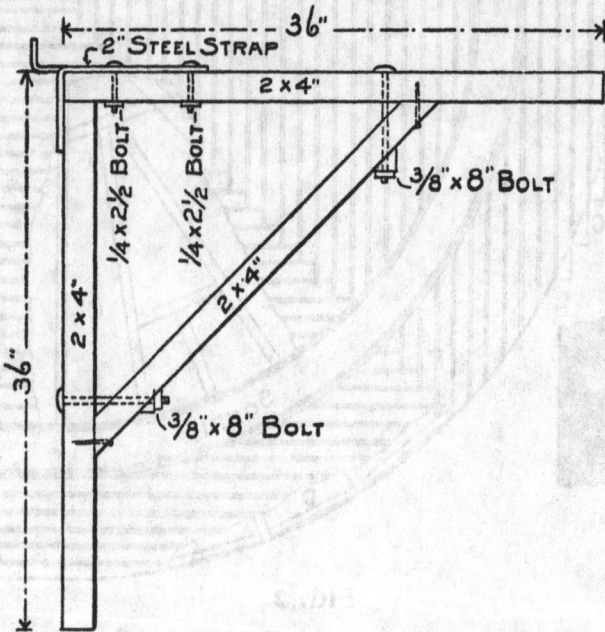

Making a Scaffold Bracket

We show herewith a sketch of a folding bracket.
The arm is made of 2 by 3-inch hard wood. The
brace is four feet and one inch long and takes up
very little space when folded. It is easy to move
from one job to another, which is a great convenience.

PERSPECTIVE VIEW OF
BRACKET IN POSITION

STUD

BRACKET IN
POSITION.

⅜″ BOLT

1½″x⅜″ IRON

RIVETED JOINT

⅜″ BOLT

BRACKET FOLDED.

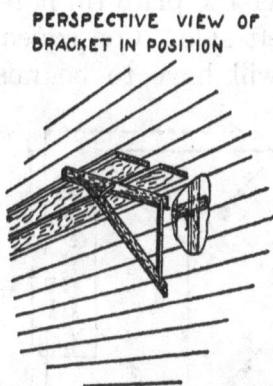

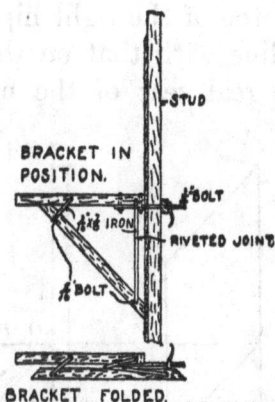

One-half Pitch Roof

I beg the privilege of submitting the enclosed drawing of the outline of a building, for which I would like to have the roof plan, or elevation for same. The roof to have one-half pitch, and hips with dormer windows in preference to gables, if possible. The building is two-story of brick, and to have shingle roof.

Answer:—While this roof is considerably cut up, there is nothing complicated about it, except at the four-foot angle on the right-hand side, which should have been opposite the angle on the left-hand side, but in this there is two feet difference. Now since the question asked for a given pitch, there are two ways to get around this irregularity. First—To make a deck, which should be placed on a level with the ridge of the highest gable and would be to the contour of the plates for the main roof and to the dimensions as shown in the small diagram. Second—To extend

the toe of the right hip over the plate till it rests
in line with that on the left side. Consequently
the real seat of the hip will have to be raised

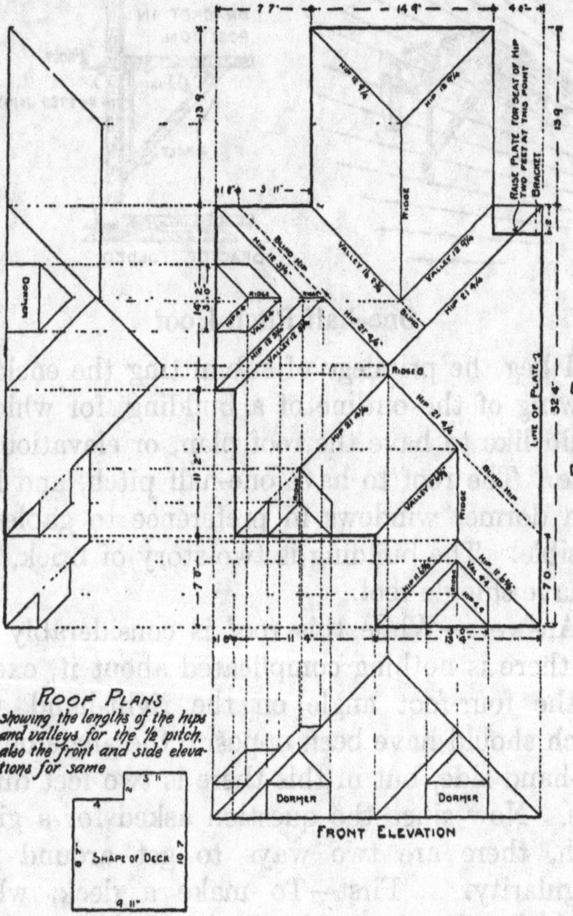

ROOF PLANS
showing the lengths of the hips
and valleys for the ½ pitch,
also the front and side elevations for same

SHAPE OF DECK

FRONT ELEVATION

above the plate as much as the common rafter
rises, in a two-foot run, which in this case, the roof
being the one-half pitch the common rafter will

rise two feet at the point where marked by the arrow. The planceer at this angle will be two feet wider than that for the regular cornice, but by placing a bracket at the corner it will relieve the wide projection. By squaring up this angle, we find the largest square contained in the roof to be twenty-four feet eight inches by twenty-five feet four inches, a difference of eight inches in the length and width and this difference will be the length of the main ridge.   In the plan, we show all of the lengths for the hips and valleys, but it must be remembered that these lengths are based on the supposition that the angles of the building are true, or square, with each other.   If not, then the irregularity is bound to show up in the framing of the rafters.   These lengths are estimated to a center line as at the center of the ridge tree.   In connection with the plan, we show a front and a side elevation and by tracing the dotted lines from the plan to the elevations, the locations of the roof lines in the plan are shown in position in the elevations and need no further explanation.

Making a Tool Chest

I send you herewith a sketch of my tool chest and work-bench combined.   It is the most useful and handiest I ever had.   It is 12 feet long, 2 feet 6 inches wide and 2 feet 10 inches up to cover, which is 6 inches high and divided into shelves as shown.   The sides and ends are covered with flooring to a depth of 1 foot 2 inches, with the front reinforced back of the pin holes.   The bottoms of

the chests are made of flooring and to the full
depth of the sides. I have them divided into
compartments to suit the tools. The top of bench
is of 1-inch select lumber. The cover is of ½-inch
lumber and covered with galvanized iron and

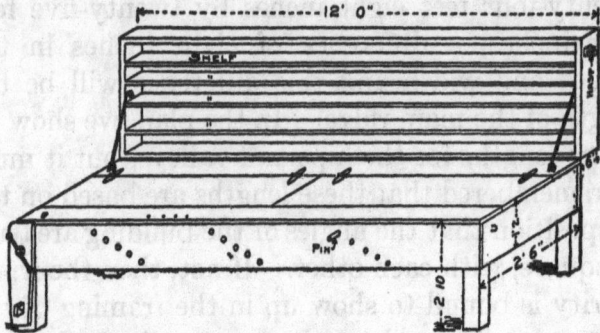

hinged to bench with heavy strap hinges. A hasp
and staple is provided at each end so that the
cover can be locked with padlocks.

Finishing White Pine

Question:—I am building a kitchen cabinet of
Georgia and White pine. We would like it a little
darker, something on the order of light cherry.
How would you finish it?

Answer:—The wood should be first given a
coat of cherry oil stain, such as can be bought at
any paint dealer's, thinning with turpentine if
necessary. Then give a coat of thin shellac and
finally give two additional coats of a first-class
interior varnish. If a dull finish is desired, the
simplest method would be to use one of the dull

drying varnishes as a last coat. Another method would be to give the wood a thin coat of grain alcohol shellac and finish with two additional coats of any of the varnish stains made with aniline color, extensively advertised by many of the varnish manufacturers under fancy trade names. These dry so quickly that great care is needed in applying them to broad surfaces, like large panels, but most of them would give good results on a piece of furniture such as is mentioned.

Cutting Siding for a Gable

How many of you ever tried cutting the siding for a gable on the ground, instead of on the tresses? It is much easier, quicker, more saving of siding and much more satisfactory every way than to cut it out on the scaffold. As soon as you are well onto the gable fit in a piece to run clear across the gable if possible. Now be sure and cut this piece to fit neatly, as it is your pattern. Now take this to your tresses and with your working gauge lay off what you want to show to the weather. Gauge all your timber for the gable this way. Now lay the piece already cut on the next piece to be cut with the gauge line at the bottom of the piece to be cut just as it will appear when put up and mark the length. Now lay your pattern on top of the piece to be cut and mark your angles same as pattern. Don't use your bevel square, for if the siding is not exactly straight your bevels will be wrong. Now proceed in like manner until all are cut; then all you need to take to the scaffold is your hammer and block plane, and you will not need your block-plane if you have been careful to cut the angles like the pattern, and you have no nails to set to hold your siding while you mark it.

Looking After Little Things

It is the little things that often escape our notice, and yet it is the little things that count, and that cause so many jobs to be a profit or a loss, just in accordance to the way we look after them. The man who looks after the little things in any calling is the successful man, and especially is this true of the carpenter trade. It should be our aim to look closely after all the minor details of a job and learn to do our work in the way that requires the least labor. Who has not been around where carpenters were at work without noticing the amount of work that is being done that does not advance the job at all, but is the result of dragging along in the same old rut and not learning the latest methods of doing things. It is the rule with carpenters in some sections to rabbet frieze lumber to receive the siding, and when asked why they do so frankly confess that they know of no other method. One way of getting around this disagreeable task is to take common lath and cut them in lengths about two inches shorter than the frieze board is wide. Nail these to the sheathing all around the house, placing them directly over each studding, where the frieze is to be nailed and when the frieze is placed over these it will project over the strips and allow the siding to slip up under just the same as if the old method has been followed; besides, you can use seven-eighths inch lumber for the frieze and when the job is done it will have all the appearance of one and one-quarter inch stuff having been used, so it will be seen that this method saves material, as well as a great amount of hard labor.

Scaffold Brackets

Thinking probably some of the craft might be interested in the scaffolding ideas we herewith pub-

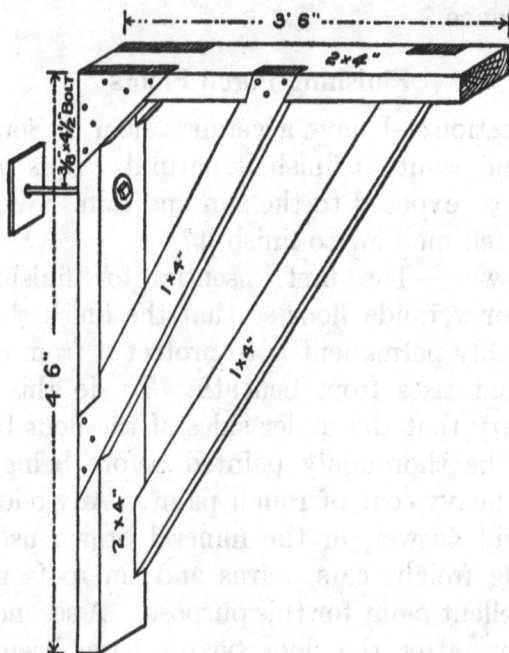

lish a sketch of one that has been very successful. The arms are made of 2 by 4 surfaced white pine and the braces of 1 by 4 as above. All parts are made exactly alike so as to be interchangeable in case of accident. The braces are let in with a shoulder to give additional strength. Bore the hole exactly the same distance from the heel and see that the bolt fits tight to prevent turning. Use a large washer under the head of the bolt. These brackets can be bolted to Byrkett sheathing the same as on the common sheathing, or can be stood on the ground with the short arm turned to the wall and make a good scaffold for siding. If they are painted it adds to their durability and appearance.

Finishing Porch Floors

Question:—I have a veranda floor of Southern pine and want to finish it natural. It is on the side most exposed to the sun and rain. Will you please tell me how to finish it?

Answer:—The first essential for finishing a porch or veranda floor so that the finish shall be reasonably permanent is to protect it from dampness that rises from beneath. To do this, it is necessary that the under sides of the floor boards should be thoroughly painted before being laid, with a heavy coat of rough paint. Any odds and ends will answer, or the mineral brown used for painting freight cars, barns and tin roofs makes an excellent paint for this purpose. A second coat of paint, after the floor boards have been laid, should also be given if there is space enough under the floor to permit of doing this. It is also advisable to lay the joints in white lead; using the ordinary pure white lead in oil, made enough thinner with pure linseed oil to permit it being pushed into the grooves of the boards with a putty knife before the boards are driven together.

If the floor is to be painted, it should be given three coats of paint, at least. The priming and second coats should be preferably pure white lead, thinned with linseed oil and turpentine, and well brushed into the wood. All cracks must be thoroughly filled with pure whiting and linseed oil putty after the first coat. The floor must be thoroughly cleaned before priming and all mud, plaster, grease or other dirt entirely removed. The surface must be thoroughly dry, and no painting should be done immediately after a frost, heavy dew or rain. The third coat may be more oily than the undercoats. Good work cannot be done with less than three coats; four are better. The last coat of paint must be brushed out smoothly and evenly so as not to leave heavy places which will dry unevenly and soon wear out.

A natural finish for a porch floor is much more difficult to make satisfactory. There is an almost unavoidable tendency to darken, especially on yellow or Southern pine. The same precaution should be taken in regard to protecting the floor from underneath as if it were to be painted. After the boards have been thoroughly cleaned they should be filled with a good paste filler (silex), applied to five or six boards at a time, allowed to set, but not to get hard, and then rubbed well into the grain of the wood, rubbing across the grain, with burlap. Any excess of filler must be wiped off with a soft rag. Many painters regard the use of paste filler on yellow pine as absurd, but it gives a solid foundation for a floor that cannot be obtained in any other way. A porch floor is obliged to stand wear and tear and exposure to the weather and for this reason the ordinary floor varnish is unsatisfactory. Three or four coats of the best exterior varnish or spar coating should be used, allowing not less than four days between the coats, and if possible, at least a week should be allowed after the last coat is applied before the floor is walked upon. When the floor becomes scratched its luster can be restored by rubbing with crude oil or an oil polish.

A Transom Window Frame

Will you please give a description and detail of a transom window frame, allowing for a three-inch bar with hinged transom light?

Answer:—The most satisfactory way we have found is to make the frame as for an ordinary two-sash window and make the meeting rail as shown in section at Fig. 1. This has the appearance of a moulded transom bar. The sash being hung the same as in the ordinary window it allows it to be lowered at will, leaving no open joints for the wind and rain to get in when closed. It is well nigh impossible to construct a transom when hung with butts and have it weather-proof. In such

cases the most satisfactory way we have found is
to hang the sash at the bottom similar to that

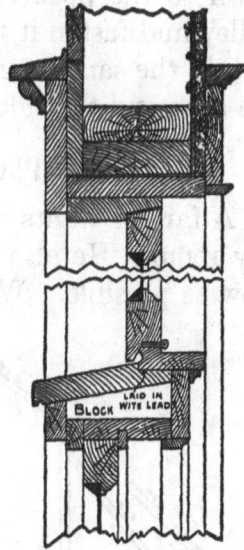

Fig. 1.

Fig. 2.

shown in Fig. 2. This makes a fairly tight job
and is simple in construction.

How to Put Ropes in Windows

To put new ropes in windows, take off the strips
and hold the lower sash in place. You will find a
pocket in the casing about ten inches from the
bottom, where the weights are concealed. Take
the old ropes from the weights and sash, taking
note of the method of fastening. Cut the rope to
be put in three-quarters as long as the window.
Take a small piece of iron that will slip through
the pulleys easily and fasten it to a piece of twine.

Tie the other end of the twine to the rope. Drop the small weight through the pulley until it comes down to the pocket. Pull the rope through the pulley and fasten it to the window weight and the sash in the same manner as the old one. Replace the sash and the side strips.

Plan of a Barn

A farmer wants a barn with an abundance of hay room. He also wants as little timber in the mow as possible. We find that a good many car-

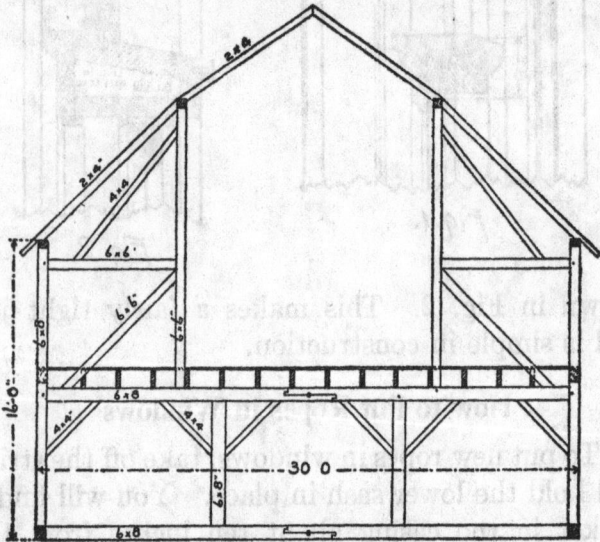

CROSS SECTION

penters are puzzled when it comes to framing a barn so as not to have much timber in the way and make a good substantial job. We print a plan of a country barn that pleases all who see it. It is the center bent to which attention is called.

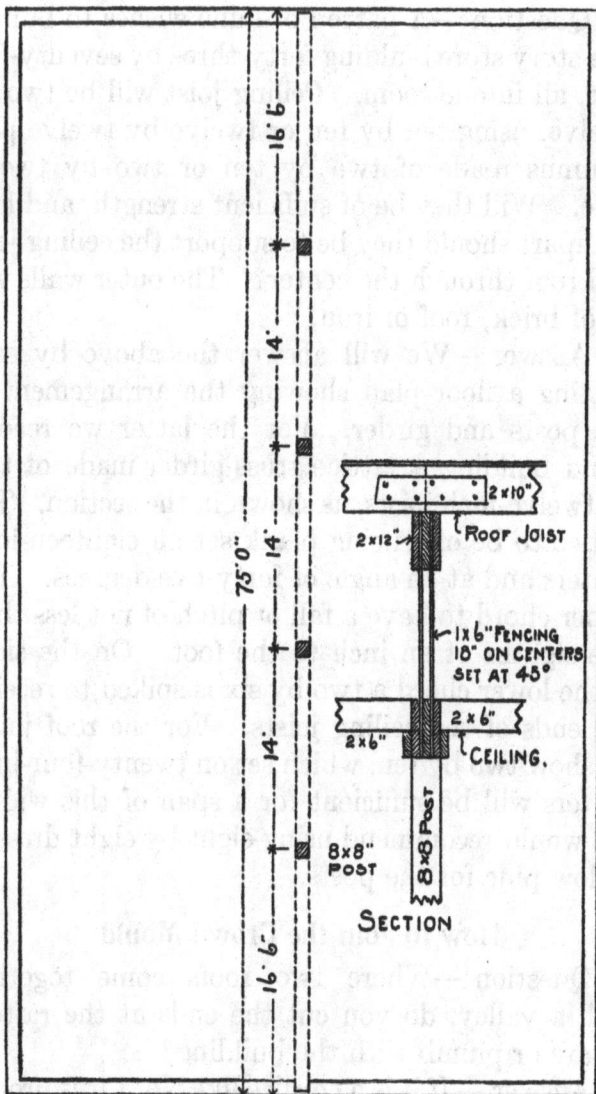

Dimensions shown on the diagram:

16' 9"
14'
14'   75'0"
14'
16' 9"

2x10"
Roof Joist
2x12"
1x6" Fencing
18" on Centers
set at 45°
2x6"
2x6"
Ceiling.
8x8" Post
8x8 Post

Section

Supporting a Ceiling

Question:—A patron of mine wishes to build a one-story store building forty-three by seventy-five feet, all in one room. Ceiling joist will be two by twelve, using ten by ten or twelve by twelve pine columns made of two by ten or two by twelve pine. Will they be of sufficient strength, and how far apart should they be to support the ceiling joist and roof through the center? The outer walls will be of brick, roof of iron.

Answer:—We will answer the above by submitting a floor plan showing the arrangement of the posts and girder. For the latter we recommend building a lattice truss girder made of two by twelve-inch joists, as shown in the section. The lattice to be of fencing plank set on eighteen-inch centers and at an angle of forty-five degrees. The upper chord to have a fall or pitch of not less than five-eighths of an inch to the foot. On the sides of the lower chord a two by six is spiked to receive the ends of the ceiling joists. For the roof joists we show two by ten, which set on twenty-four-inch centers will be sufficient for a span of this width. We would recommend using eight by eight dressed yellow pine for the posts.

How to Join the Crown Mould

Question:—Where two roofs come together with a valley, do you cut the ends at the rafters square or plumb with the building?

Answer:—If it is a cornice like No. 1 it is proper to cut the ends at the rafters square, but if it is a box cornice like No. 2, then the crown mould should be set plumb. A cornice like No. 3 will not

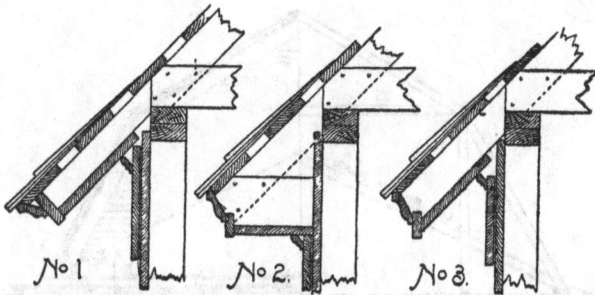

work well where there are gables, because the crown mould will not member with that of the gable.

Gutter for Gambrel Roof

Question:—I am figuring on a small cottage with a gambrel roof and I would like to know the kind of gutter which is best suited to this style of roof. The cornice is to continue all around the house, connecting with a veranda in front and a low hip roof in the rear.

Answer:—The accompanying illustration shows about all there is to be said of a combination roof and gutter for a house of this kind. The gutter is what is generally termed "Yankee gutter," and, as it is to run all around the roof, would suggest that the gutter be set near the eave with no shingles underneath, using instead a seven-eighths inch board to form the eave drip and on which to rest the brackets. Would set the gutter level and grade on the inside with false bottom. Use good quality of tin for the lining and start double course of shingles about level with top of gutter. If the

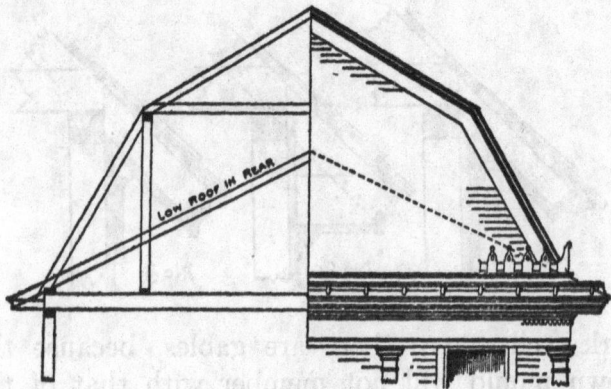

hip roof in the rear interferes with the window space of the main part it could be changed to a deck roof.

Facts About Doors

Question:—I have several questions I would like to have answered: 1. How should double sliding doors be fastened in center to prevent sliding farther than they should? 2. What kind of stop should be used? 3. Which is the most proper way to put on corner block, with grain running vertical or horizontal?

Answer:—1. The sliding is prevented by overhead track that door slides on, and also by rollers. If it is a pair of sliding doors, it should have a knuckle joint or astragal, although some sliding doors are banded, making it unnecessary to use either knuckle or astragal. 2. It is customary to use $\frac{3}{8}$ or $\frac{1}{2}$ inch stop, depending on the width.

3. Corner blocks should have the grain running horizontal, so it will match with the casing.

Constructing "Saw Horses"

In making heavy saw horses use 4 by 4, 4½ feet long, and gain in on each side 1 inch on top to nothing at the bottom. This gives about the right slope to the legs at the bottom. Use 2 by 4 for legs and a brace of 2 by 6 cut between them just underneath the top; then cut a wedge 2 by 6 and 1 foot long and nail fast against the cross piece and on under side of the top. All the rest is pinned with hardwood pins to prevent dulling of saws when you happen to cut into the horse. We find for heavy work the above will stand almost any amount of strain.

Constructing a Circular Porch

Question:—Kindly tell me how to construct a circular porch. Illustrate if possible.

Answer:—The accompanying illustration at Fig. 1 shows the method that we have used in our own work for a number of years and is probably as good as any other. The central part of the illustration shows the frame work of the floor joists with a portion of the flooring in position.

There should be supports at C, B and D. From C to D is one-quarter of a circle, and this is divided in the center, as at B, then the straight lines C–B and B–D are equal to the sides of an octagon with a circumscribed radius of seven feet and eight inches, which is the width of the frame-

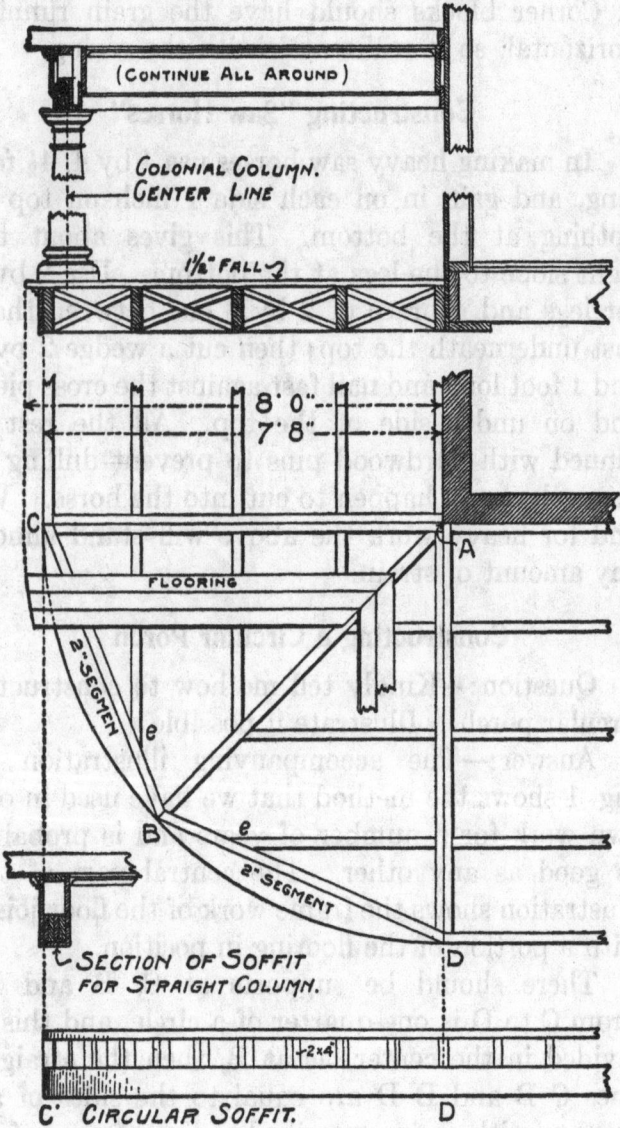

FIG. 1.

work of the porch, and the length of the sides may
be found by multiplying the radius by the decimal
9.18, which equals five feet ten and three-eighths
inches, and is the length to cut the side pieces and
is also the length of the chord of the segment to
form the circle to receive the base.

In the absence of the above decimal or in case a
person is not apt in figures, these parts may be
found as shown in Fig. 2. By placing the square
on a board, from which the segment is to be cut,
with the figures that give the octagon cuts and lay
off the raidius in line with the blade, as shown,
describe the arc, and it is ready to cut. The
figures shown on the square will give all of the
cuts required in the frame work about the octagon,
as the blade will give all of the cuts at B, also at
the other end of the side pieces at C and D. The
tongue will give the cut at e and e. The other cuts
are the square or on the 45 degree angle. Thus,
from this it will be seen that all of the pieces can
be successfully framed without first building a
part of the framework and scribing the other
pieces to it as is the general custom.

There should be four of the segment pieces
gotten out, setting one flush with the top edge and
one at the lower edge of the joists. The upper
ones should be of one and three-fourth inch stuff,
same as the joists, while seven-eighths will be sufficient for the lower member. Set blocks between
these segments, nailing them well to the joists, also
set a few blocks flush with the face of the segments
which makes an excellent form to secure the base.

The ceiling joists are usually put on the narrow way of the porch with an angle piece same as at A–B, on which to form the miter joint of the ceiling.

To form the soffit we use seven-eighths by six or eight-inch sized boards and spring them to their

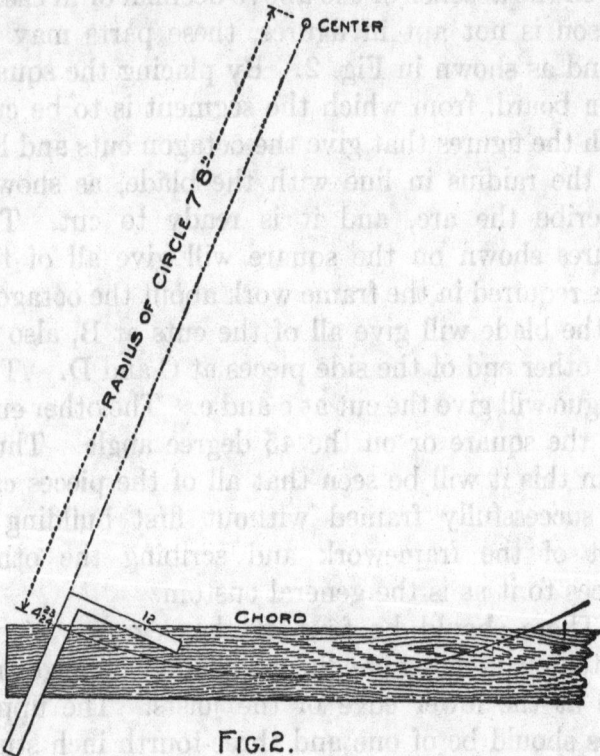

FIG. 2.

proper place just the same as building a circular girder. The first board should be sprung to a form and the next board well nailed to this one, and so on till the soffit is to the required thickness or strength, and as it is not always necessary to build to the full width desired as it can easily be furred out to the required width. The soffit should be continuous; that is, for the straight part as well as for the circle. Long boards should be used so as to lap well around the circular part, being careful not to break joints on the circular part or at C or D.

A soffit if properly built in this way will not necessarily need a column set at B, as it will be self-supporting. If straight columns are used the outer face of the framework should be flush with the framework below, but if tapered or colonial columns are to be used, then the center of the soffit should rest over the center of the column, as shown in the upper part of the illustration of Fig. 1.

In case a deep frieze is wanted, it may be had by building on top of the soffit girder with blocks, and putting a formed plate on these. For all circular mouldings, it is better to have them solid and they will then always stay in place, as there will be no kerf joints to open up after the work is completed.

Strength of Beams

Question:—Will a beam, made by spiking seven 2 by 8's together equal 14 inches by 8 inches, 50 feet long, supported in the center by post, sustain a tar and gravel roof with joists 2 feet on center and hay loft of 10 feet high with hay? Two loft posts in sum of 50 feet, the two loft posts stand out from center post 8 feet on each side.

Answer:—The width of the building is not

stated, but presume the joists are not over 24 feet. The construction indicated would be rather weak in case of a full mow of hay and the possible chance of a heavy snow on the roof. Would suggest that five 2 by 14's be used, well spiked together; this will give a girder of about 9 by 14 and not much more lumber.

Building Construction

Question:—Here is a business house 33 feet by 135 feet, five stories high. The first, second and third stories will be used for a furniture store, fourth story for a ball room, fifth story for a lodge room. Now will 3 inch by 14 inch pine placed 12 inches on center, which will be nine inches between, with four rows of bridging, carry the weight, bridging to be 2 inches by 4 inches, nailed with two 12 penny nails in each end of the bridge?

Answer:—The size 3 by 14 placed on 12 inch centers will be sufficient to carry the loads the floors will be subject to, but would recommend that the timbers be cut full size, and as these timbers will probably have to be sawed to order it would be well to have them gotten out several months in advance of using, so they will be seasoned so as to avoid shrinkage after being placed in the building. The joists should be sized to even depths with a camber of at least 1 inch. Four rows of 2 by 4 bridging is all right. Would also recommend laying a rough floor diagonally with the finished floor, the latter to be laid after all the plastering and rough work is done. Do not let the plumbers notch into the joists further out than three feet from the ends of the joists.

Cutting Rafters

We wish to make a small note right here which may set right some of the younger members of the craft and a few of the older heads who never paid any attention to it; in fact, we saw a con-

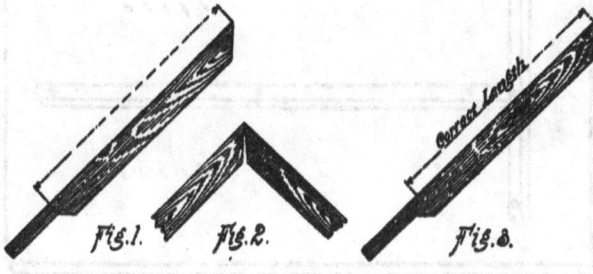

tractor cut the rafters on a three thousand dollar house in the same faulty way. It is in the manner of adding the projection for cornice. Fig. 1 is the wrong way and makes the rafters too short, causing the ridge joint to open as shown in Fig. 2. The right way is shown in Fig. 3.

Sash Pulley Gauge and Marker

We show herewith a drawing of a sash-pulley gauge and marker, such as is used. Take a piece seven-eighths by six inches wide and eight inches long. Then take a piece of parting stop one-fourth by one-half inches, nail on as shown in drawing the piece running lengthwise drops in

groove in jamb which is plowed for parting stop.
The piece running crosswise rests in the gain which
is made for the header.   The dots represent brads
which come up about one-eighth of an inch and
are filed sharp.   When you drop it in place give it
a tap with the hammer and the brads do the rest.

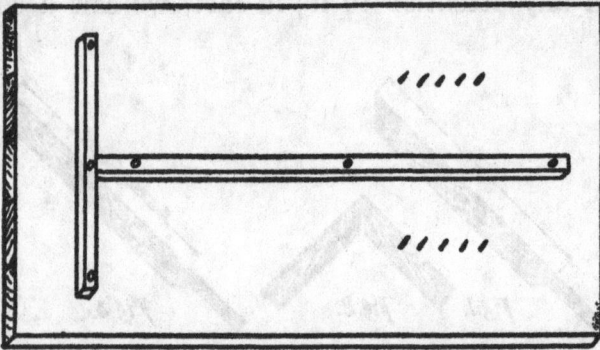

MARKER FOR SASH PULLEYS

Both pulleys are marked at once.   Set the point
of the bit where the brads marked and your
pulleys will fit as if they had grown there.   Of
course different makes of pulleys need different
kind of markers.

A Non-Freezing Potato House

Question:—I am called upon to make a plan
of a potato warehouse in which lumber will be the
main material, and, of course, it must be frost
proof as much as possible.   I have thought of
making walls like accompanying sketch.   The

studs will be 2 by 8 set on 16-inch centers. On the outside are first placed one-inch boards horizontally with the studs, and over this are one-inch boards placed perpendicularly, with building paper between them, and the cracks are covered with buttons. There will be two courses of black plaster as shown, making three dead air spaces in the width of the wall. The inside of the wall is lathed and plastered, and over this boards are placed for the protection of the plastered walls. The main object is to keep the frost out.

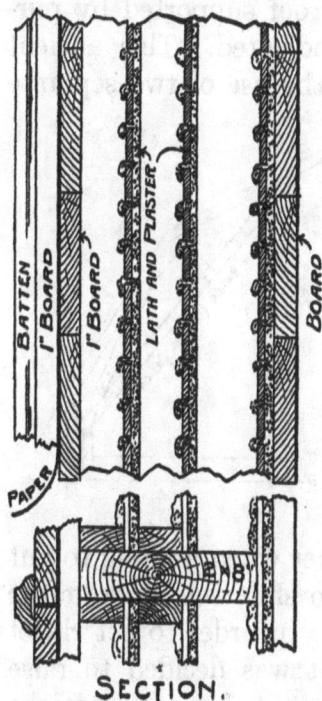

SECTION.

Answer:—We herewith produce the sketch. The plan is all right, provided all parts of the building are constructed accordingly tight. The three dead air spaces will act as a non-conductor of frost sufficiently to protect the interior from freezing in most sections of the country where potatoes are raised. However, we would suggest putting two or three thicknesses of paper between the outer boards. Would use tarred felt for the last layer.

Brace for Gambrel Roof

The accompanying sketch illustrates a brace for gambrel roof to be used instead of the purlin support.   There are two barns of different widths and heights joined end to end.   The larger barn is thirty-two by fifty feet, with regular gambrel roof. The horse-barn, which is joined on, had an old-fashioned one-third pitch roof supported by purlin plate and posts, as indicated.   This ancient arrangement necessitated the use of two separate

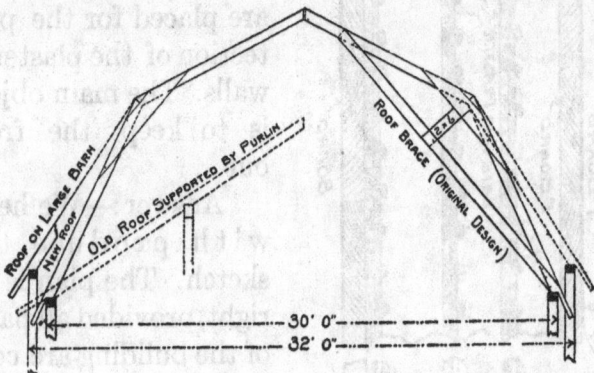

tracks, compelling the owner when wishing to put hay into the horse-barn to shift the car from the upper to the lower track.   In order to get rid of this great inconvenience it was decided to raise the roof on the horse-barn so as to have one straight ridge the entire length of the two barns, and thus secure the one track desired.   It was found after the old roof had been raised to the gambrel form that its strength was not sufficient to load and carry a hay fork.   The owner desired if possible to avoid the use of the purlin support, and the result was the brace shown in the illustration, which is constructed as follows: Take a two by six inches by twenty feet, cut it to fit on plate at bottom and strike rafter about two feet from ridge. Spike it thoroughly at top and bottom with number twenty nails. Cut a piece of two by six inches to fit snugly between ridge and top of long brace well spiked to rafter. Repeat this operation on the opposite side of the same rafter. Introduce a tie brace of two by six inches at the buckle, which is thoroughly spiked in as shown. Repeat this operation on the opposite side of the roof and you have in effect a pair of trusses where before you had only a pair of rafters. This brace would ordinarily suffice with one, two by six inches by twenty feet, instead of two placed six or eight feet apart. This plan makes a stronger roof than the purlin, and its construction requires less of both material and labor.

Trussing a Roof

Question:—What would be the proper way to truss a roof on a span of thirty-four feet, using two by six for the rafters, with roof to have a pitch of forty-five degrees, the ceiling to raise six feet above the plate? This is for a church, and would like very much to have a diagram of the trussing of the roof. Want it sufficiently strong to keep the plaster from cracking. Also what is your opinion about using two by four for the studding, which are to be fourteen feet high?

Answer:—In answer to the above we herewith print a diagram of the truss as per the dimensions given, which we presume is for a common shingle roof. The timbers required will be twenty-six feet in length for the common and tie rafters and twenty-four feet for the collar beam. At the seat of the rafters we use a two by six piece circled out to form a cove as shown. This piece should be

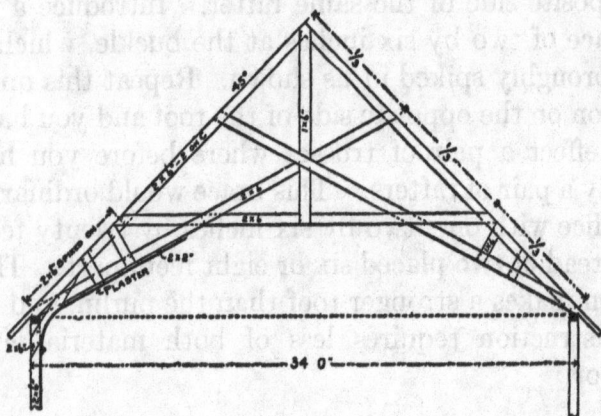

thoroughly spiked to the studding and to the side of the tie rafter and to the under edge of the common rafter. However, only one-third of these pieces will catch the rafters, owing to their being spaced on twenty-four inch centers, thus requiring the other pieces to be framed in between the rafters. Other timbers in the truss are of one by six fencing plank. All parts should be thoroughly spiked together. Cross pieces of two by two stuff are used to receive the lath, for which we would recommend using expanded metal lath. The level ceiling will lack a few inches of being six feet above the plate, but the timbers will work to a better advantage and will make a stronger job. As to the last question, we would by all means recommend using two by six studding for a building of this kind. What is called two by four is really only three and five-eighths in width, which is too narrow even if stout enough to give the proper width at the window jambs to receive the sash and other trim.

Preventing Leaky Window Frame

Question: I wish you would explain with an illustrated sketch the proper way to make a window frame, for an ordinary frame building, to gain

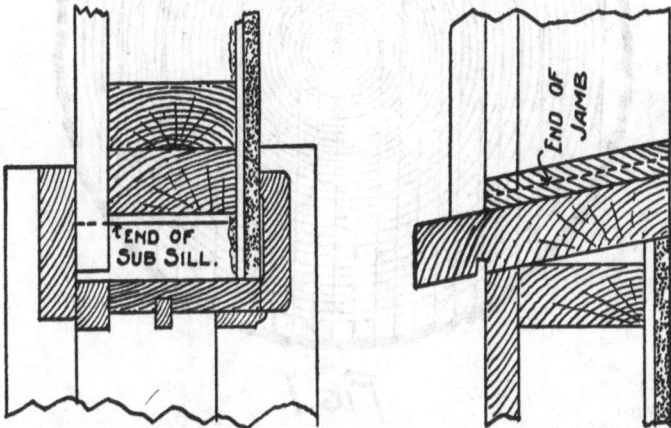

into the sill or into the jamb so as to prevent leak.

Answer: The most satisfactory way we have found is to gain the jamb into the subsill, leaving the end of this sill project same as for window sill, and only notch out enough of the back corners to fit nicely in opening for the window as shown by the sectional drawing. The joints should be set in white lead, and well painted on the outside.

Quarter Sawing Oak

Question: Would you please inform me as to what system is used in quarter sawing oak? I

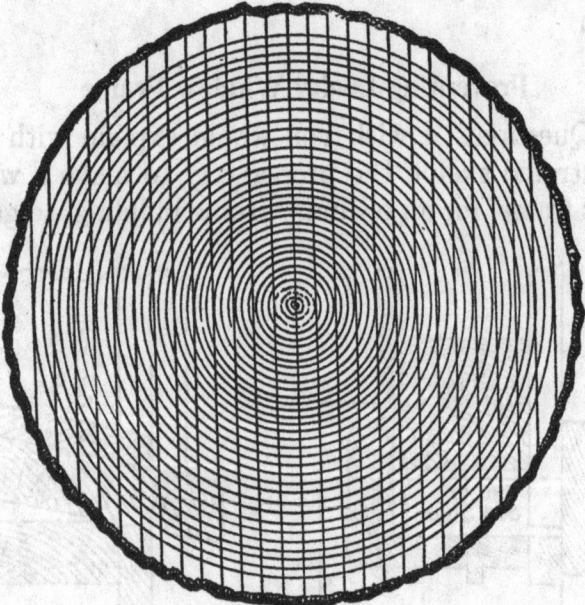

FIG. 1.

have never seen this done and hear of several ways. What I want to know is in what shape they cut the log.

Answer: Quarter sawing is simply the manip-
ulation of the log on the carriage to the saw, cutting
the same into boards so that the grain of the wood
runs from perpendicular to obliquely with the face
of the board, showing the edge instead of the flat

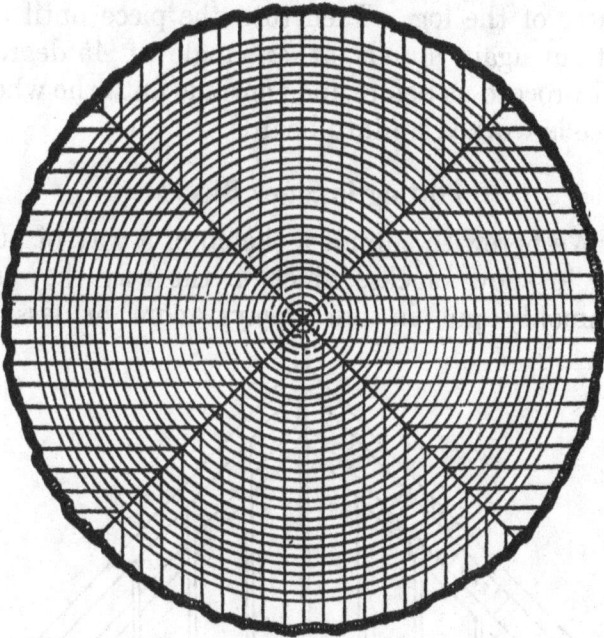

FIG.2.

grain. If the log is worked into boards parallel
with one another, as shown in Fig. 1, only a portion of the log will be cut into what is called
quarter sawed boards, but if the log is first quartered as shown in Fig. 2 (hence the name), and
these pieces again sawed into boards as shown,

then most all of the log will have been quarter
sawed. The same result may be accomplished by
first sawing the log into halves, then turn one of
these pieces on the carriage so that the sawed side
will rest at an angle of 45 degrees with the saw,
then saw the same into boards up to the original
center of the log. Then turn the piece until the
last cut again stands at an angle of 45 degrees
and proceed as before and continue until the whole
piece is worked up into boards.

Putting in Snow Blocks

We herewith submit a sketch of a way of putting in snow blocks, which in the way they are
commonly put in are always more or less of

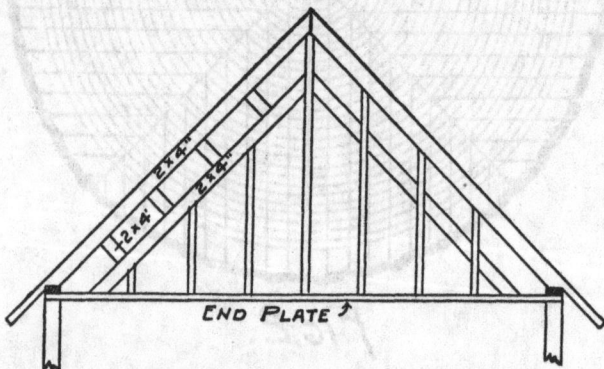

annoyance. In the sketch, both the common and
our way are shown. In the common way the
blocks are pieces of 2 by 4, cut in between the
gable studs, which if not extremely well nailed will
become loosened in nailing on the siding. In this
way the blocks are in one continual piece, fastened

n center to the gable stud just low enough to
receive the siding and cornice. While this plan
may be old to some, it may be new to many. It
is often some of the simple things that are of much
importance.

Box Sill for Frame Buildings

As we have never seen any cuts of a box sill
for frame buildings given, we print the following
plan. The advantages of a sill made this way are,
first, you get the full strength of the studding;

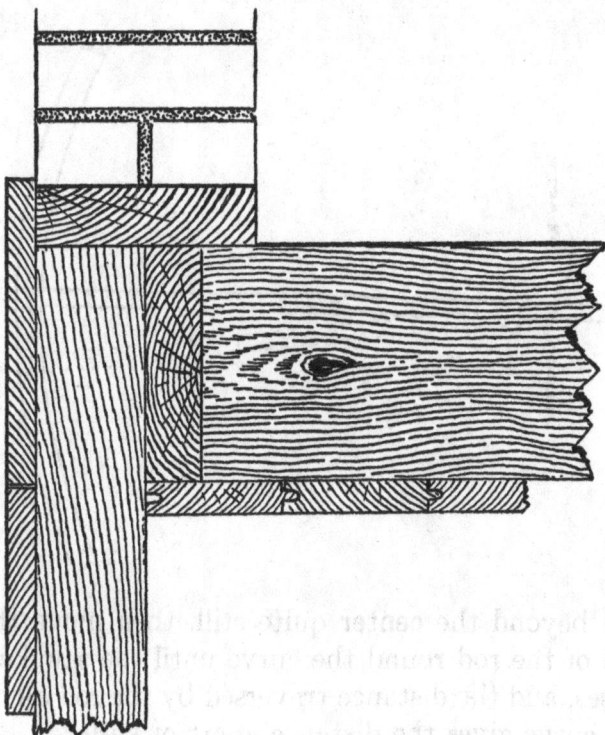

second, less work cutting same; third, it makes a tight job all around, so wind, mice or rats cannot get into the house or between the partitions; fourth, it is easier to lay floor, as you have no studding to cut around.

Sawing Kerfs

Herewith is a simple sketch of a method of finding the distance between saw kerfs for bending a board around a curve.

Take a rod and saw a kerf in it at the center, from which the curve was struck. Hold the short

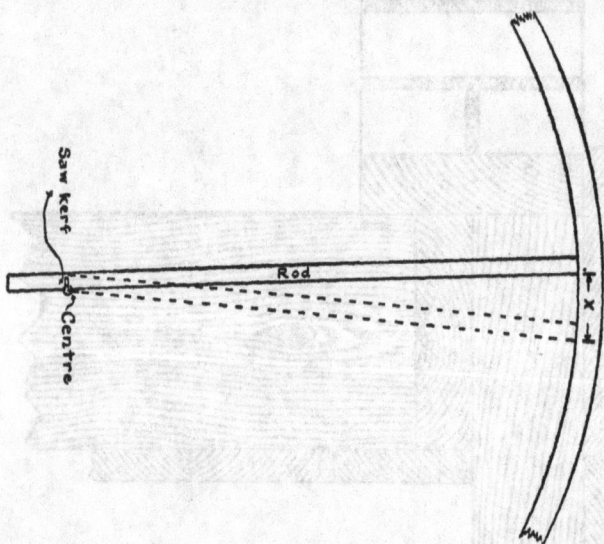

end beyond the center quite still, then move the end of the rod round the curve until the saw kerf closes, and the distance traversed by the rod along the curve gives the distance apart of kerfs.

Adjustable Trestles

We print a sketch of a pair of trestles which we think are pretty handy and save the carpenter lots of stooping. There are times when it is not necessary to have the trestles so low as we commonly use them; for instance, when working with screens, blinds and sashes. Cut a mortise about one by two and three-quarter inches in each end, say about three inches from each end of the trestle. Take a two by four the same length as the trestles

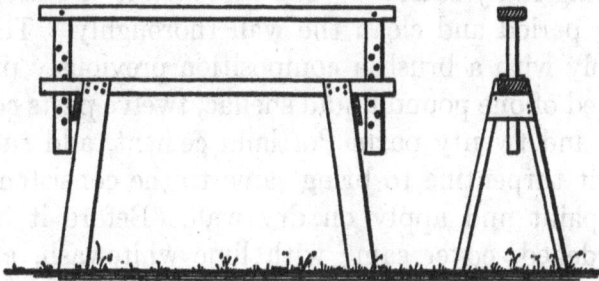

and cut a mortise same as in trestles, only not so long. Take two pieces one by two and one-half inches and two feet long. Hardwood is the best. Fit these pieces tightly into mortise in two by four and pin. Now bore holes in the one by two and one-half inch pieces in a zigzag manner and about two inches apart, and you have it complete. The holes should be a little larger than a large spike nail. You can see by having one of these for each trestle that you can raise and lower your work as it suits you best. Raise to the height wanted and slip a spike in the hole above the trestle and it will stay there until you change it.

Remedy fer Sweating Wall

Please give me some formula for dressing a basement to keep it from sweating. I have built two basements, and in damp weather the walls sweat and drip so that it becomes wet and everything kept inside inclines to mold. Is there anything that I could apply to prevent it? Will another coat of plaster stop it?

Answer: The following will greatly relieve the sweating of any basement whose wall admits water during rainy seasons. The first step is to select a dry period and clean the wall thoroughly. Then apply with a brush a composition previously prepared of one pound liquid shellac, twelve parts coal tar and twenty parts Portland cement, add sufficient turpentine to bring same to the consistency of paint and apply on dry wall. Before it has hardened, cover same with lime whitewash, and when dry apply a second coat of whitewash.

Flues in Chimneys

Question: I am going to build a chimney about twenty-four feet high. It will start on a center wall below the floor? Can I put in it a flue so as to take the cold air off in the winter?

Answer: Yes, but it should be an independent flue. In other words, it can be built alongside of the smoke flue and would, in fact, be a benefit because the warm air thrown off from same would help to create a draft in the ventilating flue. We would recommend putting in two ventilating registers. One at the floor and one at the ceiling.

The upper one to be closed in cold weather and the lower one in warm weather.

Laying a Church Floor

Question: Here is a sketch of a bowl-shaped church floor. What is the proper way of laying the flooring? We will use two floors, a rough floor and a top floor.

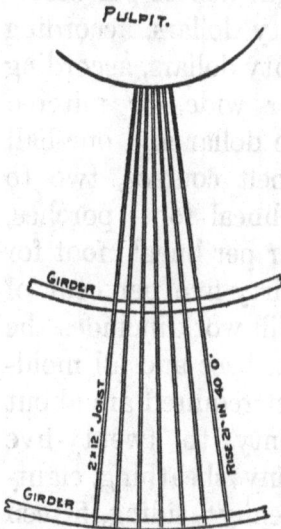

Answer: We herewith reproduce the sketch mentioned above. As the incline in this floor is only slightly over one-half inch to the foot, it will be a very easy matter to spring the boards to the required shape. The finished floor should all run one way, but the under floor should run diagonal to the upper one to prevent the shrinkage from showing up in the finished floor. The joists being laid fan shape will require the under floor to be laid at different angles. For this floor would recommend using surfaced one by six boards laid close and well nailed.

Estimated Cost of Labor in Building

Question: I would like to know how you estimate the price of labor of a house or barn and the amount of nails needed for a building.

Answer: The easiest way to estimate the labor, and one that is practical, is to estimate rough work by the thousand feet of lumber, and the finishing a per cent of the cost of the mill work, thus: Estimate the labor of framing dimension lumber at eight to ten dollars per thousand feet, board measure; sheathing, eight dollars per thousand; flooring, twelve to twenty dollars, according to kind; siding, twelve to twenty dollars, according to kind, whether narrow or wide, or mitered corners or not; shingling, one dollar and one-half per thousand; cornice and belt courses, two to four cents per member per lineal foot; porches, one to one and one-half dollar per lineal foot for labor; finishing, thirty to thirty-five per cent of the cost of the mill work. Mill work includes the sash, doors, blinds, casing and base and all moldings. Nails, per thousand feet required are about as follows: Framing, twenty to twenty-five pounds twelve to sixteen penny; sheathing, eighteen to twenty pounds eight penny; siding, fifteen to eighteen pounds six penny; flooring, twenty to twenty-five pounds eight penny nails; shingling, four pounds per thousand shingles.

How to Ceil a Circular Plancier

Question: Enclosed find a sketch of a porch that I intend running a plancier of ⅝ by 4-inch ceiling. I have a ¾ circle on one corner and want to carry the ceiling around the circle. Could I divide it into an octagon or would the pieces be too long?

Answer: We herewith reproduce the sketch showing the ¾ part of the circle divided up into octagonal parts, which is all right, but would look

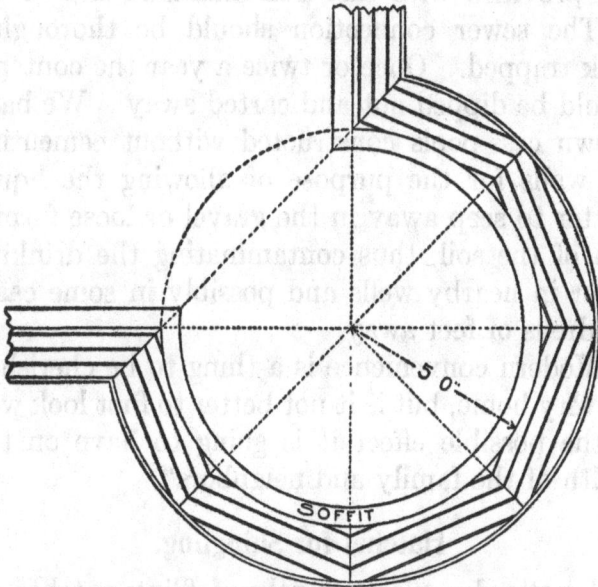

better if divided into twelve parts. This would show nine parts instead of six in the ¾ circle.

Constructing a Cess-pool

Question: Will you kindly advise me as to the best method of constructing a cess-pool; also for connecting same with bath room?

Answer: The building of cess-pools should be avoided whenever possible, but in many villages and country homes the proper drainage or sewer connections are not always to be had, and in that case the cess-pool is the last resort.

It should be built with as much care as constructing a cistern with walls of brick laid in mortar and well plastered with best quality of cement and provided with cast-iron man-hole and cover.

The sewer connection should be thoroughly back trapped. Once or twice a year the contents should be dipped out and carted away. We have known cess-pools constructed without cementing the walls for the purpose of allowing the liquid matter to seep away in the gravel or loose formation of the soil, thus contaminating the drinking water in nearby wells and possibly in some cases hundreds of feet away.

Modern convenience is a thing to be cherished in every home, but is it not better to first look well to the possible effect it is going to have on the health of the family and neighbors?

Hatchet for Shingling.

I noticed a plan recently of filing notches on the hatchet at the regular distance for courses. We use somewhat the same method here, but in place of the notches at four and one-half and five inches we drill small holes and thread them for short bolts which screw in. This leaves the end out, and in using no line is needed, as the bolt is drawn up against the course below, when the face of the hatchet gives the next course When used in this way in conjunction with the bracket shown, no chalk line or foot rest is needed. The bracket is very simple. Saw two pieces of seven-eighths stuff eight by twelve inches, and on the sloping

edge nail a strip about two inches wide; drive three
or four large nails in this strip and file to a point,
leaving them projecting about half an inch; nail a
board on top and one on front and your bracket is

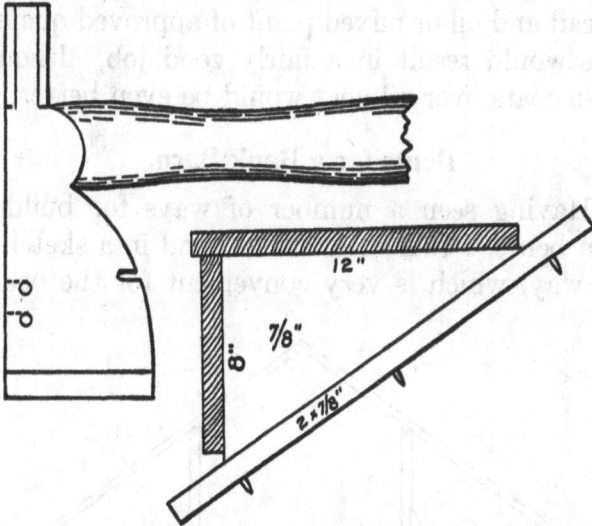

ready for use.   In drilling the holes in the hatchet
be sure to measure to the upper edge of hole and
not to center.

Repainting an Old House.

Question: What is the best way to paint an
old house from which most of the paint has gone?

Answer: If the old paint is peeling and scaling,
remove as much as possible by scraping or with
wire brushes. If powdered, brush well with
"duster." Give a coat of oil, mixed in the proportion of four gallons raw linseed oil, one gallon

turpentine, about one pint good turpentine driers. If there is still any scaling paint, scrape it off after oiling. When surface is dry, give one coat pure white lead and raw linseed oil, with necessary turpentine and driers, tinted as desired. Final coat of lead and oil or mixed paint of approved quality. This would result in a fairly good job, although three coats over oil coat would be even better.

Bents for a Bank Barn.

Having seen a number of ways for building barn bents, I thought I would send in a sketch of my way, which is very convenient for the use of

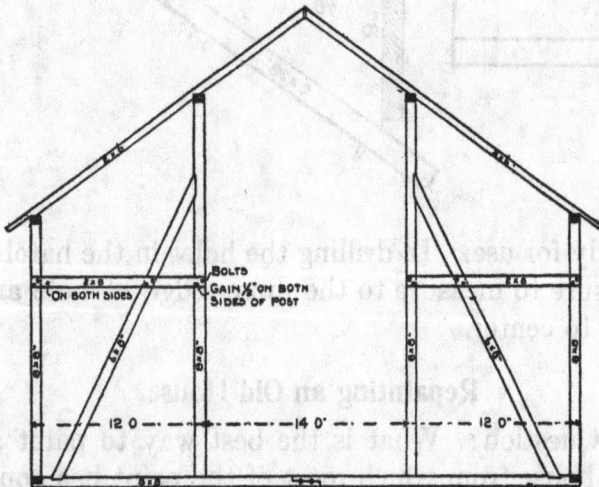

slings or hay forks. This makes a very simple, yet strong construction. A good way to build large sliding doors for barns to keep them from springing, is to cut the boards to the proper length and then stand them edgewise on the trestle and give the grooves a good coat of paint before nailing. This will prevent the siding from swelling. I have followed this way of putting doors together for several years and it has given good satisfaction.

How to Lay Out a Segment Arch

Question: Will you please publish a rule for laying out segments.

Answer: The segment is a sixth part of a circle. The radius being equal the chord, its true shape

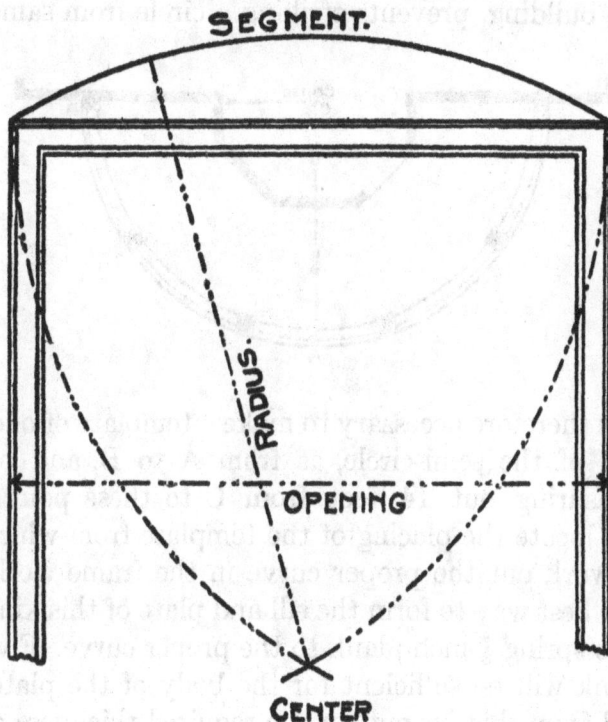

SEGMENT.

RADIUS

OPENING

CENTER

may very easily be found by letting the desired opening represent the chord. The intersection of the arcs locates the center, as shown in the illustration.

How to Build a Circular Porch

Question: Enclosed find a rough sketch of a circular porch to be built around a bay window or tower. Please give me the best method for constructing this porch. How to get the circle true and how to make the plate economical?

Answer: The center being at a point inside the building, prevents striking a circle from same.

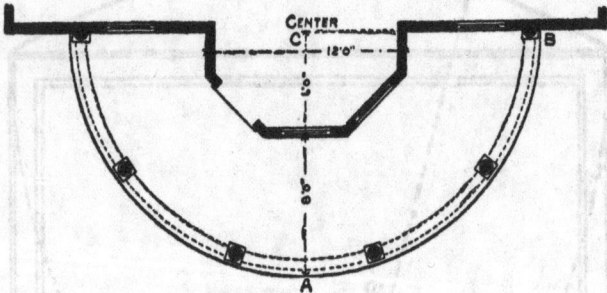

It is therefore necessary to make a template of one-half of the semi-circle, as from A to B, and by measuring out 14 feet from C to these points, will locate the placing of the template from which to work out the proper curve in the frame work. The best way to form the sill and plate of this kind is to spring ⅞-inch plank to the proper curve. Five plank will be sufficient for the body of the plate, and from this fur out for the required thickness of the soffit. It is not mentioned what kind of a roof is wanted on the porch. If it is to be a pitched roof with rafters radiating to the center, he will find that they will not strike the octagon on a level line, but will rise higher at the center of the sides and will have to be framed accordingly.

Best Method of Mixing Paint

Question: Please advise me as to the best method of mixing paints and its cost per square foot, including putting on.

Answer: Pure white lead and oil paint exclusive of labor of mixing will cost, say from one dollar and ten cents to one dollar and twenty-five cents per gallon, depending on price of oil and turpentine. Mixed paints of any good quality will range from one dollar and fifty cents to one dollar and seventy-five cents per gallon net. A gallon of paint will cover from four to seven hundred square feet of surface (perhaps even more), one coat, depending on the condition of the surface. Or it will cover from three to five hundred square feet, two coats. On an old painted surface in good condition the covering power is much greater.

Practical Method of Filing Saws

As there have been several questions asked in regard to saws, and as they have been of such a general nature, I have decided that the best way to answer them is to write a brief article covering the entire subject. I will try to be as brief as possible and cover the subject, although I fully realize it is a large one and a great deal should be written to fully give it justice.

As the saw is by far the most important of any of the carpenter's tools, yet how often we see a carpenter almost work his very life out of himself with a saw in bad shape, and yet not do even a fair day's work; again, how often we see mechanics leave an inside job (where the lumber was dry) with their tools in good order and go to framing

FIG. 1.

coarse, wet, cross-grained lumber. Their saws would cut fine until they got in a little ways, and then, as they would not be set enough for that kind of lumber, the saw would bind and it would be almost impossible to continue to push it until the piece was cut, simply because there did not happen to be a set on the job.

How easy these hard, unsatisfactory days could have been made by simply laying the saw down on some studding or joist on the trestles as shown in Fig. 1, and set with a common nail set, which would practically not dull it at all.

A common nail set makes the best saw set I know of to meet the emergency just mentioned. Many claim a hammer set is the only perfect set, while I find for general use the latest hand sets much more convenient. The ornamental nib on point of saw I am glad to say is not put on many of the best saws of to-day.

FIG. 2.

Fig. 2 shows the set in a saw which should always be just as little as possible and have the saw run free.

Fig. 3 is a rip saw, which should be filed square across for all ordinary work. The set gives all the bevel the teeth need, as rip saw teeth should march one after the other just like little chisels and

FIG. 3.

cut clear across the tooth and not simply cut on the outside edge as a cross-cut saw, which acts more like jackknife blades on each side of the saw.

Fig. 4 shows the jointing of a saw which is generally done with a flat file. A cut-off saw

FIG. 4.

should be jointed rounding, while a rip saw should be jointed perfectly straight, although many of them are jointed rounding, and some joint them hollowing.

Fig. 5 illustrates a good way to file a cut-off saw that has gotten into very bad shape.   After

FIG. 5.

the teeth are all made even and the same pitch (or rake), then give it proper bevel.

Fig. 6 shows a saw filed with about the right pitch and bevel for ordinary hardwood.

Fig. 7 shows the proper pitch, bevel and fleam, which is the bevel on the back of the tooth, for ordinary soft wood.   It also shows the file, which as you will note should point towards the point of

FIG. 6.

the saw.   Not only is that my opinion, but all the best authorities I have ever read on the subject give it the same way; still I am free to admit many good mechanics file just the other way.

Fig. 8 shows the groove in a cut-off saw. A needle should run down this groove the full length of the saw if it is well filed. To prove this state-

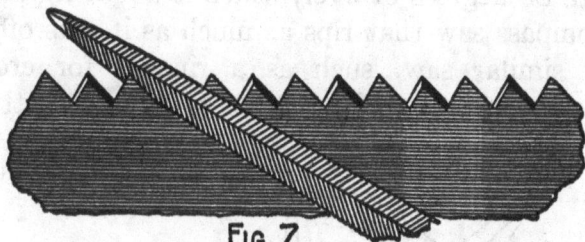

FIG. 7.

ment, I just tried my father's old course saw which is not very sharp and has had many kinks in its day. The needle went to full length. If an old saw about 40 years old, filed by a man nearly twice that proves the statement, new saws by young mechanics surely ought to.

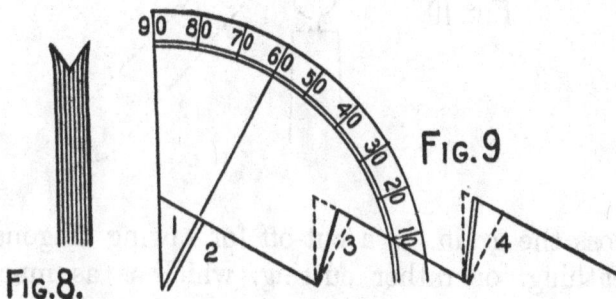

FIG. 9

FIG. 8.

Fig. 9 gives the degree of pitch. Tooth and the dotted lines show that the rip saw tooth should be on the square or at an angle of 90 degrees. I used to file even just a little sharper than that. Tooth 2, which is 60 degrees, is right for a general cut-off saw. If you wish the saw to cut fast, though possibly not quite as smooth, file it 70 degrees, or more like Tooth 3, while 4 shows a tooth 80 degrees or over, which is about right for a compass saw that rips as much as it cuts off or any similar saw, such as a rip-saw for cross-grained hardwood where it has to do some cutting

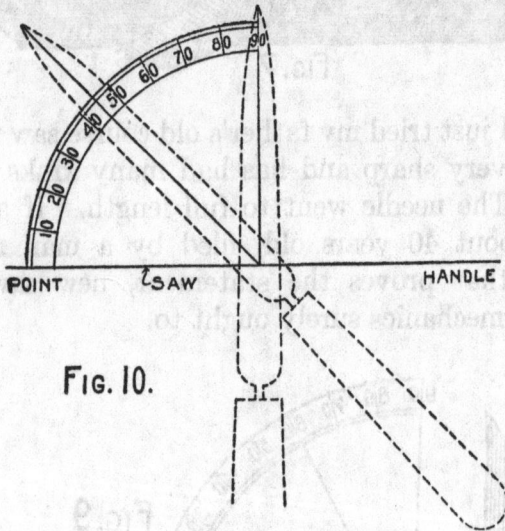

FIG. 10.

across the grain, or a cut-off for sawing diagonal sheathing, or rather cutting, which is as much ripping as cutting off.

Fig. 10 is looking right down onto the edges of the saw and shows that the rip saw should be filed square across or at an angle of 90 degrees with the saw, while a strictly cut-off saw at an angle of 45 degrees, and for the different classes of work the file should swing at different points between 90 degrees and 45 degrees.

Fig. 11 shows how the file should be held level for rip saws, and some even hold it level for all saws, and others drop the handle so the point is raised to about 30 degrees. I seldom raise the point of my file more than 10 degrees.

Anyone who is willing to give the time required to keep a saw in good order (and that time is time well spent) ought to be interested enough in his

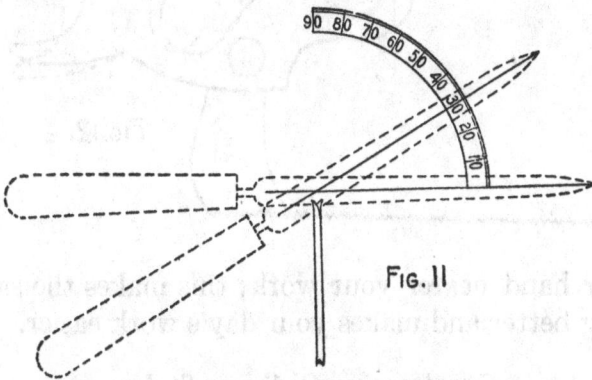

Fig. 11

saw to secure a good one, even if it does cost a little more. A cheap saw is a poor investment at any price, for the files and time it takes to keep it in order would soon pay for the very best.

While there have been no radical changes in saws in my time, yet there have been some improvements, the main improvement being in the perfecting of the steel, until to-day we have silver steel, which stands at the head. The perfection handle which is shown in Fig. 12 by the main lines

is also an improvement, while the dotted lines
show the old style.   The perfection handle is hung
as you will notice, more onto the saw and places

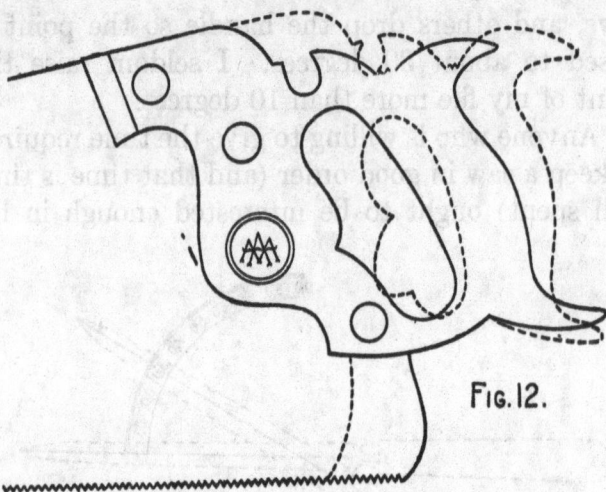

FIG. 12.

your hand nearer your work; this makes the saw
hang better and makes your day's work easier.

Constructing Ordinary Stairs

Our object will be to consider carefully the
various details of construction, but in order not to
be too general in the discussion we must limit ourselves somewhat.   A very satisfactory plan is to
take up the different kinds of stairs successively,
and as the housed string stair is one of the simplest,
and at the same time an important one, we will
begin with it.   This class of stairs may be divided
into two kinds.   First, where the stair is between
walls, that is, both strings are fastened to and

supported by the walls; and second, where only one of the strings is fastened to the walls, and the other, the face, or outside string, is free.

The first is the cheaper, and is used very much in small cottages, and also as a rear stair in the better grade of houses. Of course, we find very often both of these stairs without the housed

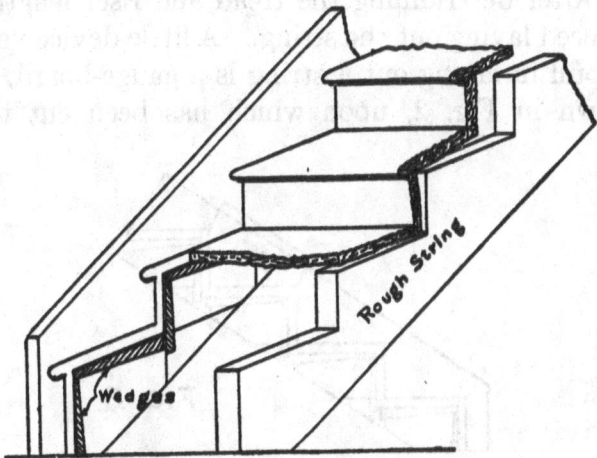

FIG. 1

string. The treads are carried on a rough string, and the finished string is fastened to the treads and risers by nailing through it into the treads and risers, but, as this is very poor construction, we will not discuss it.

As some may not understand the term "Housed String," we will explain. By housed string we mean that the string is notched out to receive the ends of the treads and risers. An examination of Fig. 1 will show clearly what is meant.

In the stair between two walls, which we are now considering, rough strings are unnecessary, unless the stair is over 2 feet 6 inches wide, when a rough string must be provided under the middle of the stair. The finished strings are fastened to the walls, and are more rigid than if a rough string was the means of support.

After determining the tread and riser lengths, proceed laying out the string. A little device very helpful in laying out a string is a gauge-board, as shown in Fig. 2, upon which has been cut the

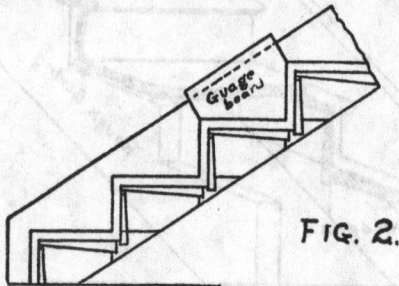

FIG. 2.

proper length of tread and riser to the pitch of the stair. In notching out the treads and risers the notches should be cut large enough to receive a small wedge below the tread and back of the riser. See Fig. 2. These are used to make a tight fit in front, where the treads and risers come against the edge of the notches. When the stairs is put together the wedges are covered with glue before being driven into place.

A closed string stair, that is, one between two walls, is built up against one of the walls before the second wall is built, which, when the stair is in place, is set up against it, and the string nearest to this wall is then fastened to the studding. If the stair is put in and lathing done afterwards pieces of inch stuff will have to be cut between the studding along the string to receive the ends of the lath. A better way, however, is to lath wall No. 1, before building the stair, then put in the stair, and when placing the studding of wall No. 2, leave the thickness of a lath clearance between the studs and the near string. Then lath this wall, shoving the lath through behind the string and nailing them below and above the stair. When this is done the string may be fastened to the studding by nailing through the string beneath the treads and risers.

Care must be taken, however, that the string lies well against the studding—even if necessary to put in thin blocking strips at each nailing place. Otherwise, there is danger of breaking the glue joint, where the treads and risers fit into the string.

A stair of this kind must always be put in before plastering and the plastering is finished down upon the string. The stair should be covered with paper and small strips of boards on the treads so as to preserve the unfinished wood work from the injurious effects of the plaster which might be spattered upon it. It is also a good idea to give the stair a coat of linseed oil before plastering as the wood work is then less easily affected by plaster.

There are several methods of joining the risers and treads. In Fig. 3 are shown the various methods of doing this. Some are considered better than others, but that is much a matter of opinion. At A in Fig. 3 is shown how the risers come down upon the tread and the tongue on the front side of the riser fits into a groove in the tread. If the fit is not good there will be a crack in the front of the riser, which will become more apparent as the stair becomes older. At G is another method. This is the very way when there is a

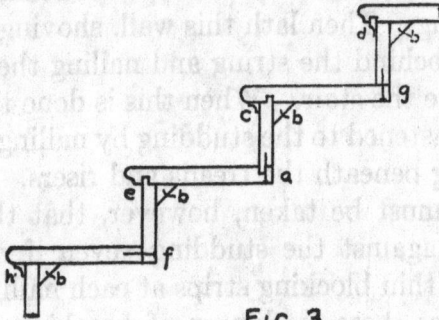

FIG. 3

rough string in under the middle of the stair. At F is a similar arrangement except that the riser goes down behind the tread. All of these joints should be nailed or fastened with wood screws.

At E and D, Fig. 3, are shown two ways of joining the tread and riser at the nosing. They are practically the same, the one at D having the moulding set into the groove, while at E a tongue is cut on the front of the riser to fit into the groove in the tread. In D and E there is the danger of having the nosing break off, because the tread, unless made of a rather thick piece of wood, may

crack over the groove. In C and H are two more
satisfactory methods. At C a tongue is cut on
the back edge of the riser and the groove in the
tread is as a result farther back from the front of
the tread than in E and D. A small moulding is
put under the nosing and there is little danger of
the nosing breaking off. In G no groove is cut in
the tread, so that the full strength of the tread is
preserved. In this case, however, it is almost
necessary to put a triangular strip, B, as shown, so
as to fasten the tread and riser together. This

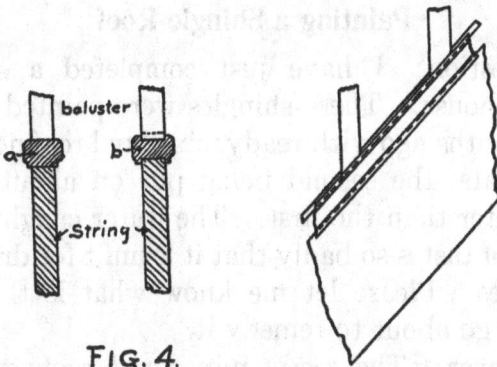

FIG. 4.

strip may be put on with good results in all the
cases, as it will stiffen up the work considerably.

Now we will take up the stair which has one
side open; that is, the string farthest from the wall,
which is the face string. Fig. 4 shows a section
and side elevation of this string. The treads and
risers are housed the same as in the other stair,
but the string must be made somewhat differently
to fit the conditions. The strip A is placed upon

the string to give it a better finish and may also be used to receive the lower ends of the balusters. These are beveled to fit the pitch of the string and fastened to the strip A. A better method is shown in B. The strip is grooved below to fit over the string, and above a groove is cut as wide as the balusters. These are set into the groove and a small strip as wide as the groove in B and a little thicker than the groove is deep, is fitted between the balusters. This gives a better finish to the joint between the string and balusters.

Painting a Shingle Roof

Question: I have just completed a shingle roofed house. These shingles were painted about two months ago with ready mixed red roofing paint two coats, the second being put on about three days later than the first. The water caught from this roof tastes so badly that it is unfit for drinking purposes. Please let me know what is the best way to go about to remedy it.

Answer: The ready mixed red roofing paint referred to is in all probability a paint made by mixing a mineral red or metallic paint, red oxide of iron, with linseed oil (probably more or less adulterated) with resin oil or mineral oil, since this is the usual composition of such paints. While such a paint would undoubtedly give a disagreeable taste to the water for some time and would discolor it to a certain extent, there is nothing poisonous about it, such as there would be in any white lead paint that could be used. As a rule, it is better to avoid painting a shingle roof if the water from it is to be used for drinking purposes. Any additional coats of paint applied to this roof would add to the difficulty, and moreover the paint would find its way into the crevices between the shingles, causing little dams, which would hold back the water and rot the shingles. The present condition will probably disappear in the course of a month or two at most. When the paint becomes powdered on the surface it may be given a coat of hot linseed oil, but other than that we should not advise any treatment. Care should be taken in heating the oil to avoid fire. The best way is to put the can containing it into a large kettle of water, which is brought to a gentle boil over a slow fire. Cold raw linseed oil will answer the purpose, but will not penetrate the wood as well as hot oil. It might be well to add here that while dipping the shingles in creosote stain is found to preserve them, painted shingles do not last any longer than unpainted. Creosote, however, will give a very disagreeable taste to water taken from a roof where such stains are used.

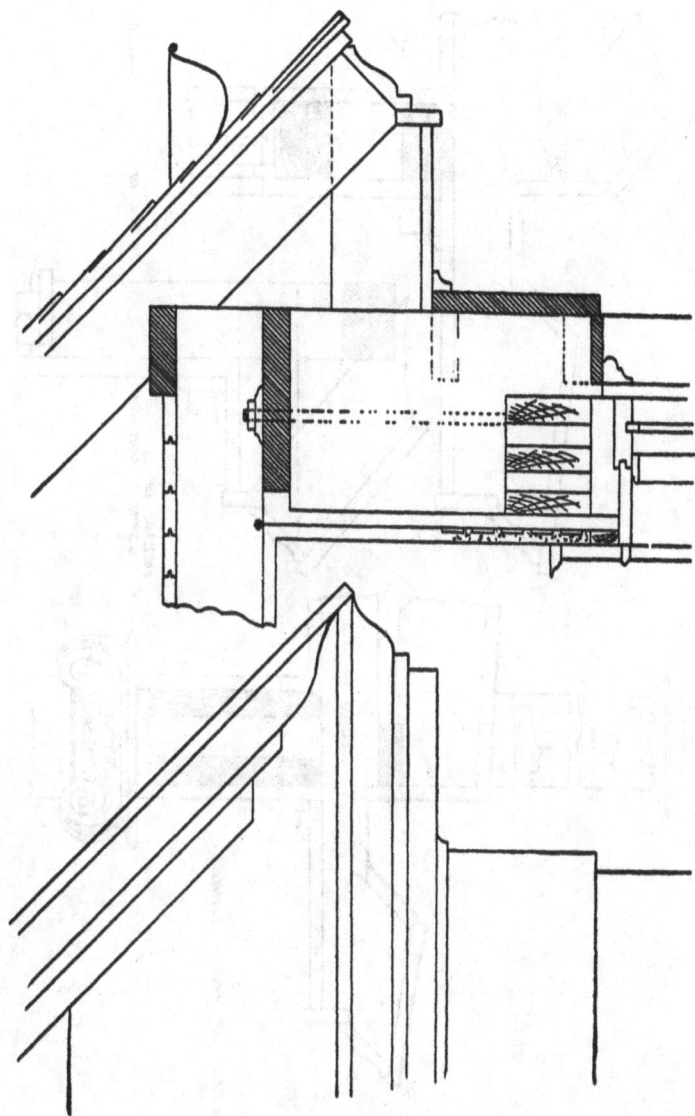

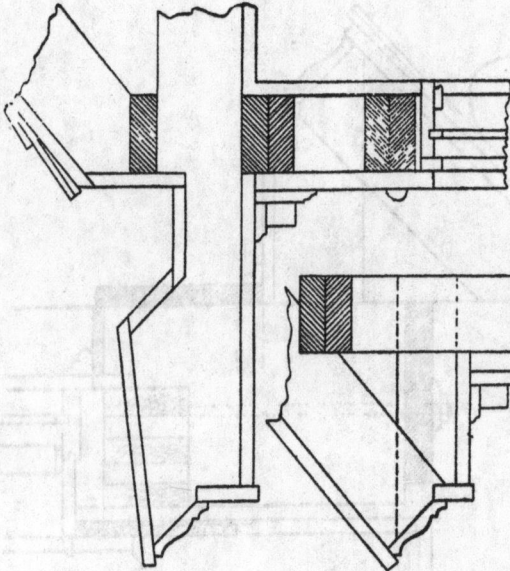

DETAIL OF MAIN CORNICE

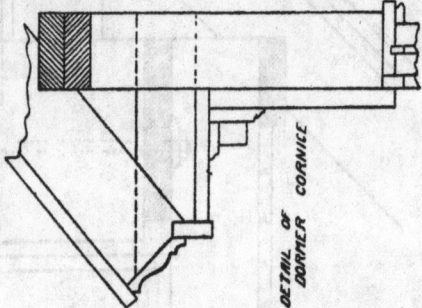

DETAIL OF DORMER CORNICE

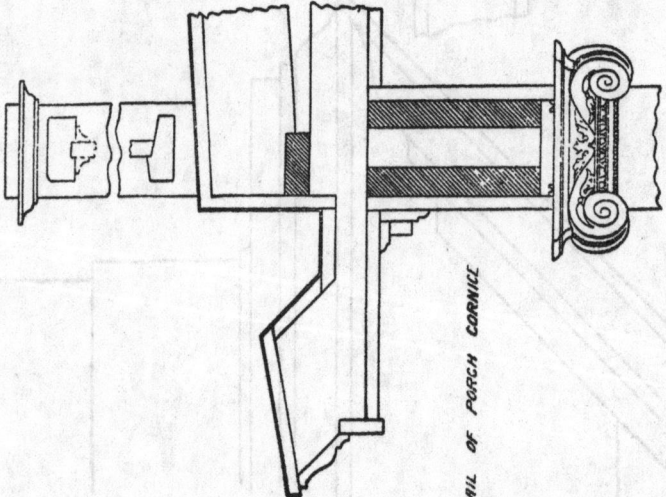

DETAIL OF PORCH CORNICE

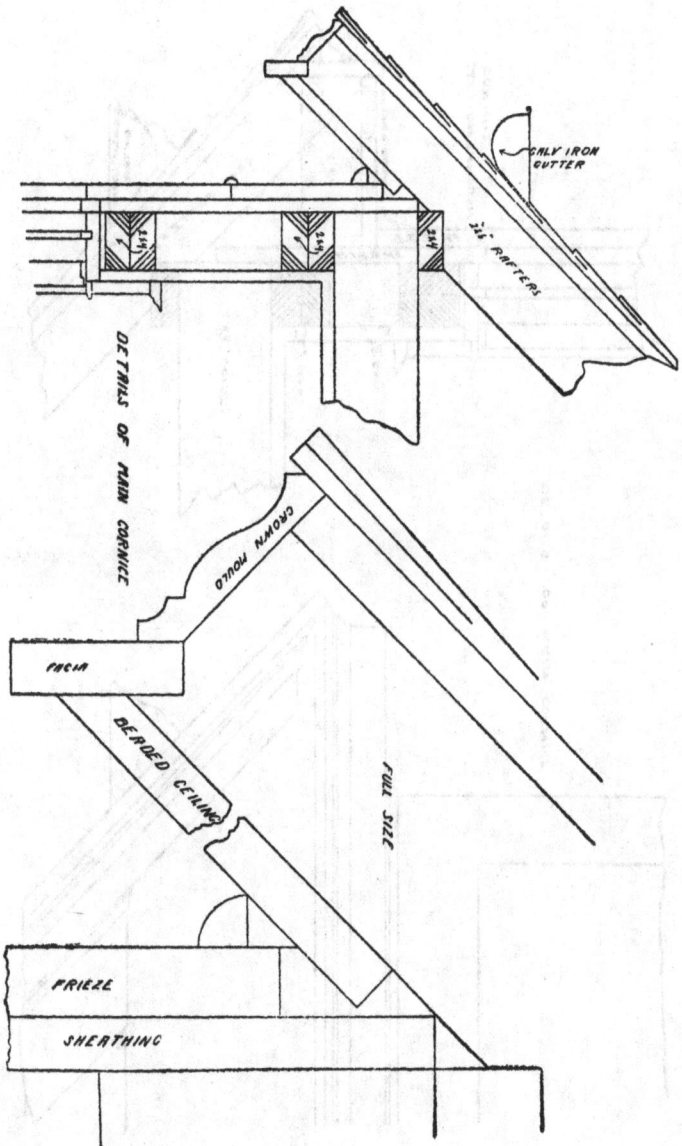

DETAILS OF MAIN CORNICE

GALV IRON
GUTTER

3x5" RAFTERS

4x6"

4x6"

4x4"

CROWN MOULD

FACIA

BEADED CEILING

FULL SIZE

FRIEZE

SHEATHING

SECTION THROUGH CORNICE

EARMS CEILING

2 x 4

2 x 6" RAFTERS

SHEATHING

GALV. IRON
RAFTER

FRIEZE

SHEATHING

2 x 4

2 - x 4

2 x 8" JOIST

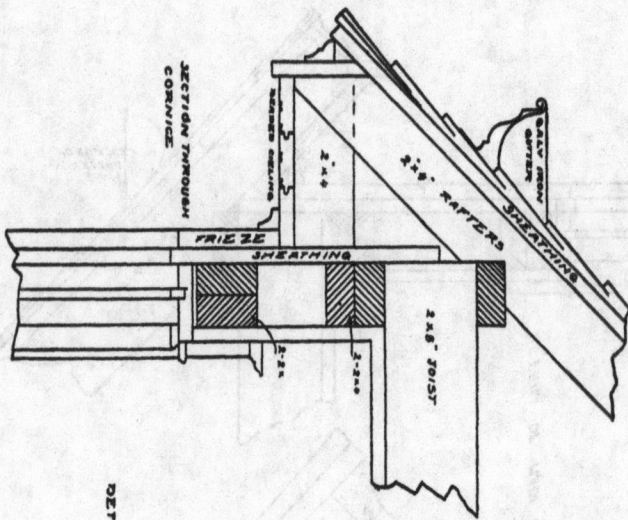

DETAILS OF MAIN CORNICE

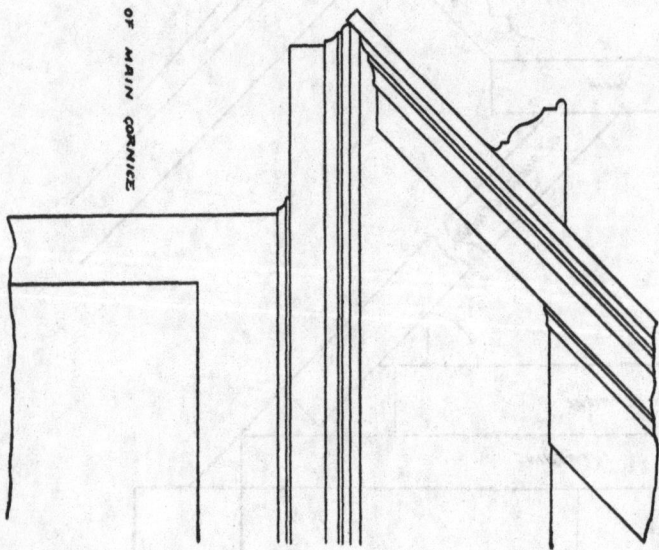

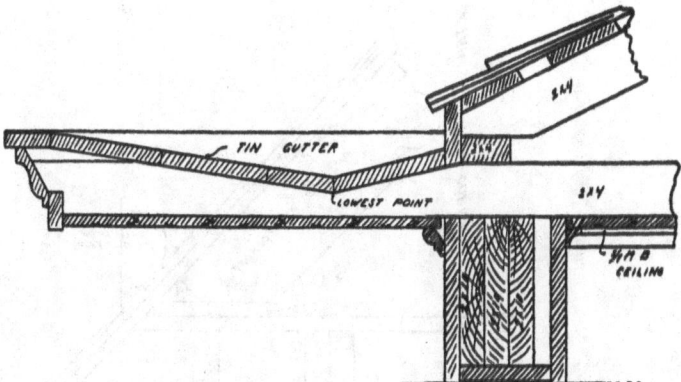

TIN GUTTER

LOWEST POINT

2x4

2x4

W H B
CEILING

1'

1'-0"

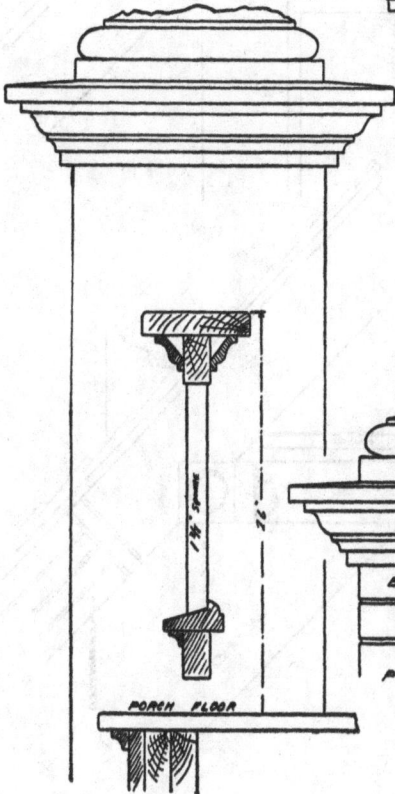

1 1/4" SQUARE

72"

PORCH FLOOR

10

BRICK

PORCH DETAIL

DETAIL OF MAIN CORNICE.

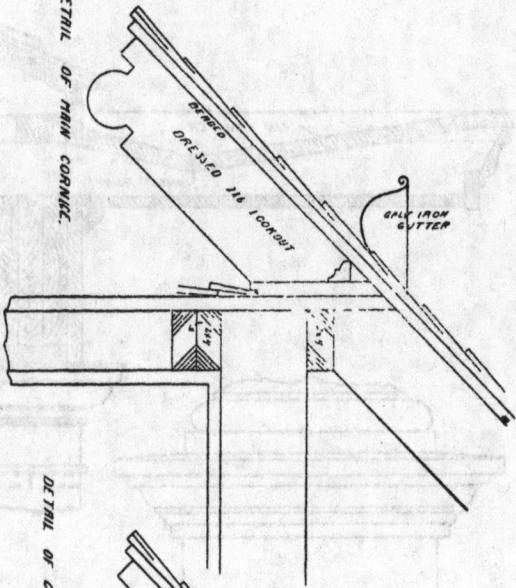

BEADED

DRESSED 1x6 CONDUIT

GALV IRON GUTTER

DETAIL OF GABLE

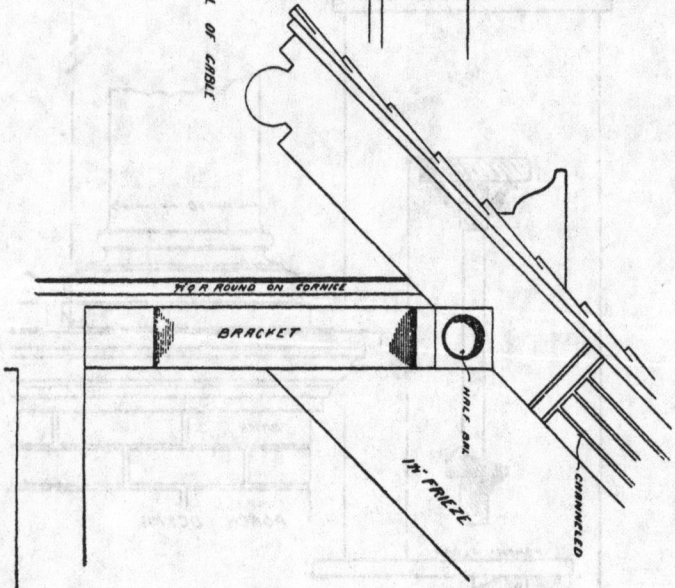

WG ¾ ROUND ON CORNICE

BRACKET

HALF BRK

CHANNELLED

14" FRIEZE

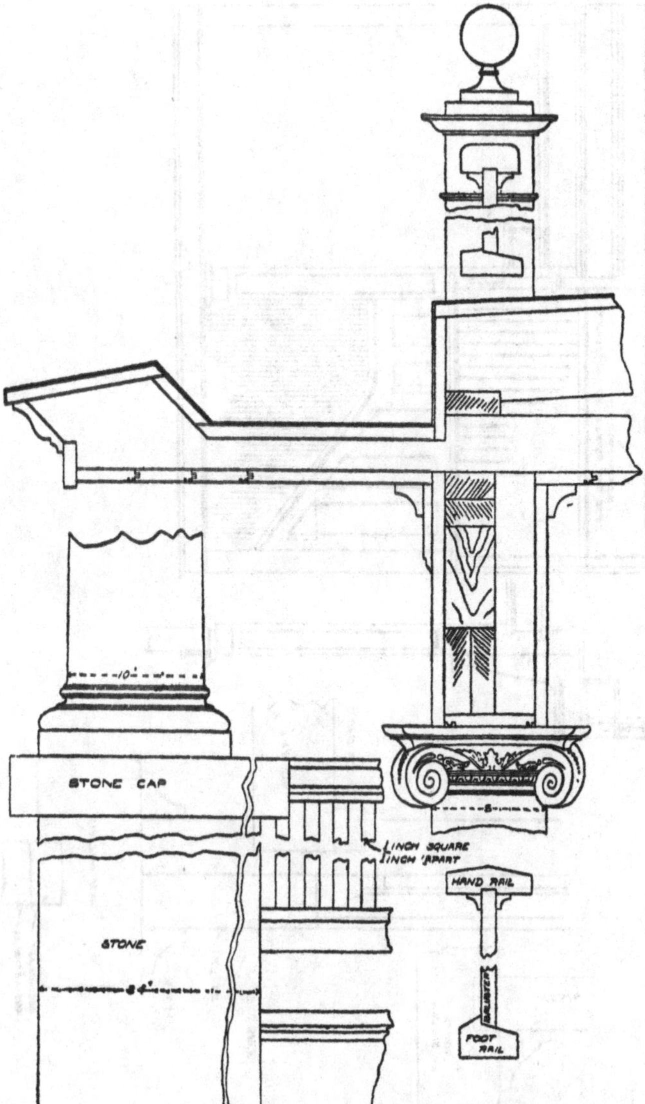
STONE CAP
LINCH SQUARE
LINCH APART
HAND RAIL
BALUSTER
FOOT RAIL
STONE

DETAILS OF PORCH

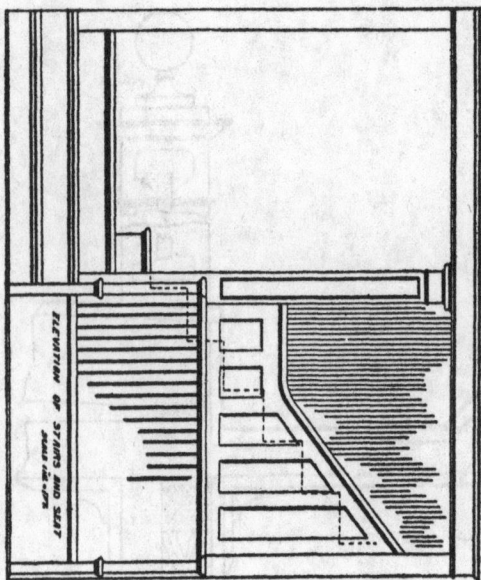

ELEVATION OF STAIRS AND SEAT

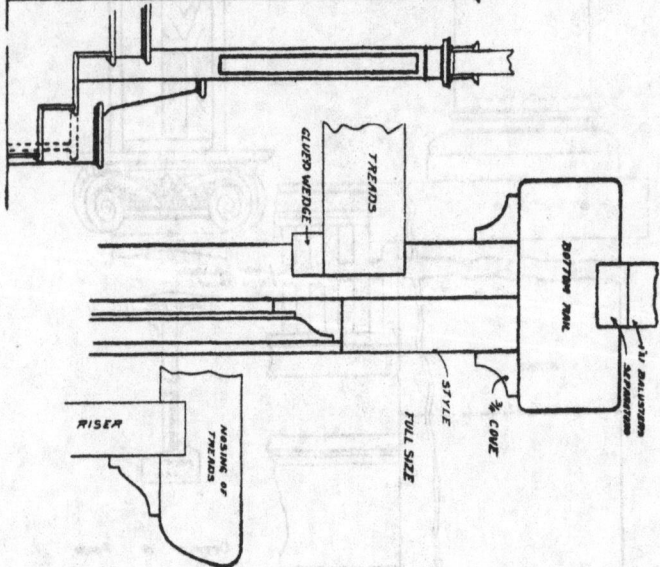

GLUED WEDGE

TREADS

RISER

NOSING OF TREADS

BOTTOM RAIL

STYLE

FULL SIZE

3/8 COVE

1/2 BALUSTERS

SEPARATIONS

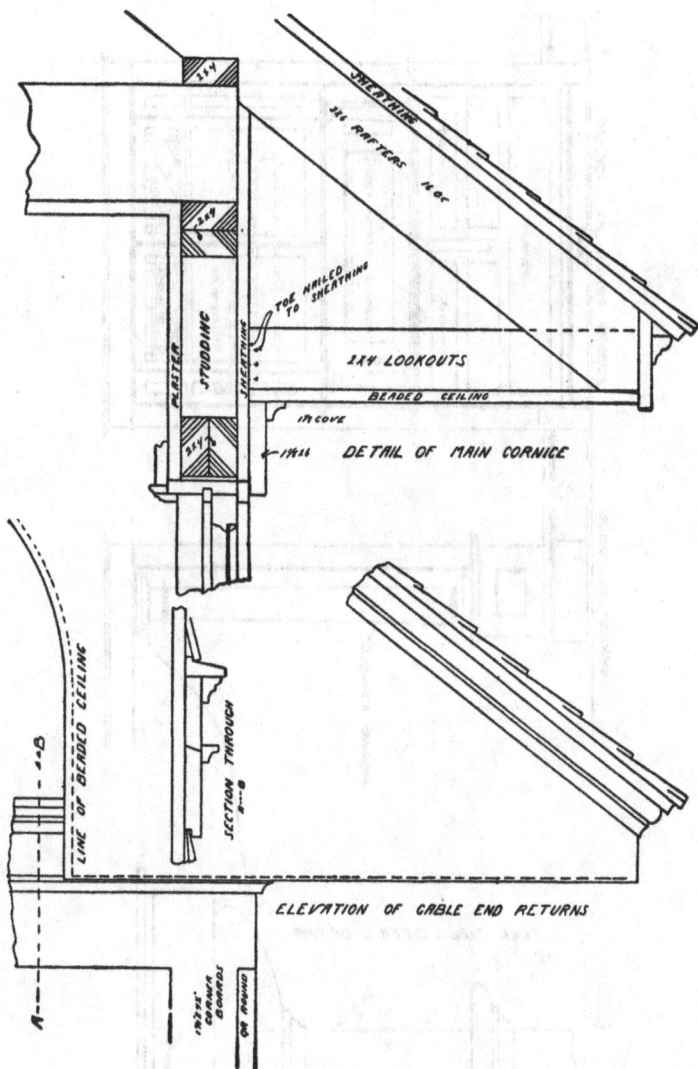

2 x 4

2 x 4

PLATE

STUDDING

TOE NAILED
TO SHEATHING

SHEATHING

2 x 4 LOOKOUTS

BEADED CEILING

IN COVE

2 x 4 ⅛

⅞ x 2⅛

1 x 1 DETAIL OF MAIN CORNICE

SECTION THROUGH A—B

LINE OF BEADED CEILING

A—B

A

13 x 2¼
CORNER
BOARDS

OR ROUND

ELEVATION OF GABLE END RETURNS

2 x 4 RAFTERS 16 OC

SHEATHING

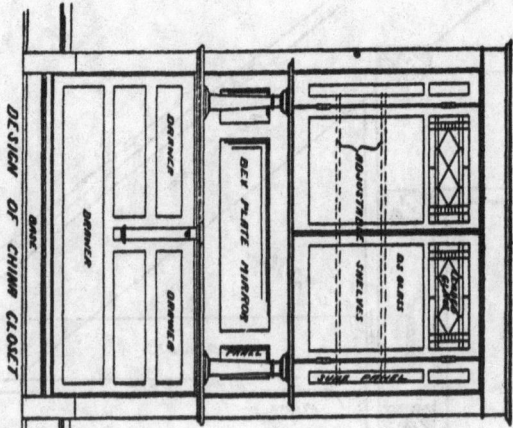

DESIGN OF CHINA CLOSET

BAR

DRAWER

DRAWER

DRAWER

BEV PLATE MIRROR

PANEL

ADJUSTABLE SHELVES

ART GLASS

TURNED DOWEL

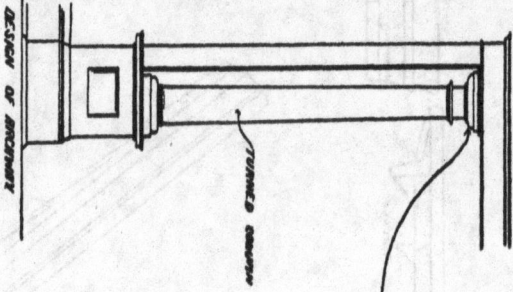

DESIGN OF BROOKERY

TURNED COLUMN

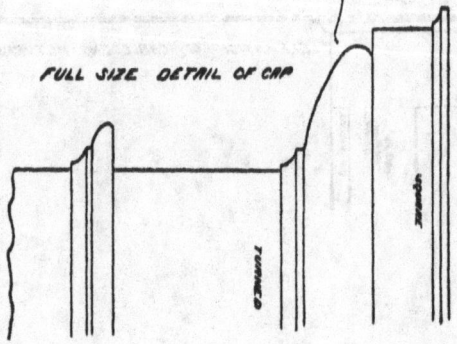

FULL SIZE DETAIL OF CAP

TURNED

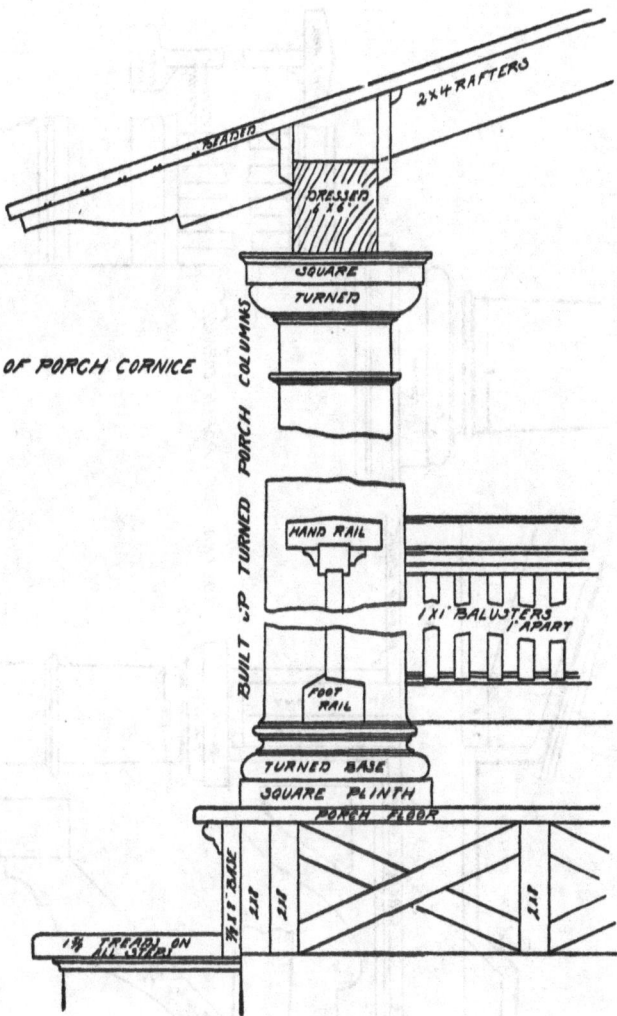

DETAILS OF PORCH CORNICE

2X4 RAFTERS

SHEATHED

DRESSED 6 X 6

SQUARE

TURNED

BUILT UP TURNED PORCH COLUMNS

HAND RAIL

1X1" BALUSTERS 1" APART

FOOT RAIL

TURNED BASE

SQUARE PLINTH

PORCH FLOOR

3/4 X 12" BASE

2 X 4

2 X 4

2 X 8

1 3/4 TREADS ON ALL STEPS

HAND RAIL

ROOF RAIL

1" SQUARE 1" APART

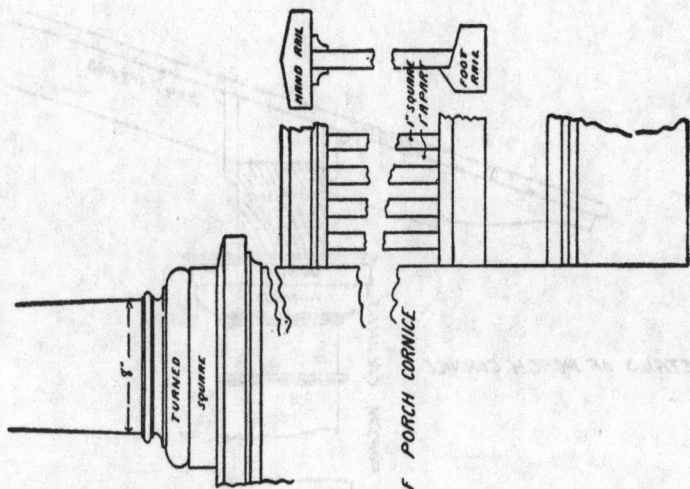

DETAIL OF PORCH CORNICE

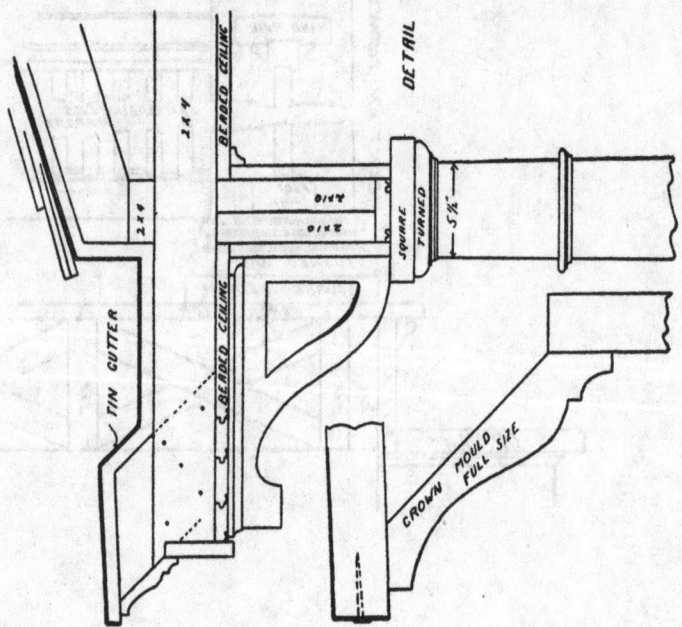

5"

TURNED

SQUARE

TIN GUTTER

2 x 4

2 x 4

BEADED CEILING

2 x 10

2 x 10

BEADED CEILING

SQUARE

TURNED

5 ½"

CROWN MOULD FULL SIZE

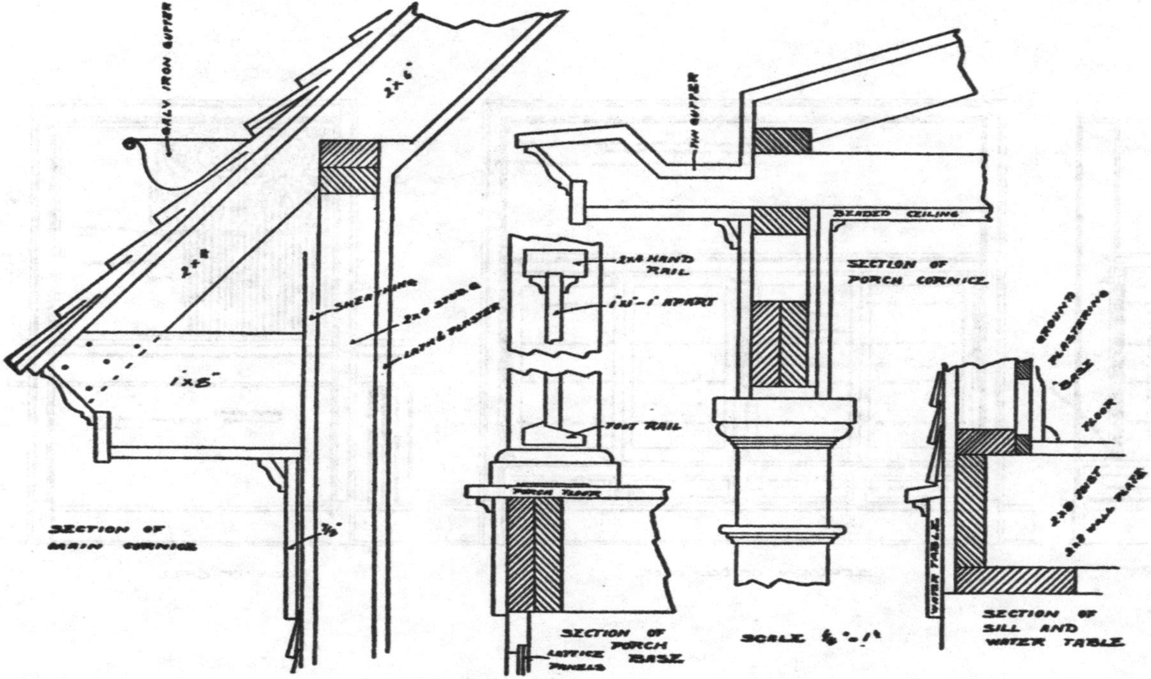

SECTION OF MAIN CORNICE

GALV IRON GUTTER

1 x 8

1 x 8

1 x 6

SHEATHING

1 x 8 FASCIA

SECTION OF OFFICE PORCH RAIL

SECTION OF PORCH CORNICE

1½" - 1" APART

2 x 4 HAND RAIL

TIN GUTTER

BEADED CEILING

SCALE ¾" = 1'-0"

SECTION OF SILL AND WATER TABLE

WATER TABLE

2 x 8 JOIST

2 x 4 WALL PLATE

FLOOR

SILL

GROUND

PLASTERING

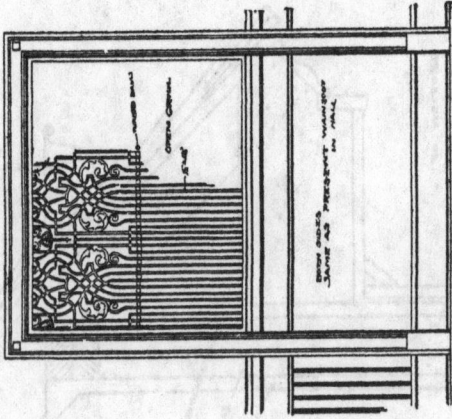

DESIGN OF GRILL

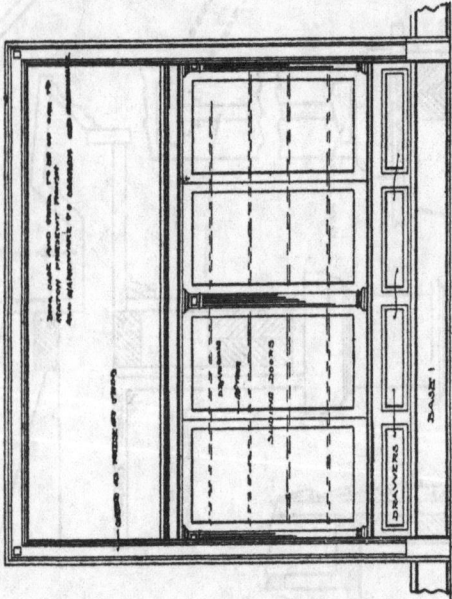

ELEVATION OF BOOK CASE

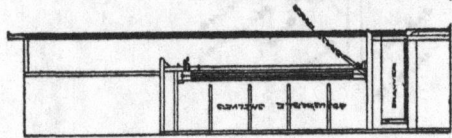

SECTION

DRAWERS

BASE 1

Index to Volume 1

Practical House Plans

WE ILLUSTRATE IN THIS BOOK the perspective view and floor plans of 50 low and medium-priced houses. In the preparation of this work great care has been exercised in the selection of original, practical and attractive house designs, such as seventy-five to ninety per cent of the people to-day wish to build. In drawing these plans special effort has been made to provide for the MOST ECONOMICAL CONSTRUCTION, thereby giving the home builder and contractor the benefit of the saving of many dollars; for in no case have we put any useless expense upon the building simply to carry out some pet idea. Every plan illustrated will show, by the complete working plans and specifications, that we give you designs that will work out to the best advantage and will give you the most for your money; besides every bit of space has been utilized to the best advantage.

$50.00 PLANS FOR ONLY $5.00 This department has for its foundation the best equipped architectural establishment ever maintained for the purpose of furnishing the public with complete working plans and specifications at the remarkably low price of only $5.00 per set. Every plan we illustrate has been designed by a licensed architect, who stands at the head of his profession in this particular class of work and has made a specialty of low and medium-priced houses. The price usually charged for this work is from $50.00 to $75.00.

WHAT WE GIVE YOU The first question you will ask is, "What do we get in these complete working plans and specifications? Of what do they consist? Are they the cheap printed plans on tissue paper without details or specifications?" We do not blame you for wishing to know what you will get for your money.

BLUE PRINTED WORKING PLANS

The plans we send out are the regular blue printed plans, drawn one-quarter inch scale to the foot, showing all the elevations, floor plans and necessary interior details. All of our plans are printed by electricity on an electric circular blue-printing machine, and we use the very best grade of electric blue-printing paper; every line and figure showing perfect and distinct.

FOUNDATION AND CELLAR PLANS

This sheet shows the shape and size of all walls, piers, footings, posts, etc., and of what materials they are constructed; shows the location of all windows, doors, chimneys, ash-pits, partitions, and the like. The different wall sections are given, showing their construction and measurements from all the different points.

FLOOR PLANS

These plans show the shape and size of all rooms, halls and closets; the location and size of all doors and windows; the position of all plumbing fixtures, gas lights, registers, pantry work, etc., and all the measurements that are necessary are given.

ELEVATIONS

A front, right, left and rear elevation are furnished with all the plans. These drawings are complete and accurate in every respect. They show the shape, size and location of all doors, windows, porches, cornices, towers, bays, and the like; in fact, give you an exact scale picture of the house as it should be at completion. Full wall sections are given showing the construction from foundation to roof, the height of stories between the joists, height of plates, pitch of roof, etc.

ROOF PLAN This plan is furnished where the roof construction is at all intricate. It shows the location of all hips, valleys, ridges, decks, etc. All the above drawings are made to scale one-quarter inch to the foot.

DETAILS All necessary details of the interior work, such as door and window casings and trim, base, stools, picture moulding, doors, newel posts, balusters, rails, etc., accompany each set of plans. Part is shown in full size, while some of the larger work, such as stair construction, is drawn to a scale of one and one-half inch to the foot. These blue prints are substantially and artistically bound in cloth and heavy water-proof paper, making a handsome and durable covering and protection for the plans.

SPECIFICATIONS The specifications are typewritten on Lakeside Bond Linen paper and are bound in the same artistic manner as the plans, the same cloth and water-proof paper being used. They consist of from about sixteen to twenty pages of closely typewritten matter, giving full instructions for carrying out the work. All directions necessary are given in the clearest and most explicit manner, so that there can be no possibility of a misunderstanding.

BASIS OF CONTRACT The working plans and specifications we furnish can be made the basis of contract between the home builder and the contractor. This will prevent mistakes, which cost money, and they will prevent disputes which are unforeseen and never settled satisfactorily to both parties. When no plans are used the contractor is often obliged to do some work he did not figure on, and the home builder often does not get as much for his money as he expected, simply because there was no basis on which to work and upon which to base the contract.

NO MISUNDERSTANDING CAN ARISE when a set of our plans and specifica-

tions is before the contractor and the home builder, showing the interior and exterior construction of the house as agreed upon in the contract. Many advantages may be claimed for the complete plans and specifications. They are time savers and, therefore, money savers. Workmen will not have to wait for instructions when a set of plans is left on the job. They will prevent mistakes in cutting lumber, in placing door and window frames, and in many other places where the contractor is not on the work and the men have received only partial or indefinite instructions. They also give instructions for the working of all material to the best advantage.

FREE PLANS FOR FIRE INSURANCE ADJUSTMENT You take

every precaution to have your house covered by insurance; but do you make any provision for the adjustment of the loss, should you have a fire? There is not one man in ten thousand who will provide for this embarrassing situation. You can call to mind instances in your own locality where settlements have been delayed because the insurance companies wanted some proof which could not be furnished. They demand proof of loss before paying insurance money, and they are entitled to it. We have provided for this and have inaugurated the following plan, which cannot but meet with favor by whoever builds a house from our plans.

IMMEDIATELY UPON RECEIPT OF INFORMATION from you

that your house has been destroyed by fire, either totally or partially, we will forward you, free of cost, a duplicate set of plans and specifications, and in addition we will furnish an affidavit giving the number of the design and the date when furnished, to be used for the adjustment of the insurance.

WITHOUT ONE CENT OF COST TO YOU

and without one particle of trouble. We keep a record of the number of the house design and the date it was furnished, so that, in time of loss, all it will be necessary for you to do is to drop us a line and we will furnish the only reliable method of getting a speedy and satisfactory adjustment. This may be the means of saving you hundreds of dollars, besides much time and worry.

OUR LIBERAL PRICES

Many have marveled at our ability to furnish such excellent and complete working plans and specifications at such low prices. We do not wonder at this, because we charge but $5.00 for a more complete set of working plans and specifications than you would receive if ordered in the ordinary manner, and when drawn especially for you, at a cost of from fifty to seventy-five dollars. On account of our large business and unusual equipment, and owing to the fact that WE DIVIDE THE COST of these plans among so many, it is possible for us to sell them at these low prices. The margin of profit is very close, but it enables us to sell thousands of sets of plans, which save many times their cost to both the owner and the contractor in erecting even the smallest dwelling.

OUR GUARANTEE

Perhaps there are many who feel that they are running some risk in ordering plans at a distance. We wish to assure our customers that there is no risk whatever. If, upon receipt of these plans, you do not find them exactly as represented, if you do not find them complete and accurate in every respect, if you do not find them as well prepared as those furnished by any architect in the country, or any that you have ever seen, we will refund your money upon the return of the plans from you in perfect condition. All of our plans are prepared by architects standing at the head of their profession, and the standard of their work is the very highest We could not afford to make this guarantee if we were not positive that we were furnishing the best plans put out in this country, even though our price is not more than one-seventh to one-tenth of the price usually charged.

BILL OF MATERIAL We do not furnish a bill of material. We state this here particularly, as some people have an idea that a bill of material should accompany each set of plans and specifications. In the first place, our plans are gotten up in a very comprehensive manner, so that any carpenter can easily take off the bill of materia' without any difficulty. We realize that there are hardly two sections of the country where exactly the same kinds of materials are used, and, moreover, a bill which we might furnish would not be applicable in all sections of the country. We furnish plans and specifications for houses which are built as far north as the Hudson Bay and as far south as the Gulf of Mexico. They are built upon the Atlantic and Pacific Coasts, and you can also find them in Australia and South Africa. Each country and section of a country has its peculiarities as to sizes and qualities; therefore, it would be useless for us to make a list that would not be universal. Our houses, when completed, may look the same whether they are built in Canada or Florida, but the same materials will not be used, for the reason that the customs of the people and the climatic conditions will dictate the kind and amount of materials to be used in their construction.

ESTIMATED COST It is impossible for anyone to estimate the cost of a building and have the figures hold good in all sections of the country. We do not claim to be able to do it. The estimated cost of the houses we illustrate is based on the most favorable conditions in all respects and includes everything but the plumbing and heating. We are not familiar with your local conditions, and, should we claim to know the exact cost of a building in your locality, a child would know that our statement was false. We leave this matter in the hands of the reliable contractors, for they, and they alone, know your local conditions.

WE WISH TO BE FRANK WITH YOU

and therefore make no statement that we cannot substantiate in every respect. If a plan in this book pleases you; if the arrangement of the rooms is satisfactory, and if the exterior is pleasing and attractive, then we make this claim—that it can be built as cheaply as if any other architect designed it, and we believe cheaper.

WE HAVE STUDIED ECONOMY

in construction, and our knowledge of all the material that goes into a house qualifies us to give you the best for your money. We give you a plan that pleases you, one that is attractive, and one where every foot of space is utilized at the least possible cost. Can any architect do more, even at seven to ten times the price we charge you for plans?

REVERSING PLANS

We receive many requests from our patrons for plans exactly according to the designs illustrated, with the one exception of having them reversed or placed in the opposite direction. It is impossible for us to make this change and draw new plans, except at a cost of about eight times our regular price. We see no reason why our regular plans will not answer your purpose. Your carpenter can face the house exactly as you wish it, and the plans will work out as well facing in one direction as in another. We can, however, if you wish, and so instruct us, make you a reversed blue print and turnish it at our regular price; but in that case all the figures and letters will be reversed and, therefore, liable to cause as much confusion as if your carpenter reversed the plan himself while constructing the house.

WE WOULD ADVISE however, in all cases where the plan is to be reversed, and there is the least doubt about the contractor not being able to work from the plans as we have them, that two sets of blue prints be purchased, one regular and the other reversed, and in such cases we will furnish two sets of blue prints and one set of specifications for only fifty per cent added to the regular cost, making the $5.00 plan cost only $7.50.

Design No. 2015

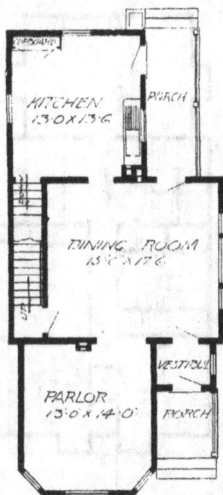

First Floor Plan

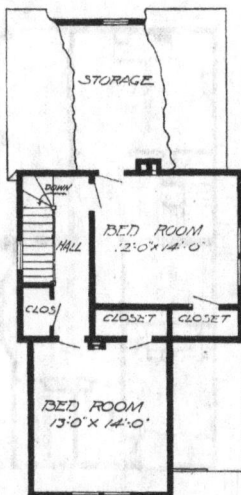

Second Floor Plan

PRICE
of Blue
Prints, to-
gether with
a complete
set of type-
written
specifica-
tions is

ONLY

$5.00

We mail
Plans and
Specifica-
tions the
same day
order is re-
ceived.

Size: Width, 22 feet; length, 44 feet 6 inches

Blue prints consist of basement plan; roof plan; first and second floor plans; front, rear, two side elevations, wall sections and all necessary interior details. Specifications consist of about twenty pages of typewritten matter.

Full and complete working plans and specifications of this house will be furnished for $5.00. Cost of this house is from about $1,550.00 to about $1,750.00, according to the locality in which it is built.

Design No. 2036

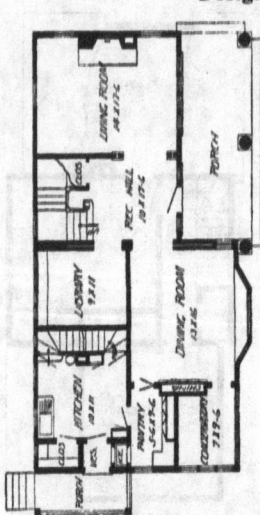

First Floor Plan

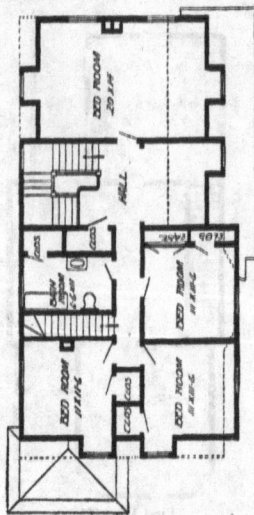

Second Floor Plan

Size: Width, 52 feet 6 inches; length, 27 feet, exclusive of porches

Blue prints consist of basement plan; roof plan; first and second floor plans; front, rear, two side elevations; wall sections and all necessary interior details. Specifications consist of about twenty pages of typewritten matter.

Full and complete working plans and specifications of this house will be furnished for $5.00. Cost of this house is from about $3,000.00 to about $3,250.00 according to the locality in which it is built.

Design No. 2003

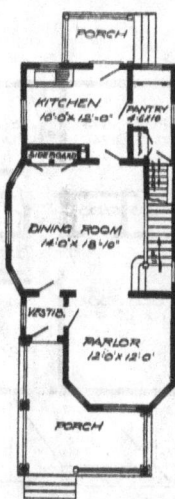

First Floor Plan

KITCHEN 10'0"x12'0"

PANTRY 4'6"x10'

SIDE BOARD

DINING ROOM 14'0"x18'10"

VESTIB.

PARLOR 12'0"x12'0"

PORCH

PORCH

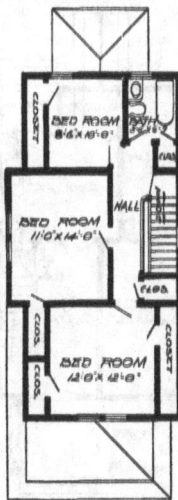

Second Floor Plan

CLOSET

BED ROOM 8'6"x18'0"

BATH

HALL

BED ROOM 11'0"x14'0"

CLOS.

CLOS.

CLOS.

BED ROOM 12'0"x12'6"

CLOSET

PRICE

of Blue Prints, together with a complete set of typewritten specifications is

ONLY

$5.00

We mail Plans and Specifications the same day order is received.

Size: Width, 19 feet 10 inches; length, 38 feet, exclusive of porches

Blue prints consist of basement plan; first and second floor plans; front, two side elevations; wall sections and all necessary interior details. Specifications consist of about twenty pages of typewritten matter.

Full and complete working plans and specifications of this house will be furnished for $5.00. Cost of this house is from about $2,550.00 to about $2,800.00, according to the locality in which it is built.

Design No. 2009

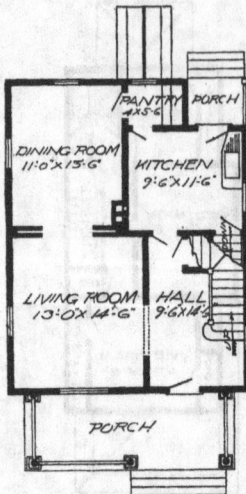

First Floor Plan:

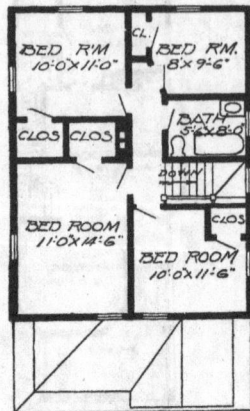

DINING ROOM 11'0"X15'6"

PANTRY 4'X5'6"

PORCH

KITCHEN 9'6"X11'6"

LIVING ROOM 13'0"X14'6"

HALL 9'6"X14'6"

PORCH

Second Floor Plan:

BED R'M 10'0"X11'0"

CL.

BED R'M 8'X9'6"

CLOS

CLOS

BATH 5'6"X8'0"

BED ROOM 11'0"X14'6"

CLOS

BED ROOM 10'0"X11'6"

First Floor Plan   **Second Floor Plan**

Size: Width, 24 feet; length, 30 feet, exclusive of porch

Blue prints consist of cellar and foundation plan; first and second floor plans; front, rear, two side elevations; wall sections and all necessary interior details. Specifications consist of about twenty pages of typewritten matter.

Full and complete working plans and specifications of this house will be furnished for $5.00. Cost of this house is from about $1,950.00 to about $2,150.00, according to the locality in which it is built.

Design No. 2049

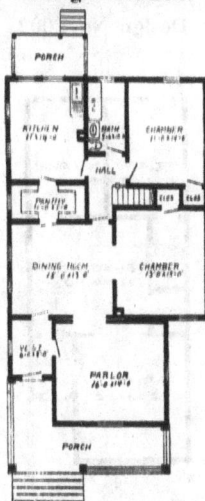

PRICE

of Blue Prints, together with a complete set of typewritten specifications is

ONLY

$5.00

We mail Plans and Specifications the same day order is received.

Floor Plan

Size: Width, 29 feet 6 inches; length, 49 feet, exclusive of porches

Blue prints consist of basement plan; roof plan; floor plan; front, rear, two side elevations; wall sections and all necessary interior details. Specifications consist of about fifteen pages of typewritten matter.

Full and complete working plans and specifications of this house will be furnished for $5.00. Cost of this house is from about $1,650.00 to about $1,800.00, according to the locality in which it is built.

Design No. 2002

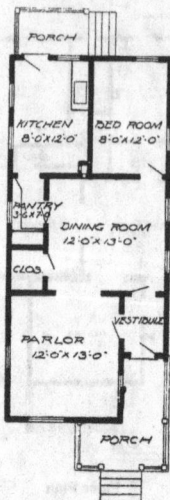

Floor Plan

Size: Width, 18 feet; length, 39 feet, exclusive of porehes

Blue prints consist of foundation plan; roof plan; floor plan; front, rear, two side elevations; wall sections and all necessary interior details. Specifications consist of about fifteen pages of typewritten matter.

Full and complete working plans and specifications of this house will be furnished for $5.00. Cost of this house is from about $850.00 to about $1,000.00, according to the locality in which it is built.

Design No. 2033

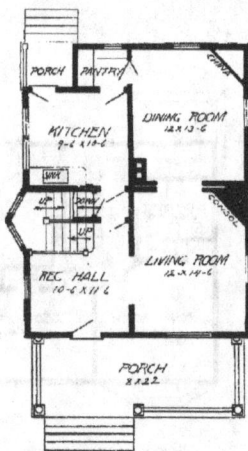

First Floor Plan

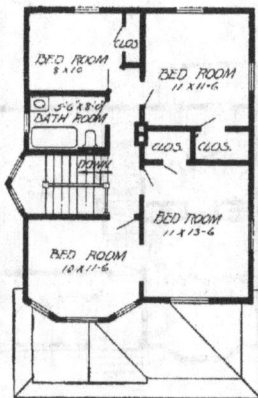

Second Floor Plan

Size: Width, 24 feet; length, 30 feet, exclusive of porch

Blue prints consist of cellar and foundation plan; first and second floor plans; front, two side elevations; wall sections and all necessary interior details. Specifications consist of about twenty pages of typewritten matter.

Full and complete working plans and specifications of this house will be furnished for $5.00. Cost of this house is from about $1,550.00 to about $1,750.00, according to the locality in which it is built.

Design No. 2039

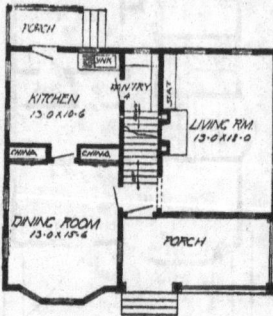

First Floor Plan

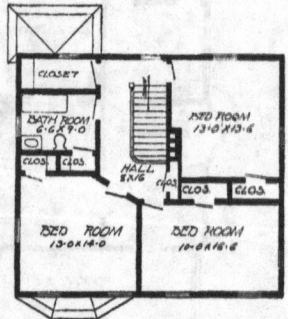

Second Floor Plan

Size: Width, 32 feet; length, 27 feet 6 inches, exclusive of porches

Blue prints consist of basement plan; roof plan; first and second floor plans; front, rear, two side elevations; wall sections and all necessary interior details. Specifications consist of about twenty pages of typewritten matter.

Full and complete working plans and specifications of this house will be furnished for $5.00. Cost of this house is from about $1,750.00 to about $1,950.00, according to the locality in which it is built.

Design No. 2022

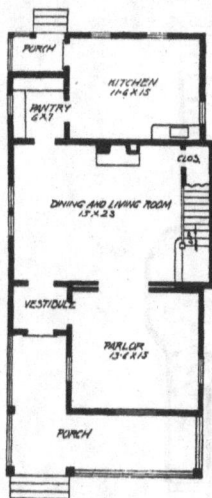

First Floor Plan

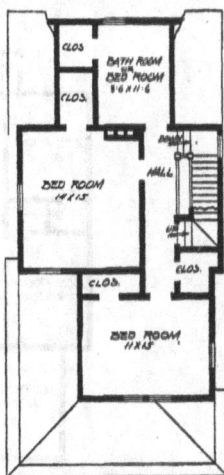

Second Floor Plan

Size: Width, 24 feet; length, 43 feet, exclusive of porches

Blue prints consist of cellar and foundation plan; first and second floor plans; front, two side elevations; wall sections and all necessary interior details. Specifications consist of about twenty pages of typewritten matter.

Full and complete working plans and specifications of this house will be furnished for $5.00. Cost of this house is from about $1,850.00 to about $2,100.00, according to the locality in which it is built.

Design No. 2016

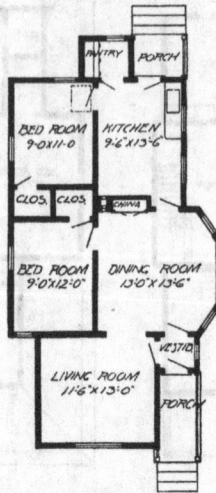

Floor Plan

Size: Width, 24 feet; length, 44 feet 6 inches, exclusive of porch

Blue prints consist of foundation plan; roof plan; floor plan; front, rear, two side elevations; wall sections and all necessary interior details. Specifications consist of about fifteen pages of typewritten matter.

Full and complete working plans and specifications of this house will be furnished for $5.00. Cost of this house is from about $1,250.00 to about $1,400.00, according to the locality in which it is built.

Design No. 2005

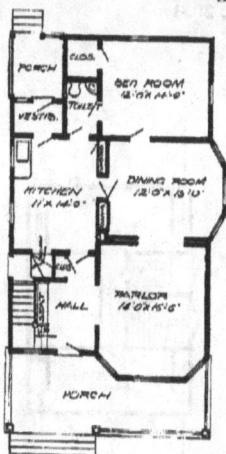

First Floor Plan

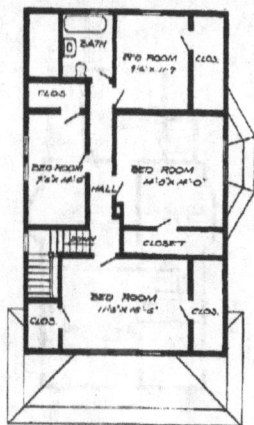

Second Floor Plan

Size: Width, 28 feet 6 inches; length, 43 feet, exclusive of porch

Blue prints consist of basement plan; first and second floor plans; front, rear, two side elevations; wall sections and all necessary interior details. Specifications consist of about twenty pages of typewritten matter.

Full and complete working plans and specifications of this house will be furnished for $5.00. Cost of this house is from about $2,250.00 to about $2,500.00, according to the locality in which it is built.

Design No. 2031

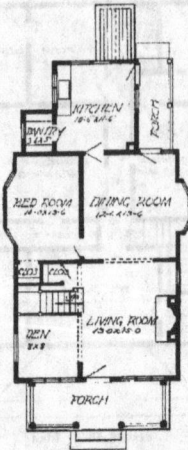

KITCHEN

PANTRY

PORCH

BED ROOM

DINING ROOM

DEN

LIVING ROOM

PORCH

First Floor Plan

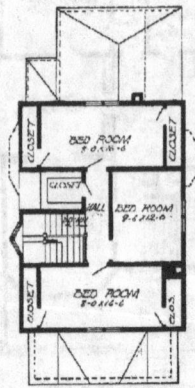

BED ROOM

CLOSET

CLOSET

CLOSET

HALL BED ROOM

BED ROOM

CLOSET

CLOSET

Second Floor Plan

Size: Width, 26 feet 2 inches; length, 42 feet 3 inches, exclusive of porch

Blue prints consist of basement plan; first and second floor plans; front, rear, two side elevations; wall sections and all necessary interior details. Specifications consist of about twenty pages of typewritten matter.

Full and complete working plans and specifications of this house will be furnished for $5.00. Cost of this house is from about $2,450.00 to about $2,650.00, according to the locality in which it is built.

Design No. 2004

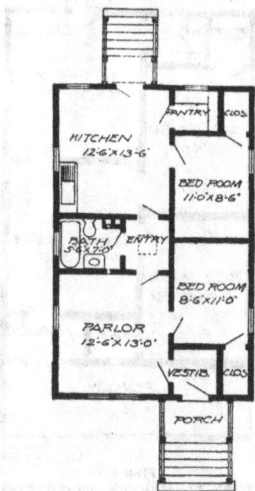

Floor Plan

KITCHEN
12'-6" X 13'-6"

PANTRY   CLOS

BED ROOM
11'-0" X 8'-6"

BATH
5'-3" X 7'-0"   ENTRY

BED ROOM
8'-6" X 11'-0"

PARLOR
12'-6" X 13'-0"

VESTIB   CLOS

PORCH

Size: Width, 22 feet 6 inches; length, 34 feet, exclusive of porch

Blue prints consist of foundation plan; floor plan; front, rear, two side elevations; wall sections and all necessary interior details. Specifications consist of about fifteen pages of type-written matter.

Full and complete working plans and specifications of this house will be furnished for $5.00. Cost of this house is from about $850.00 to about $1,000.00, according to the locality in which it is built.

Design No. 2011

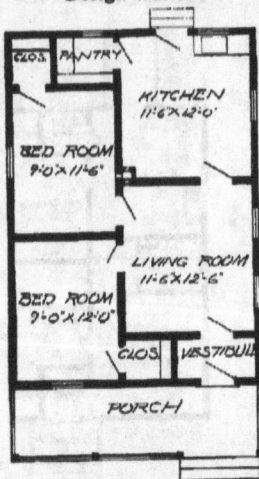

Floor Plan

Size: Width, 22 feet; length, 30 feet, exclusive of porch

Blue prints consist of foundation plan; roof plan; floor plan; front, rear, two side elevations; wall sections and all necessary interior details. Specifications consist of about fifteen pages of typewritten matter.

Full and complete working plans and specifications of this house will be furnished for $5.00. Cost of this house is from about $750.00 to about $900.00, according to the locality in which it is built.

Design No. 2047

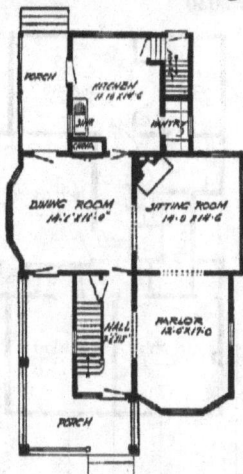

First Floor Plan

Second Floor Plan

Size: Width, 30 feet; length, 46 feet, exclusive of porches

B ue prints consist of basement plan; roof plan; first and second floor plans; front, rear, two side elevations; wall sections and all necessary interior details. Specifications consist of about twenty pages of typewritten matter.

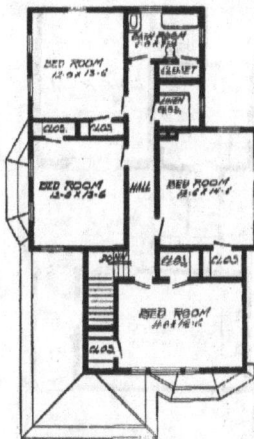

Full and complete working plans and specifications of this house will be furnished for $5.00. Cost of this house is from about $2,750.00 to about $3,000.00, according to the locality in which it is built.

Design No. 2030

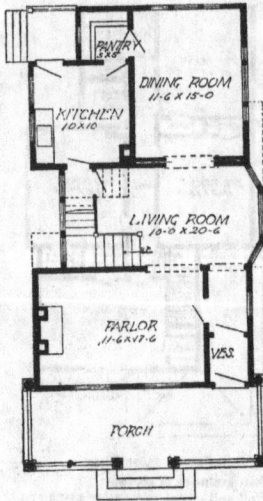

PN'TRY
5x8

DINING ROOM
11-6 X 15-0

KITCHEN
10x10

LIVING ROOM
10-0 X 20-6

PARLOR
11-6 X 17-6

VEST.

PORCH

First Floor Plan

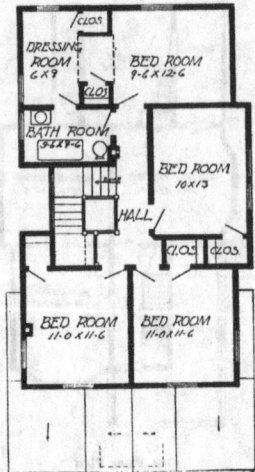

CLOS

DRESSING
ROOM
6 X 9

CLOS

BED ROOM
9-6 X 12-6

BATH ROOM
5-6 X 7-6

BED ROOM
10 X 13

HALL

CLOS   CLOS

BED ROOM
11-0 X 11-6

BED ROOM
11-0 X 11-6

Second Floor Plan

Size: Width, 24 feet 6 inches; length, 39 feet, exclusive of porch

Blue prints consist of basement plan;
roof plan; first and second floor plans;
front, rear, two side elevations wall sections and all necessary interior details.
Specifications consist of about twenty
pages of typewritten matter.

Full and complete working plans and
specifications of this house will be furnished for $5.00.   Cost of this house is
from about $1,950.00 to about $2,100.00,
according to the locality in which it is
built.

Design No. 2010

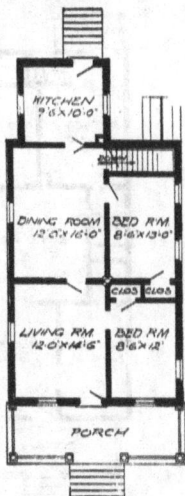

Floor Plan

Size: Width 22 feet 8 inches; length, 42 feet 8 inches, exclusive of porch

Blue prints consist of basement plan; roof plan; floor plan; front, rear, two sides elevations; wall sections and all necessary interior details. Specifications consist of about fifteen pages of type-written matter.

Full and complete working plans and specifications of this house will be furnished for $5.00. Cost of this house is from about $850.00 to about $1,000.00, according to the locality in which it is built.

Design No. 2001

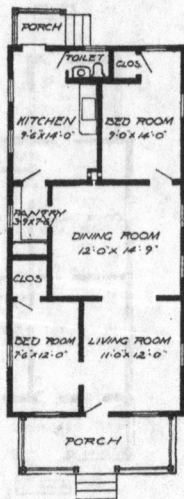

Floor Plan

Size: Width, 20 feet; length, 40 feet, exclusive of porches

Blue prints consist of foundation plan; roof plan; floor plan; front, rear, two side elevations; wall sections and all necessary interior details. Specifications consist of about fifteen pages of typewritten matter.

Full and complete working plans and specifications of this house will be furnished for $5.00. Cost of this house is from about $1,150.00 to about $1,300.00, according to the locality in which it is built.

Design No. 2032

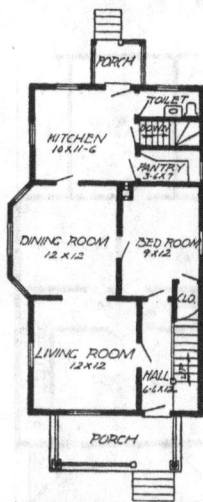

First Floor Plan

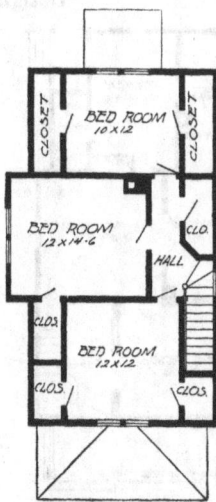

Second Floor Plan

Size: Width, 22 feet 6 inches; length, 36 feet, exclusive of porches

Blue prints consist of basement plan; of plan; first and second floor plans; nt, rear, two side elevations; wall ions and all necessary interior de- ls. Specifications consist of about enty pages of typewritten matter.

Full and complete working plans and specifications of this house will be fur- nished for $5.00. Cost of this house is from about $1,650.00 to about $1,800.00, according to the locality in which it is built.

Design No. 2013

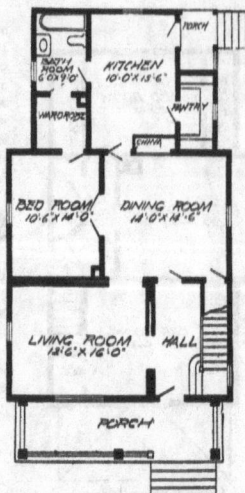

BATH
ROOM
6'0"X9'0"

KITCHEN
10'0"X13'6"

PORCH

PANTRY

WARDROBE

CHIM

BED ROOM
10'6"X14'0"

DINING ROOM
14'0"X14'6"

LIVING ROOM
12'6"X16'0"

HALL

PORCH

BED ROOM
13'0"X15'6"

BED ROOM
10'0"X13'6"

BED ROOM
10'0"X11'6"

CLOS CLOS

HALL

BED ROOM
12'6"X21'6"

CLOS

First Floor Plan   **Second Floor Plan**

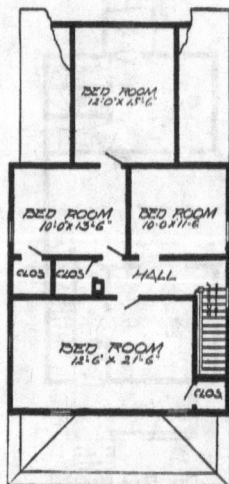

Size: Width, 26 feet; length, 44 feet 6 inches, exclusive of porch

Blue prints consist of cellar and foundation plan; roof plan; first and second
floor plans; front, rear, two side elevations; wall sections and all necessary interior details. Specifications consist of about
twenty pages of typewritten matter.

Full and complete working plans and
specifications of this house will be furnished for $5.00. Cost of this house is
from about $2,050.00 to about $2,300.00,
according to the locality in which it is
built.

Design No. 2041

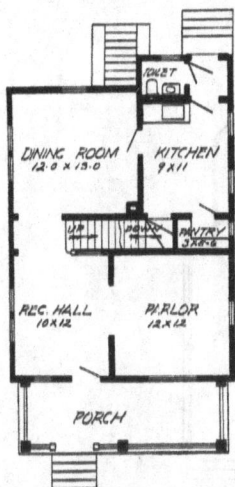

First Floor Plan

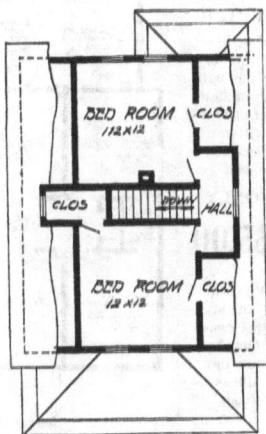

Second Floor Plan

Size: Width, 23 feet 6 inches; length, 32 feet exclusive of porch

Blue prints consist of basement plan; roof plan; first and second floor plans; front, rear, two side elevations; wall sections and all necessary interior details. Specifications consist of about twenty pages of typewritten matter.

Full and complete working plans and specifications of this house will be furnished for $5.00. Cost of this house is from about $1,050.00 to about $1,200.00, according to the locality in which it is built.

Design No. 2028

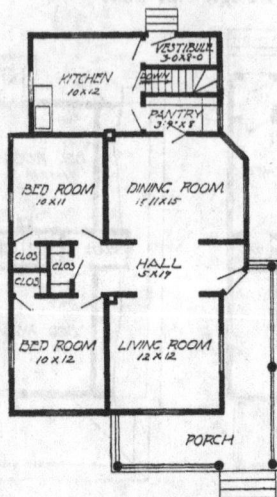

Floor Plan

Size: Width, 26 feet 6 inches; length, 40 feet 6 inches exclusive of porches

Blue prints consist of basement plan; roof plan; floor plan; front, rear, two side elevations; wall sections and all necessary interior details. Specifications consist of about fifteen pages of typewritten matter.

Full and complete working plans and specifications of this house will be furnished for $5.00. Cost of this house is from about $1,550.00 to about $1,750.00, according to the locality in which it is built.

Design No. 2044

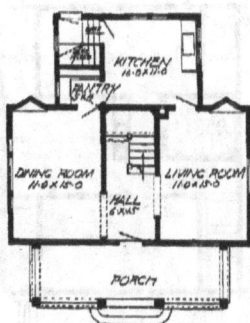

First Floor Plan

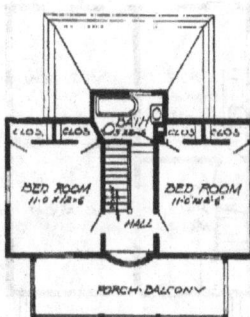

Second Floor Plan

Size: Width, 30 feet; length, 26 feet 6 inches, exclusive of porch

Blue prints consist of basement plan; roof plan; first and second floor plans; front, rear and two side elevations; wall sections and all necessary interior details. Specifications consist of about twenty pages of typewritten matter.

Full and complete working plans and specifications of this house will be furnished for $5.00. Cost of this house is from about $1,650.00 to about $1,850.00, according to the locality in which it is built.

Design No. 2021

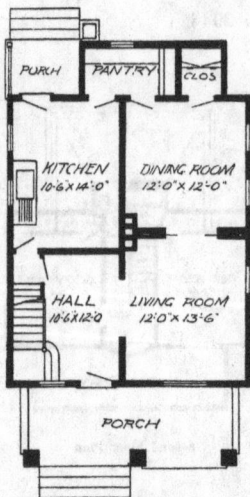

PORCH PANTRY CLOS

KITCHEN
10'-6"x14'-0"

DINING ROOM
12'-0"x12'-0"

HALL
10'-6"x12'-0"

LIVING ROOM
12'-0"x13'-6"

PORCH

First Floor Plan

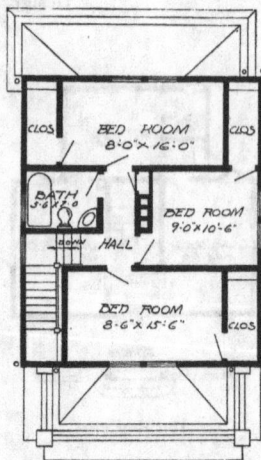

CLOS

BED ROOM
8'-0"x16'-0"

CLOS

BATH
5'-6"x6'-0"

BED ROOM
9'-0"x10'-6"

HALL

BED ROOM
8'-6"x15'-6"

CLOS

Second Floor Plan

Size: Width, 24 feet; length, 32 feet, 6 inches, exclusive of porch

Blue prints consist of basement plan; first and second floor plans; front, rear, two side elevations; wall sections and all necessary interior details. Specifications consist of about twenty pages of typewritten matter.

Full and complete working plans and specifications of this house will be furnished for $5.00. Cost of this house is from about $1,950.00 to about $2,200.00 according to the locality in which it is built.

Design No. 2020

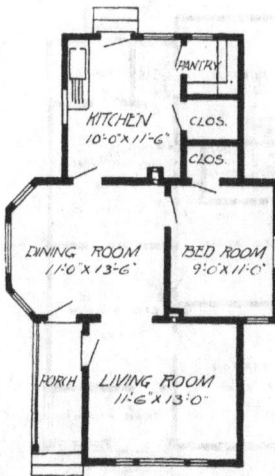

Floor Plan

Size: Width, 24 feet; length, 36 feet

e prints consist of foundation plan;)lan: front, rear, two side elevawall sections and all necessary)r deatils. Specifications consist)ut fifteen pages of typewritten r.

Full and complete working plans and specifications of this house will be furnished for $5.00. Cost of this house is from about $950.00 to about $1,100.00, according to the locality in which it is built.

Design No. 2038

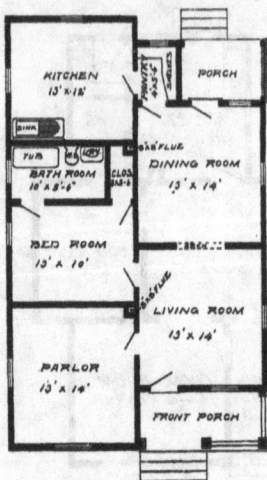

Floor Plan

Size: Width, 27 feet, 6 inches; length, 44 feet

Blue prints consist of basement plan; roof plan; floor plan; front, rear, two side elevations; wall sections and all necessary interior details. Specifications consist of about fifteen pages of typewritten matter.

Full and complete working plans and specifications of this house will be furnished for $5.00. Cost of this house is from about $1,750.00 to about $1,900.00, according to the locality in which it is built.

Design No. 2023

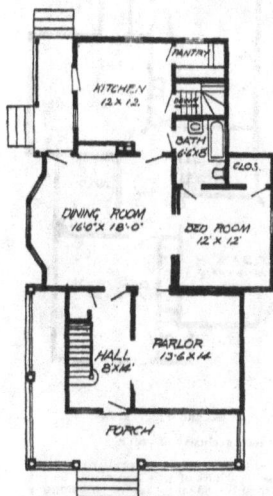

First Floor Plan

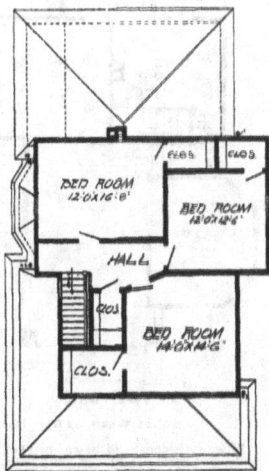

Second Floor Plan

PRICE
of Blue
Prints, to-
gether with
a complete
set of type-
written
specifica-
tions is

ONLY

$5.00

We mail
Plans and
Specifica-
tions the
same day
order is re-
ceived.

Size: Width, 32 feet; length, 46 feet, exclusive of porches

Blue prints consist of basement plan; first and second floor plans; front, rear, two side elevations; wall sections and all necessary interior details. Specifications consist of about twenty pages of typewritten matter.

Full and complete working plans and specifications of this house will be furnished for $5.00. Cost of this house is from about $2,350,00 to about $2,600,00, according to the locality in which it is built.

Design No. 2012

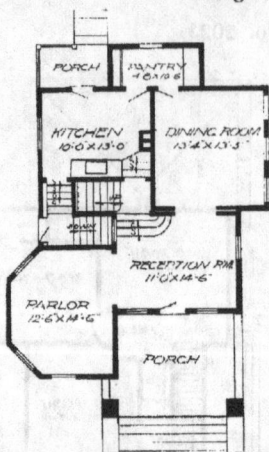

First Floor Plan

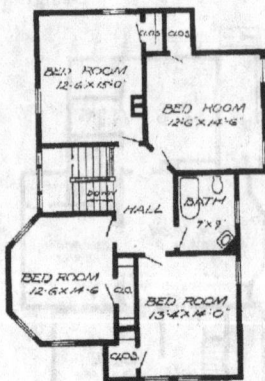

PORCH

PANTRY
4'6"x8'6"

KITCHEN
10'0"x13'0"

DINING ROOM
13'4"x13'5"

RECEPTION RM
11'0"x14'6"

PARLOR
12'6"x14'6"

PORCH

Second Floor Plan

CLOS CLOS

BED ROOM
12'6"x15'0"

BED ROOM
12'6"x14'6"

HALL

BATH
7'x9'

BED ROOM
12'6"x14'6"

BED ROOM
13'4"x14'0"

CLOS

First Floor Plan   **Second Floor Plan**

Size: Width, 32 feet; length, 39 feet, exclusive of porch

Blue prints consist of basement plan; roof plan; first and second floor plans; front, rear, two side elevations; wall sections and all necessary interior details. Specifications consist of about twenty pages of typewritten matter.

Full and complete working plans and specifications of this house will be furnished for $5.00. Cost of this house is from about $2,750.00 to about $3,000.00, according to the locality in which it is built.

Design No. 2018

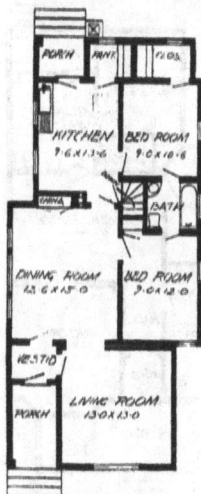

Floor Plan

Size: Width, 25 feet; length, 48 feet 6 inches

Blue prints consist of foundation plan; roof plan; floor plan; front, rear, two side elevations; wall sections and all necessary interior details. Specifications consist of about fifteen pages of typewritten matter.

Full and complete working plans and specifications of this house will be furnished for $5.00. Cost of this house is from about $1,250.00 to about $1,500.00, according to the locality in which it is built.

Design No. 2027

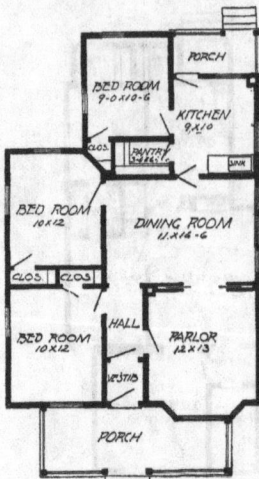

PRICE

of Blue Prints, together with a complete set of typewritten specifications is

ONLY

$5.00

We mail Plans and Specifications the same day order is received.

Floor Plan

Size: Width, 28 feet; length, 41 feet, exclusive of porch

Blue prints consist of basement plan; floor plan; front, rear, two side elevations; wall sections and all necessary interior details. Specifications consist of about fifteen pages of typewritten matter.

Full and complete working plans and specifications of this house will be furnished for $5.00. Cost of this house is from about $1,150.00 to about $1,300.00, according to the locality in which it is built.

Design No. 2007

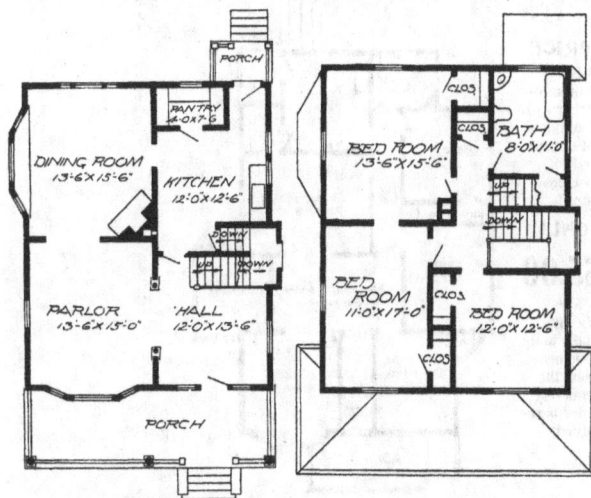

First Floor Plan

Second Floor Plan

Size: Width, 29 feet 6 inches; length, 33 feet, exclusive of porches

Blue prints consist of basement plan; first and second floor plans; front, rear, two side elevations; wall sections and all necessary interior details Specifications consist of about twenty pages of typewritten matter.

Full and complete working plans and specifications of this house will be furnished for $5.00. Cost of this house is from about $2,300.00 to about $2,550.00, according to the locality in which it is built

Design No. 2029

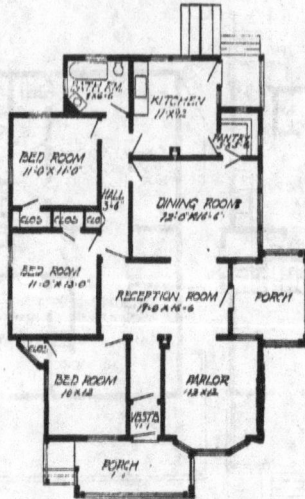

Floor Plan

Size: Width, 32 feet 6 inches; length, 47 feet 6 inches, exclusive of porch

Blue prints consist of basement plan; roof plan; floor plan; front, rear, two side elevations; wall sections and all necessary interior details. Specifications consist of about fifteen pages of type-written matter.

Full and complete working plans and specifications of this house will be furnished for $5.00. Cost of this house is from about $1,650.00 to about $1,850.00, according to the locality in which it is built.

Design No. 2040

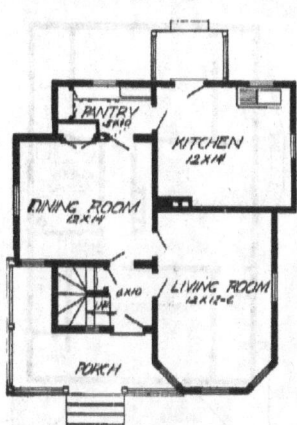

First Floor Plan

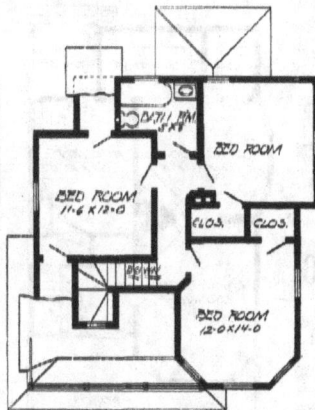

Second Floor Plan

Size: Width, 29 feet 6 inches; length, 31 feet, exclusive of porches

Blue prints consist of basement plan; roof plan; first and second floor plans; front, rear, two side elevations; wall sections and all necessary interior details. Specifications consist of about twenty pages of typewritten matter.

Full and complete working plans and specifications of this house will be furnished for $5.00. Cost of this house is from about $1,750.00 to about $1,000.00, according to the locality in which it is built.

Design No. 2017

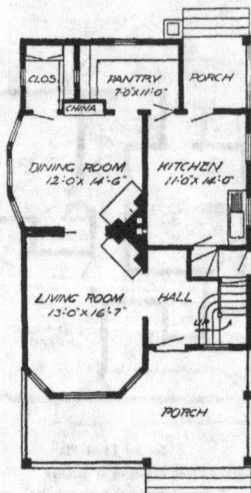

First Floor Plan

Second Floor Plan

Size: Width, 27 feet; length, 38 feet, exclusive of porch

Blue prints consist of basement plan; roof plan; first and second floor plans; front, rear, two side elevations; wall sections and all necessary interior details. Specifications consist of about twenty pages of typewritten matter.

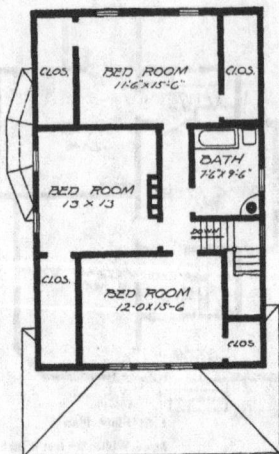

Full and complete working plans and specifications of this house will be furnished for $5.00. Cost of this house is from about $1,850.00 to about $2,250.00, according to the locality in which it is built.

Design No. 2042

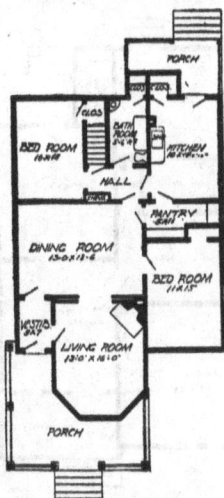

Floor Plan

Size: Width, 31 feet; length, 45 feet 6 inches, exclusive of porches

Blue prints consist of basement plan; roof plan; first and second floor plans; front, rear, two side elevations; wall sections and all necessary interior details. Specifications consist of about twenty pages of typewritten matter.

Full and complete working plans and specifications of this house will be furnished for $5.00. Cost of this house is from about $1,750.00 to about $1,900.00, according to the locality in which it is built.

Design No. 2048

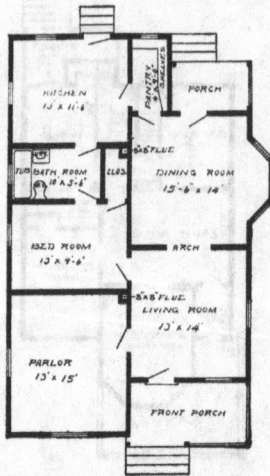

Floor Plan

Size: Width, 27 feet 6 inches; length, 44 feet, exclusive of porch

Blue prints consist of basement plan roof plan; floor plan; front, rear, two ide elevations; wall sections and all ecessary interior details. Specifications consist of about fifteen pages of ypewritten matter.

Full and complete working plans and specifications of this house will be furnished for $5.00. Cost of this house is from about $1,450.00 to about $1,600.00, according to the locality in which it is built.

Design No. 2026

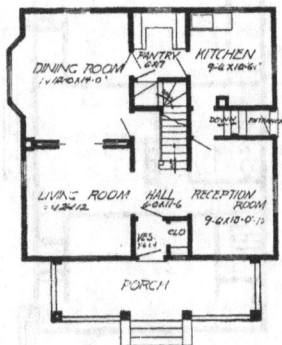

First Floor Plan

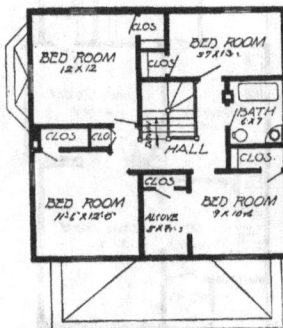

Second Floor Plan

PRICE

of Blue Prints, together with a complete set of typewritten specifications is

ONLY

$5.00

We mail Plans and Specifications the same day order is received.

Size: Width, 29 feet 6 inches; length, 28 feet exclusive of porch

Blue prints consist of basement plan; roof plan; first and second floor plans; front, rear, two side elevations; wall sections and all necessary interior details. Specifications consist of about twenty pages of typewritten matter.

Full and complete working plans and specifications of this house will be furnished for $5.00. Cost of this house is from about $1,650.00 to about $1,850.00, according to the locality in which it is built.

Design No. 2024

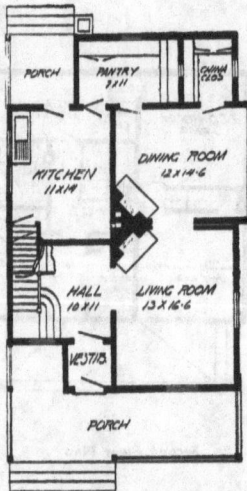

First Floor Plan

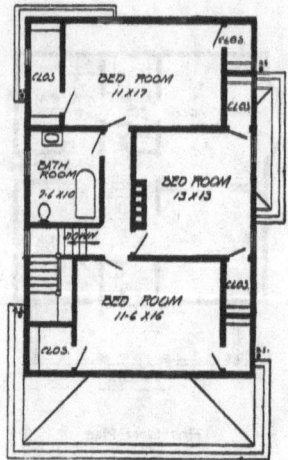

Second Floor Plan

Size: Width, 26 feet 9 inches; length, 44 feet 6 inches, exclusive of porch

Blue prints consist of basement plan; first and second floor plans; front, rear, two side elevations; wall sections and all necessary interior details. Specifications consist of about twenty pages of typewritten matter.

Full and complete working plans and specifications of this house will be furnished for $5.00. Cost of this house is from about $1,850.00 to about $2,100.00, according to the locality in which it is built.

Design No. 2043

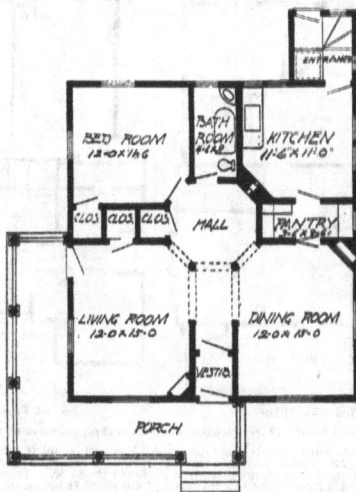

Floor Plan

Size: Width, 30 feet; length, 38 feet 6 inches, exclusive of porches

Blue prints consist of basement plan; floor plan; front, rear, two side eleva-tions; wall sections and all necessary interior details. Specifications consist of about fifteen pages of typewritten matter.

Full and complete working plans and specifications of this house will be fur-nished for $5.00. Cost of this house is from about $950.00 to about $1,100.00, according to the locality in which it is built.

Design No. 2045

First Floor Plan

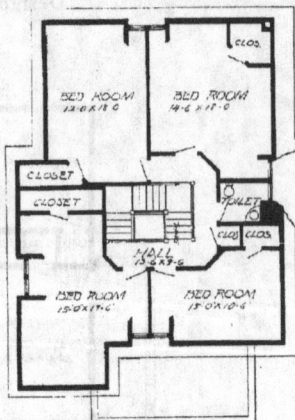

Second Floor Plan

Size: Width, 32 feet; length, 42 feet 6 inches, exclusive of porches

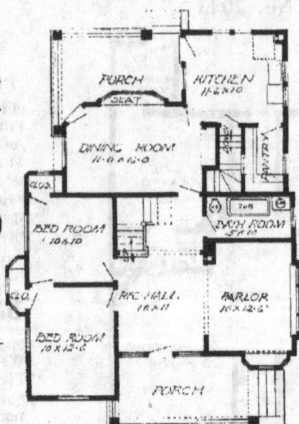

Blue prints consist of basement plan; roof plan; first and second floor plans; front, rear, two side elevations; wall sections and all necessary interior details. Specifications consist of about twenty pages of typewritten matter.

Full and complete working plans and specifications of this house will be furnished for $5.00. Cost of this house is from about $1,850.00 to about $2,000.00, according to the locality in which it is built.

Design No. 2014

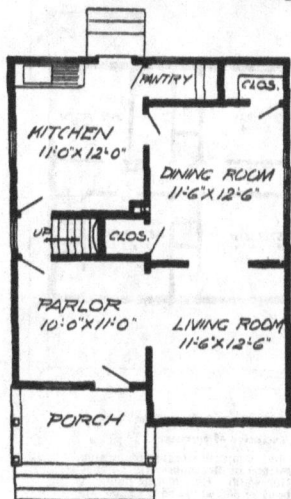

First Floor Plan

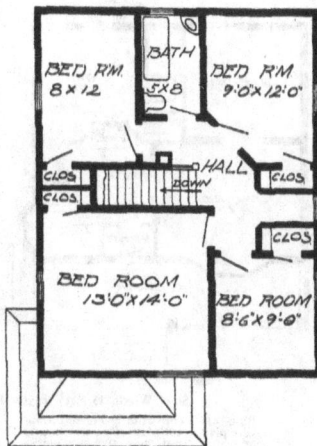

Second Floor Plan

Size. Width, 24 feet; length, 30 feet, exclusive of porch

Blue prints consist of basement plan; first and second floor plans; front, rear, two side elevations; wall sections and all necessary interior details. Specifications consist of about twenty pages of typewritten matter.

Full and complete working plans and specifications of this house will be furnished for $5.00. Cost of this house is from about $1,750.00 to about $1,950.00, according to the locality in which it is built.

Design No. 2034

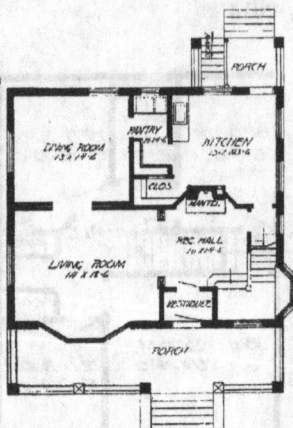

First Floor Plan

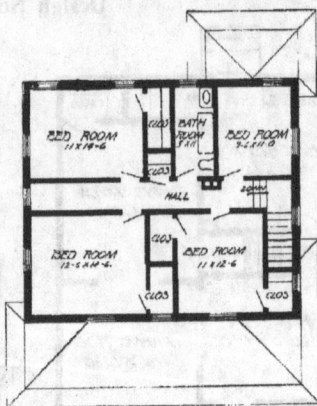

Second Floor Plan

Size: Width, 35 feet; length, 31 feet, exclusive of porches

Blue prints consist of cellar and foundation plans; roof plan; first and second floor plans; front, rear, two side elevations; wall sections and all necessary interior details. Specifications consist of about twenty pages of typewritten matter.

Full and complete working plans and specifications of this house will be furnished for $5.00. Cost of this house is from about $2,550.00 to about $2,800.00, according to the locality in which it is built.

Design No. 2008

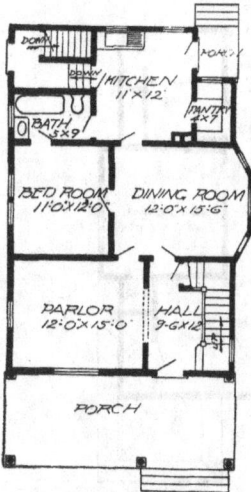

First Floor Plan

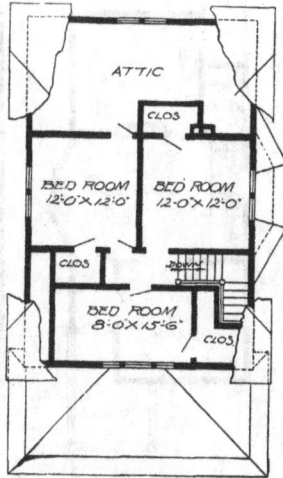

Second Floor Plan

DOWN

KITCHEN
11'x12'

POR..

PANTRY
4'x7'

BATH
5x9

BED ROOM
11'0"x12'0"

DINING ROOM
12'0"x15'-6"

PARLOR
12'-0"x15'-0"

HALL
9'-6x12'

PORCH

ATTIC

CLOS

BED ROOM
12'0"x12'0"

BED ROOM
12'-0"x12'-0"

CLOS

DOWN

BED ROOM
8'-0"x15'-6"

CLOS

Size: Width, 28 feet; length, 38 feet, exclusive of porch

Blue prints consist of foundation plan; first and second floor plans; front, two ide elevations; wall sections and all necessary interior details. Specifications consist of about fifteen pages of typewritten matter.

Full and complete working plans and specifications of this house will be furnished for $5.00. Cost of this house is from about $2,000.00 to about $2,250.00, according to the locality in which it is built.

Design No. 2006

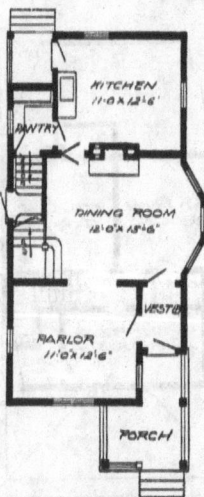

KITCHEN
11'0"x12'6"

PANTRY

DINING ROOM
12'0"x15'6"

VESTIB.

PARLOR
11'0"x12'6"

PORCH

BATH
5x7

BED ROOM
7'6"x11'0"

CLOS.

CLOS.

BED ROOM
12'0"x12'0"

CLOS.

CLOS.

BED ROOM
11'0"x11'0"

CLOS.

First Floor Plan   **Second Floor Plan**

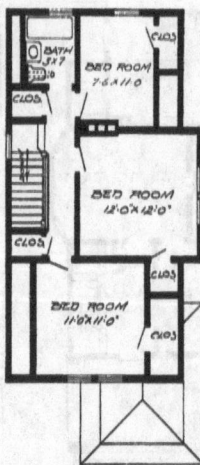

Size: Width, 22 feet; length, 36 feet, exclusive of porch

Blue prints consist of cellar and foundation plan; first and second floor plans;
front, rear, two side elevations; wall sections and all necessary interior details.
Specifications consist of about twenty
pages of typewritten matter.

Full and complete working plans and
specifications of this house will be furnished for $5.00. Cost of this house is
from about $1,850.00 to about $2,050.00
according to the locality in which it is
built.

Design No. 2046

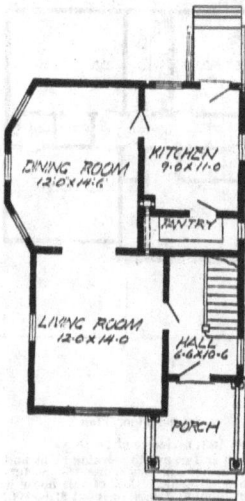

First Floor Plan

DINING ROOM
12·0 X 14·6

KITCHEN
9·0 X 11·0

PANTRY

LIVING ROOM
12·0 X 14·0

HALL
6·6 X 10·6

PORCH

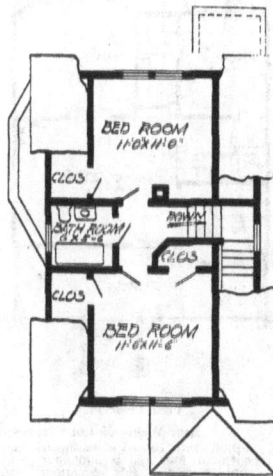

Second Floor Plan

BED ROOM
11·6 X 11·0

CLOS

BATH ROOM
8 X 7·0

ROOM

CLOS

CLOS

BED ROOM
11·6 X 11·6

Size: Width, 22 feet 6 inches; length, 30 feet, exclusive of porches

Blue prints consist of basement plan; roof plan; first and second floor plans; front, rear, two side elevations, wall sections and all necessary interior details. Specifications consist of about twenty pages of typewritten matter.

Full and complete working plans and specifications of this house will be furnished for $5.00. Cost of this house is from about $950.00 to about $1,100.00, according to the locality in which it is built.

Design No. 2037

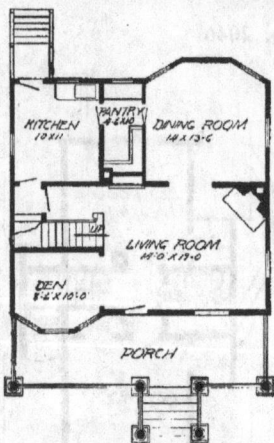

KITCHEN
10x11

PANTRY
4x6-6

DINING ROOM
14 x 13-6

LIVING ROOM
14'-0 x 13'-0

DEN
8-6 x 10-0

PORCH

BED ROOM
10x11

BATH ROOM

BED ROOM
11x12

HALL

CLOS. CLOS.

BED ROOM
13 x10

BED ROOM
8-6 x 12-6

CLOS

First Floor Plan

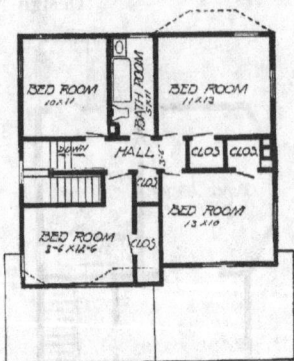

Second Floor Plan

Size: Width, 30 feet 6 inches; length, 31 feet, exclusive of porch

Blue prints consist of basement plan; roof plan; first and second floor plans; front, rear, two side elevations; wall sections and all necessary interior details. Specifications consist of about twenty pages of typewritten matter.

Full and complete working plans and specifications of this house will be furnished for $5.00. Cost of this house is from about $1,750.00 to about $1,960.00, according to the locality in which it is built.

Design No. 2050

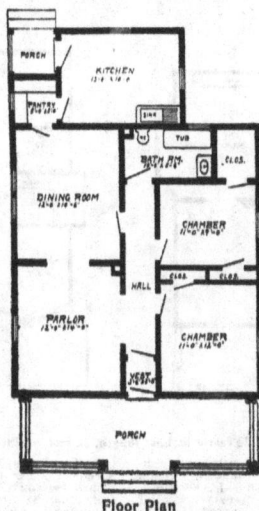

Floor Plan

Size: Width, 28 feet 6 inches; length, 40 feet 6 inches, exclusive of porch

Blue prints consist of basement plan; roof plan; floor plan; front, rear, two side elevations; wall sections and all necessary interior details. Specifications consist of about fifteen pages of typewritten matter.

Full and complete working plans and specifications of this house will be furnished for $5.00. Cost of this house is from about $1,250.00 to about $1,400.00, according to the locality in which it is built.

Design No. 2035

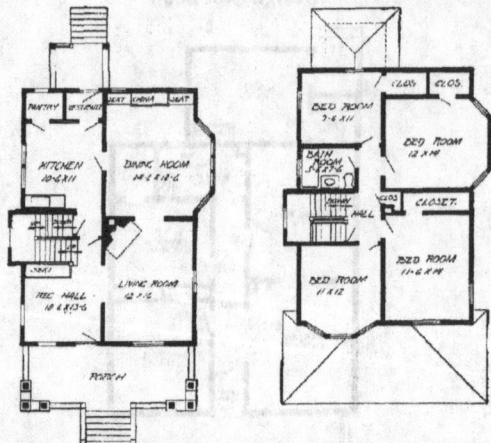

Size: Width, 28 feet 4 inches; length, 34 feet, exclusive of porches

Blue prints consist of cellar and foundation plan; roof plan; first and second floor plans; front, two side elevations; wall sections and all necessary interior details. Specifications consist of about twenty pages of typewritten matter.

Full and complete working plans and specifications of this house will be furnished for $6.00. Cost of this house is from about $2,750.00 to about $3,000.00, according to the locality in which it is built.

Design No. 2019

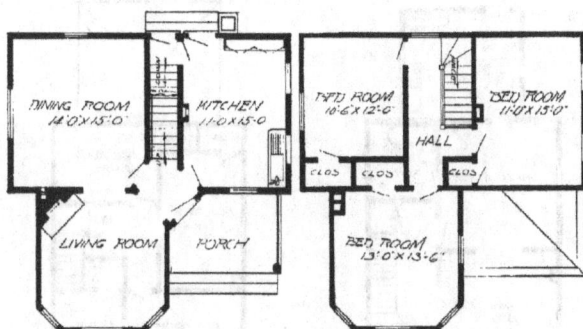

First Floor Plan

Second Floor Plan

Size: Width, 30 feet; length, 30 feet

Blue prints consist of basement plan; first and second floor plans; front, rear, two side elevations; wall sections and all necessary interior details. Specifications consist of about twenty pages of typewritten matter.

Full and complete working plans and specifications of this house will be furnished for $5.00. Cost of this house is from about $1,250.00 to about $1,400.00, according to the locality in which it is built.

Design No. 2025

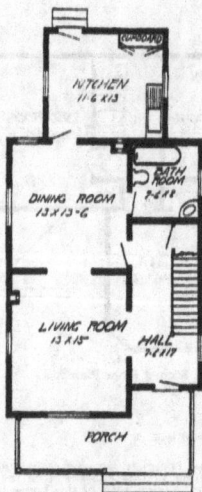

First Floor Plan

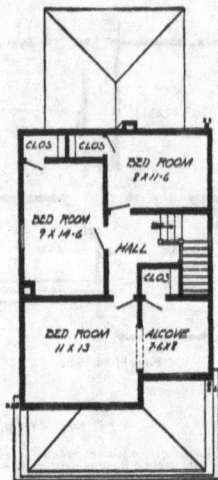

Second Floor Plan

Size: Width, 22 feet; length, 52 feet, exclusive of porch

Blue prints consist of foundation plan; first and second floor plans; front, rear, two side elevations; wall sections and all necessary interior details. Specifications consist of about twenty pages of typewritten matter.

Full and complete working plans and specifications of this house will be furnished for $5.00. Cost of this house is from about $1,950.00 to about $2,200.00, according to the locality in which it is built.

Practical Carpentry

Being a Complete, Up-to-Date Explanation of Modern Carpentry and An Encyclopedia on the Modern Methods Used in the Erection of Buildings, From the Laying of the Foundation to the Delivery of the Building to the Painter

IN TWO VOLUMES

EDITED UNDER THE SUPERVISION OF

WILLIAM A. RADFORD

Editor-in-Chief of the "American Carpenter and Builder," President of "The Radford Architectural Co.," Author of "The Steel Square and its Uses" and the Best Authority in the Country on All Things Pertaining to the Building Trade

ASSISTED BY

ALFRED W. WOODS and WILLIAM REUTHER

VOLUME II.

CONTENTS

Part I

Part II

Part III

Part IV

JOINTS, STRAPS AND OTHER FASTENINGS. Lengthen-

Part V

STAIR BUILDING.   To cut a pitch-board — Manner of apply-

Part VI

PREFACE

In preparing the second volume of the study in carpentry, it has been my endeavor to take up the subject where it was left off in volume one. Special chapters are devoted to shingling, and other methods of covering roofs, mouldings for interior finish, joinery, mitering and dovetailing, lengthening ties, stair building and questions and answers. This last part will be found to be of special interest, as it covers the entire building field in a way that could not be done in a regular article. The first chapter on Building Construction will be found very valuable as the illustrations are all large and easily understood, and the treatment of window construction is the most complete ever published.

I have followed the plan of illustrating every example given, as experience has taught me that an illustration is of more help than twice the amount of description.

WILLIAM A. RADFORD,
Chicago, Ill.

VOL. II

Practical Carpentry

Part I

BUILDING CONSTRUCTION. Foundations—Materials commonly used in walls of buildings—Bonding stretchers—Lime, cement and mortar—An arch—Brick arches—Foundation construction—Cellar windows—Framing—Joints used in framing—Framing around a fireplace—Girders—The fireplace—Windows in frame walls—Double hung sash windows—Double hung sash frame in a double plastered wall—Storm resisting windows—Inside blinds siding in grooves

The object of a foundation is to form a solid base, arranged to distribute the weight of the superstructure over a large area of ground and so reduce the inevitable "settlement" to a minimum and to provide for a uniform movement in settling, so that the framework will not be strained and the plaster cracked.

We are therefore dwelling briefly upon the subject of foundations, as it should be the duty of every carpenter to become familiar with the construction and principles which govern foundations.

The method of constructing such foundations

is largely determined by the nature of the soil and may be classed as follows:

Class I.—Foundations where the soil is firm enough to bear the weight of the building.

Class II.—Foundations on marshy grounds.

In Class I the foundation may be formed—

Firstly, by brick footings formed by spreading the wall by means of off-sets as shown in Figs. 1, 2 and 3.

Secondly, by laying down rough thick stones of width equal to twice the thickness of the wall, and then forming brick off-sets to distribute the weight of the wall on the stones, as in Fig. 4.

Thirdly, by concrete footings composed of Portland cement, broken stones and sand mixed together with water; a good proportion being, one part of cement, two of sand, and four or five of broken stone. The material thus produced becomes a solid mass as hard as stone. This method as shown in Fig. 5, is preferable to the preceding ones and is being generally adopted. The success of it depends upon the proportion of cement, sand and stone, as above given, being strictly adhered to, and care should be taken that the mixing is inspected, as the tendency is to economize on the cement to the detriment of the concrete.

In Class II, a solid bed is formed for the foundation by driving wooden piles into the marshy soil as shown in Fig. 6. Oak, yellow pine, spruce and hemlock are the woods commonly used. The piles are driven through the soft soil to the firm bed beneath. The heads are then cut off at a certain

level and a timber capping put on them. This
capping is commonly of yellow pine and serves as a

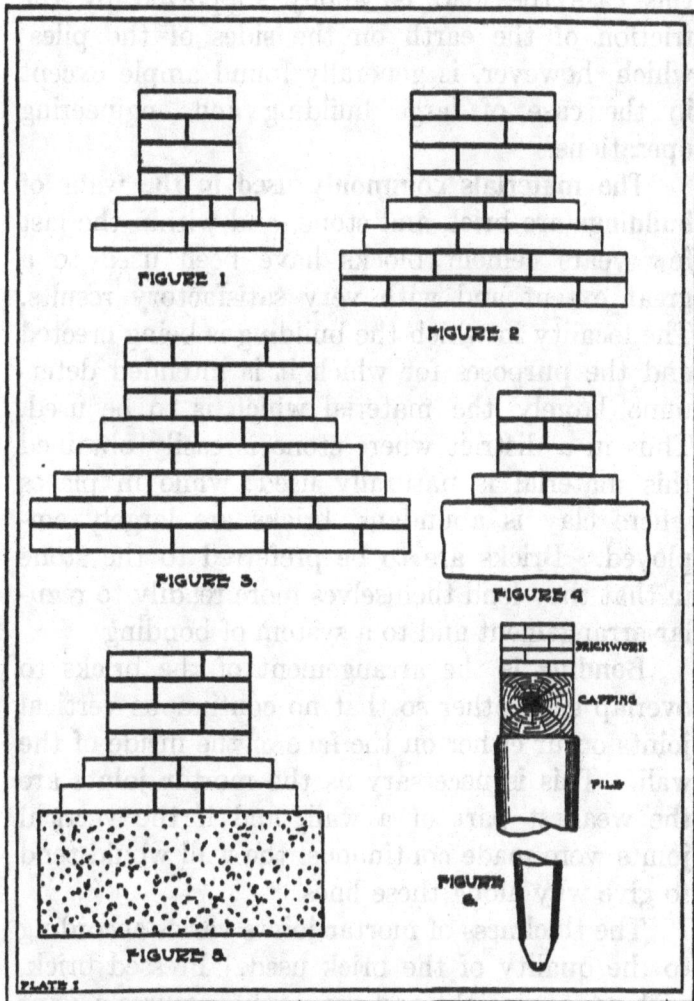

FIGURE 1

FIGURE 2

FIGURE 3.

FIGURE 4

FIGURE 5

BRICKWORK

CAPPING

PILE

FIGURE 6.

PLATE I

support for the foundation above and at the same
time ties the piles together.

It frequently happens that the piles do not reach the firm soil owing to its great depth. In this case the load is wholly supported by the friction of the earth on the sides of the piles; which, however, is generally found ample except in the case of large building and engineering operations.

The materials commonly used in the walls of buildings are brick and stone, and within the last few years cement blocks have been used to a great extent and with very satisfactory results. The locality in which the building is being erected and the purposes for which it is intended determine largely the material which is to be used. Thus in a district where stone is easily obtained this material is naturally used; while in places where clay is abundant, bricks are largely employed. Bricks are to be preferred to the stone in that they lend themselves more readily to regular arrangement and to a system of bonding.

Bonding is the arrangement of the bricks to overlap each other so that no continuous vertical joints occur either on the face or the inside of the wall. This is necessary as the mortar joints are the weakest part of a wall, and if the vertical joints were made continuous, the wall would tend to give way along these lines.

The thickness of mortar joints varies according to the quality of the brick used. Pressed brick, with edges straight and true, only requires a joint one-eighth of an inch thick; ordinary brick joints at from one-quarter to three-eighths of an inch;

while common brick frequently have as much as five-eighths of an inch at the joint.

Stretchers are the bricks laid with their lengths in the direction of the length of the wall. Headers are the brick laid with their lengths across the wall. See Fig. 7.

All bricks, to be laid in dry weather, should be wetted before being used, in order to wash off any dust and to prevent too rapid absorption of the moisture of the mortar. Whenever new brickwork is joined to old, the old work should be thoroughly wetted to insure proper adhesion. All foundation brickwork should be started well below the lowest frost line.

In the erection of brickwork, all the walls should rise at about the same rate; no part being carried more than three feet above the rest, or unequal settlement is likely to occur with the result that the wall soon shows signs of fracture. If it is not possible to carry all the walls up simultaneously, the portion first built should be "stepped back" rather than "toothed." See Fig. 7.

Fig. 7 shows the bond commonly used in brickwork, with headers every sixth course.

Fig. 8 shows what is known as English bond. The plans show the method of laying the bricks in the two courses.

Fig. 9 shows the arrangement of bricks to form the Flemish bond. In this and also the English bond, particular care should be taken to keep each vertical joint in any one course directly over the corresponding vertical joint in the course next but one below. A neglect of this precaution detracts considerably from the appearance of the finished work.

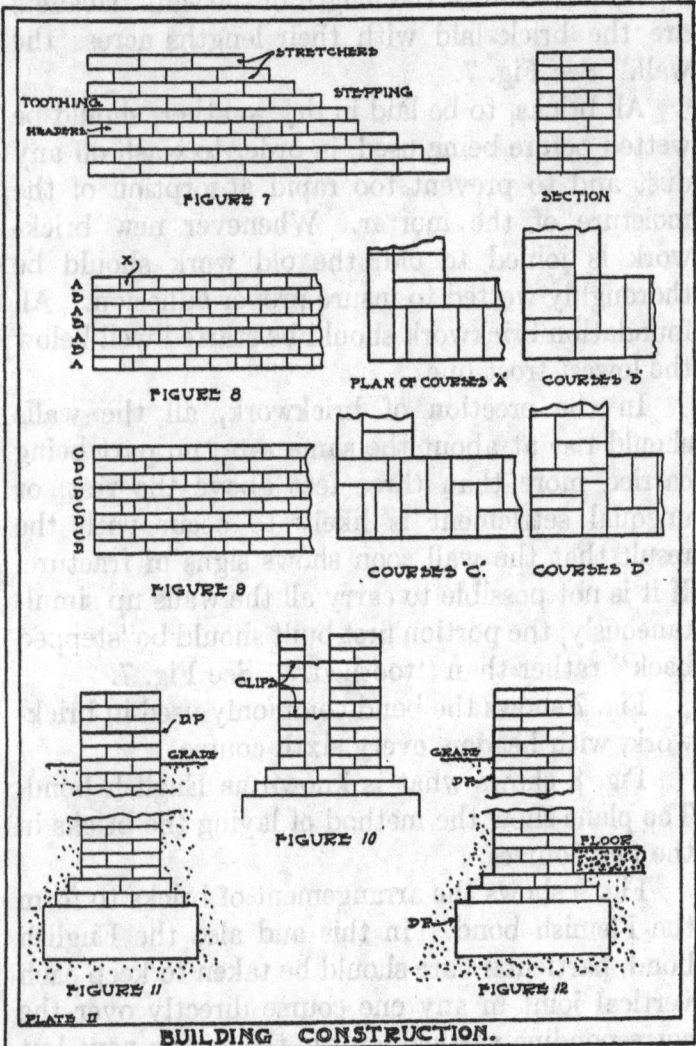

PLATE II

BUILDING CONSTRUCTION.

Fig. 10 shows the construction of a hollow wall, which, with the same amount of material, is more stable than a solid wall and possesses many other advantages. It consists of two separate walls, with an airspace of four inches between them, tied together with bonding irons or "clips" every few feet. A wall of this kind prevents dampness from penetrating to the inside.

Fig. 11 shows a damp-proof course, marked D P., which should never be omitted in important work. It consists of a layer of impervious material laid on the walls just above the ground and below the floor beams; its object being to prevent dampness from rising from the ground and getting into the building. Materials suitable for damp-proofing courses are: Asphalt, pitch, slate, damp-resisting paints and cements, or any material that does not allow moisture to pass through it.

In buildings having a cellar below the ground, this damp-proofing material is applied to the outside of the cellar walls from a little above the ground level, well down to the under side of the footing course. See Fig. 12.

Lime, Cement and Mortar

This is a subject to which considerable space might well be devoted, but, as it is not the purpose of this series to go into the subject so deeply, only such information will be given as may be required in the ordinary erection of cottages and the smaller buildings.

All lime should be freshly burned and thoroughly slacked.

All cement should be finely ground and free from lumps.

All sand should be clean, sharp, free from loam and salt, properly screened and washed.

Lime mortar is usually composed of three parts of sand to one of lime, but two parts of sand to one of lime makes a much better material.

Lime.—Rosendale cement mortar is mixed one part of Rosendale cement, one part of lime and five parts of sand and should be well mixed before the water is added.

Lime.—Portland cement mortar is mixed one part of Portland cement, one part of lime and six parts of sand. All should be well mixed before the water is added.

Rosendale cement mortar is mixed one part of Rosendale cement to two parts of sand.

Portland cement mortar is mixed one part of Portland cement to three of sand, for ordinary use, and for important work one part of Portland cement to two of sand, thoroughly mixed dry, adding only enough water to render mortar of good working consistency.

A little lime should be added to cement mortar to be used in freezing weather. The mortar should not be made up in greater quantities than required for the work on hand, and no excess that may have been left over night should be used in any way.

An arch is the arrangement of bricks, stone or other materials to span an opening. They are named from the outline of their soffits, as seg-mental (Fig. 13), semicircular (Fig. 14), or flat (Fig. 17). The terms of the various parts of the arch are shown in Fig. 21. The soffit or under side

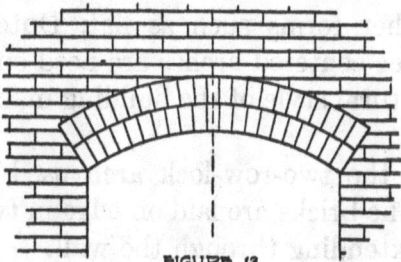 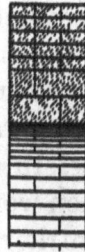

FIGURE 13   SECTION.

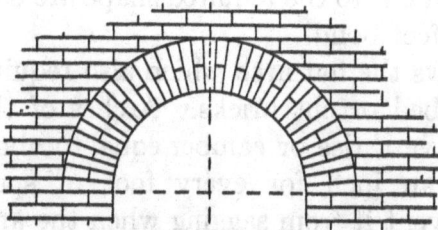 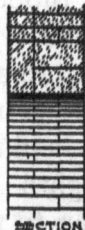

FIGURE 14   SECTION.

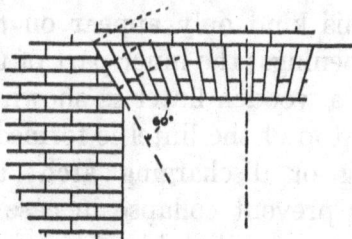 

FIGURE 15   SECTION.

PLATE III

BRICK ARCHES.

of the arch is also called the intrados. The back or upper side is also called the extrados. The pieces marked "B" of which the arch is composed are called voussoirs. The skewback is also known as the springing line.

Semicircular and segmental arches are the best as far as strength is concerned and are the simplest to construct. Other forms such as flat, Dutch, elliptical, and three-centered arches are used only where the architectural style of the building makes them preferable.

Fig. 13 shows the two-row-lock arch used in common work. The bricks are laid on edge in two concentric rings extending through the wall.

Fig. 14 shows a segmental arch. Ordinary bricks rubbed or cut to the required shape are used and form a perfect bond.

Fig. 15 shows the flat arch which also requires the use of rubbed or cut bricks. Arches of this form should have a rise or camber equal to about one-eighth of an inch for every foot of span, in order to prevent it from sagging when the arch settles. The skewback is usually made to an angle of 60°. Arches of this kind only appear on the exterior face of the openings; the inner part of the wall being carried on a wooden lintel as shown in Figs. 16 and 17. On top of the lintel is formed a rough brick relieving or discharging arch; the object of which to is prevent collapse in case of the destruction of the wooden lintel by fire or rot. The brickwork on top of the lintel and under the relieving arch is called the core (Fig. 16). The construction when this core is omitted is shown in Fig. 17.

Figs. 18 and 19 illustrate the temporary wooden structure required to support the bricks of an

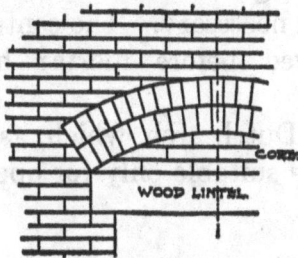

FIGURE 16.

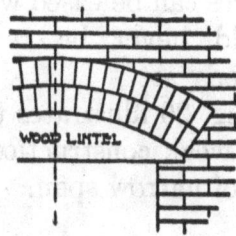

FIGURE 17.

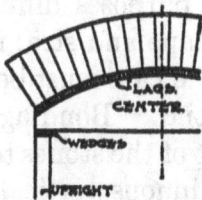

FIGURE 18.

FIGURE 19.

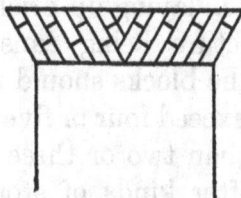

FIGURE 22.

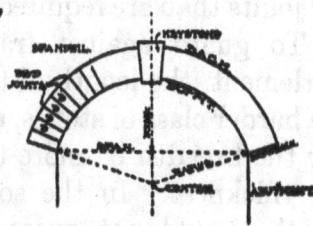

FIGURE 21.

PLATE IV

BRICK ARCHES.

arch while the arch is being built. The upper surface of a center corresponds in outline to the soffit of the arch. They consist of two parallel boards cut to the required curvature and covered with narrow wooden strips called "lags" for supporting the bricks. The centers are supported on wooden uprights with wedges as shown. These wedges can be eased when necessary. The centers should never be removed before mortar has properly set.

Fig. 20 illustrates the Dutch arch, which, as it is of weak construction, is suitable only for openings of narrow span.

Foundation Construction

Stones used for building purposes differ from bricks in being of unequal shape and size; in consequence of which considerable care must be taken in order to obtain a good bond. Bonding, as in brickwork, is the arrangement of the stones to overlap each other so that no continuous vertical joints occur. The unevenness of the stones, except in the case of cut stonework, also necessitates thicker mortar joints than are required in laying up brickwork.

To guard against fracture being caused by settlement, the length of the blocks should not, in the harder class of stones, exceed four or five times, nor the breadth be more than two or three times, the thickness. In the softer kinds of stone the length should not exceed three times, nor the breadth be more than one and one-half times, the thickness.

All stones should be laid on their natural bed; that is, laid on that plane of division along which the stones are spilt when obtained from the quarry. A neglect of this precaution very often causes the stones to split.

Fig. 22 shows a wall with the stones laid up random rubble. Stones of all shapes and sizes are used and for that reason require considerable skill in laying, as their irregular surfaces make them difficult to bed and bond. Proper bonding requires the insertion of through stones (marked T. S.) at intervals of four or five feet in the length of the wall, and about every eighteen inches in the height of the wall. An equally good substitute for a through stone, is a stone extending from the inside of the wall, three-quarters of the thickness of the wall, and overlapping another stone extending from the outside of the wall three-quarters the thickness of the wall. Through stones and three-quarter stones should be of sufficient thickness to prevent fracture through settlement of the wall. Fig. 23 shows a section through the wall.

Fig. 24, at A, shows uncoursed square rubble, and, at B, shows squared rubble built up to courses. Both of these arrangements are superior to random rubble. In both cases the wall is built of rectangular stones with squared ends. In the latter case, the wall is brought to a level every fifteen to eighteen inches.

Fig. 25 shows a wall built up of field stones. This kind of a wall is used principally in the picturesque cottages and country homes.

Fig. 26 shows an elevation and section of coursed ashlar. Ashlar is a facing of stones, never less than four inches thick, used to cover walls of

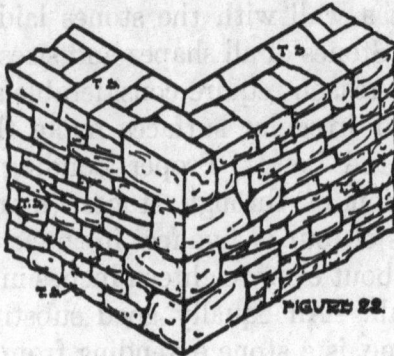

FIGURE 23.

FIGURE 23.

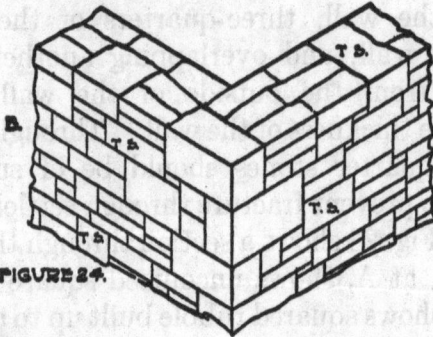

FIGURE 24.

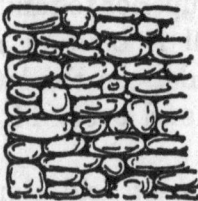

FIGURE 25.

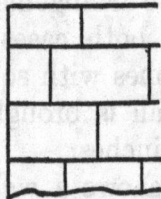

FIGURE 26.

SECTION.

PLATE V

BUILDING CONSTRUCTION.

brick or rough stone, to which it should be securely anchored. Foundation walls, where ashlar is used, should be made of sufficient thickness to carry superstructure, independent of the ashlar unless the ashlar be at least eight inches thick and bonded into the backing, in which case it may be counted as part of the thickness of the wall.

Cellar Windows

Plate VI takes up the construction of an ordinary cellar window in a stone wall. The sash is hinged at the top with heavy wrought iron butts and arranged to swing in. It is secured in place, when closed, by a catch on bottom rail, and, when open is secured with a hook on cellar ceiling or beams.

Fig. 27 shows the elevation; Fig. 28, the section, and Fig. 29, the plan.

Fig. 30 is a section through the head of the frame. The staff bead is sometimes omitted, but, as it makes a better finish at the junction between wood and masonry, hiding the roughness of the stonework where it takes up with the wood, it is desirable to use it.

Fig. 31 is a section through the jamb. The "lug" is a piece left on the ends of head and sill and built into the wall to secure the frame in place.

Fig. 32 is a section through the sill. Stone sills near grade do not require a drip. The inside of the wall is finished in cement.

Figs. 33 and 34 show the head and sill of a

window with an iron guard and a window screen outside of the sash. The iron bars are let into the head and sill about an inch.

FIGURE 27

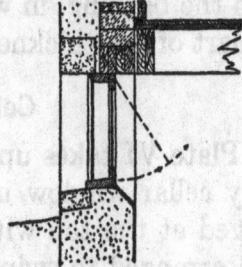

FIGURE 28

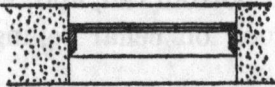

FIGURE 29

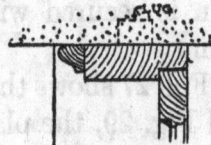

FIGURE 31

STAFF BEAD

FIGURE 30

SASH BAR

FIGURE 32

CEMENT

PLUG

FIGURE 33

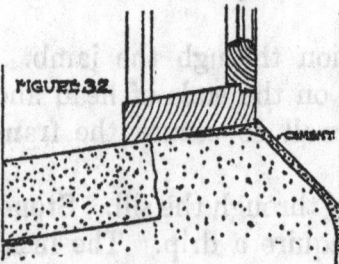

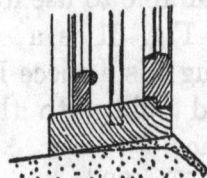

FIGURE 34

PLATE VI.

BUILDING CONSTRUCTION.

Building Construction

In the East, spruce and hemlock are commonly used in the framework of cottages, and occasionally oak in the better class of buildings. In the South and West, other materials indigenous to the particular locations are employed.

Fig. 35 shows the use of shingles for the exterior covering of a house, and Fig. 36 illustrates the use of clapboards.

Fig. 37 is a section showing the construction of the framework at the first tier of beams. The main sill is the first piece of timber to be put in place, and should be well bedded in mortar on the walls so that it may have an even bearing at every point. It should have a halved joint at all corners, and if splicing is necessary, it should be done by means of a scarfed joint. This cut also shows a base course of shingles. Three forms of base courses, where clapboards are used, are shown in Figs. 38, 39 and 41.

At "A," Fig. 39, is shown a fire stop of bricks laid between the floor beams, which should never be omitted. It also serves as a stop for wind and vermin. It is frequently built on the sill as indicated by the dotted lines at "B."

Fig. 40 shows an isometric view of the base of the framework.

Fig. 41 shows the cross bridging between the floor beams. There should be a row of cross bridging for every eight feet of span. This cut also shows the studs mortised into the sill, a constructive feature only used in the best grade of work.

Fig. 42 shows an elevation at the corner of a building, from the sill to the rafters. The girts

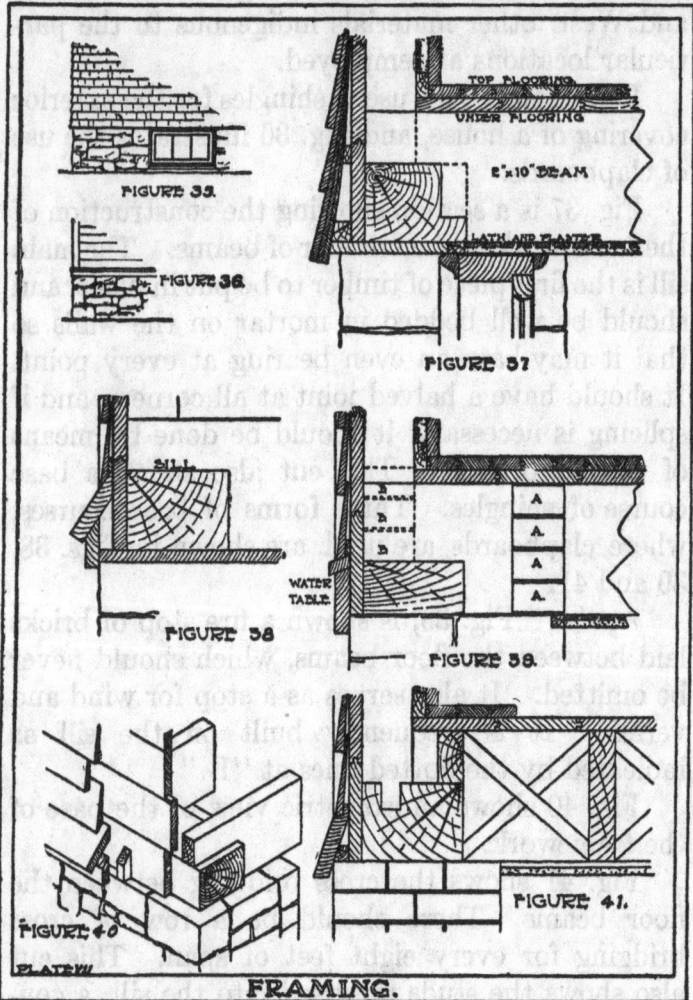

FIGURE 35.

FIGURE 36.

FIGURE 37.

FIGURE 38.

FIGURE 39.

FIGURE 40.

FIGURE 41.

FRAMING.

are mortised and tenoned into the corner posts and pinned with hardwood pins. The braces are mortised and tenoned into the corner posts, sills, girts, and plates, and are pinned with hardwood pins.

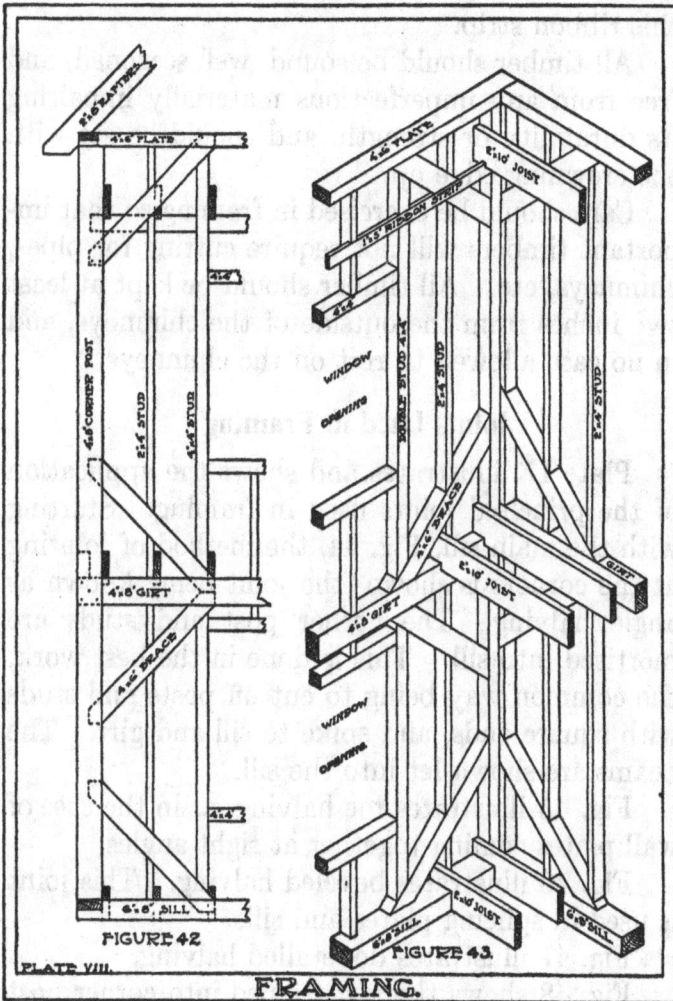

FIGURE 42

FIGURE 43.

PLATE VIII.

FRAMING.

The second tier of beams are notched over the girts, the ribbon strips, or ledger boards, as they are also called, are notched one inch into the studs, braces, and posts, and should be well spiked. The third tier of beams should be notched over this ribbon strip.

All timber should be sound, well seasoned, and free from any imperfections materially impairing its durability or strength, and should be set with the crowning edge up.

Care should be exercised in framing so that important timbers will not require cutting for pipes, chimneys, etc. All timber should be kept at least two inches from the outside of the chimneys, and in no case allowed to rest on the chimneys.

Joints Used in Framing

Plate IX illustrates and shows the application of the principal joints used in framing. Starting with the main sill, Fig. 44, the method of jointing at the corners is shown; the joint being known as angle halving. The corner post and studs are mortised into sill. This is done in the best work, the common way being to cut off posts and studs with square ends, and spike to sill and girt. The beams are shown let into the sill.

Fig. 45 illustrates tee halving, as in the case of wall plates coming together at right angles.

Fig. 46 illustrates beveled halving. This joint is used in splicing plates and sills.

Fig. 47 illustrates dovetailed halving.

Fig. 48 shows the girt framed into corner post

with a mortise and tenon joint, pinned with hard-wood pins or well spiked.

Fig. 49 shows the girt framed into corner post

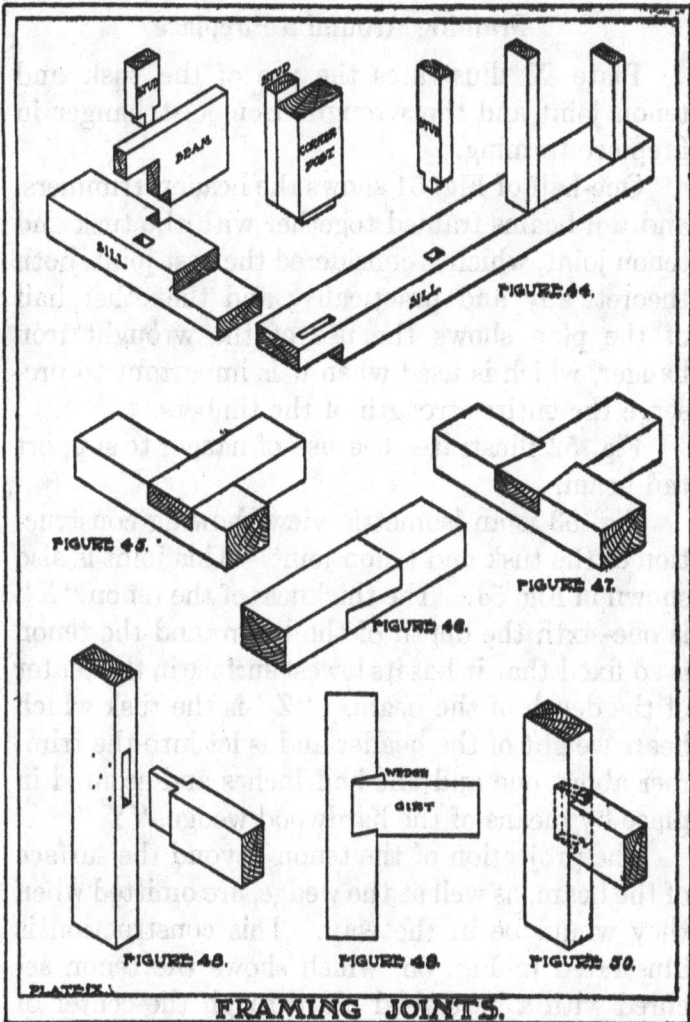

FRAMING JOINTS.

with a dovetailed tenon joint. The girt is secured firmly in place by driving the wedge shown.

Fig. 50 shows an isometric view of the dove-tailed tenon joint.

Framing Around a Fireplace

Plate X illustrates the use of the tusk and tenon joint and the wrought iron joist hanger in fireplace framing.

One-half of Fig. 51 shows the header, trimmers, and tail beams framed together with the tusk and tenon joint, which is considered the best joint, both theoretically and practically; and the other half of the plan shows the use of the wrought iron hanger, which is used when it is important to preserve the entire strength of the timbers.

Fig. 52 illustrates the use of hanger to support tail beam.

Fig. 53 is an isometric view showing construction of the tusk and tenon joint. This joint is also shown in Fig. 54. The thickness of the tenon "X" is one-sixth the depth of the beam, and the tenon is so fixed that it has its lowest surface in the center of the depth of the beam. "Z" is the tusk which bears weight of the header and is let into the trimmer about one and one-half inches and secured in place by means of the hardwood wedge, "Y."

The projection of the tenon beyond the surface of the beam, as well as the wedge, are omitted when they would be in the way. This construction is illustrated in Fig. 55, which shows the tenon secured with a hardwood pin through the center of

the header; and is also shown in Fig. 56, where it
is secured in place by means of a three-quarter-
inch bolt. A hole is cut in the beam as shown, to

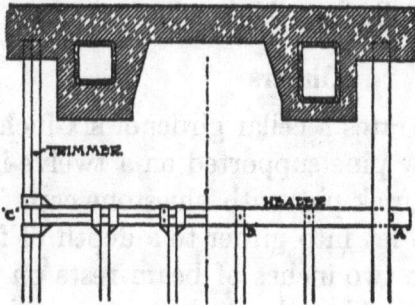

TRIMMER.

HEADER.

FIGURE 51

FIGURE 57.

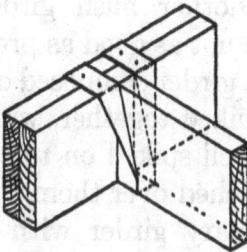

FIGURE 52.

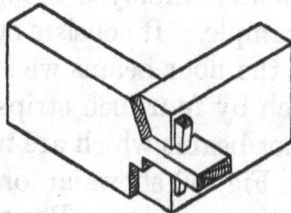

FIGURE 53.

FIGURE 54.

FIGURE 55.

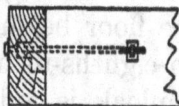

FIGURE 56.

PLATE X.

FRAMING.

receive the nut, and is made large enough so that nut may be turned.

Fig. 57 shows a cheaper method of framing‘ which is used to a considerable extent, though not with as satisfactory results as the previous examples. The tail beam is supported on a "two-by four" joist, spiked to the header.

Girders

Fig. 58 illustrates a cellar girder of six-inch by eight-inch yellow pine supported on a twelve-inch by twelve inch brick pier with bluestone cap. The floor beams are let into girder to a depth of four inches. The top two inches of beam rests on top of girder, thus making the under side of girder flush with the under side of floor beam.

Fig. 59 illustrates another flush girder frequently employed though not as good as preceding example. It consists of a girder composed of three of the floor beams well spiked together, with two-inch by four-inch strips well spiked on to support floor beams which are notched over them.

Fig. 60 shows an ordinary girder with beams resting on top. Beams should be lapped over girder and spiked together as shown.

Fig. 61 illustrates the construction of an interior partition running at right angles to the direction of the floor beams. A rough flooring, usually of seven-eighths-inch by eight-inch matched spruce or hemlock, is laid diagonally over floor beams and well nailed to them. On top of this floor, the partition is erected by putting down a two-inch by four-inch partition sill or shoe and then raising the studs, which are usually of two-inch by four-inch stuff set sixteen inches on centers. A seven-eighths-inch by two-inch ground is nailed at the

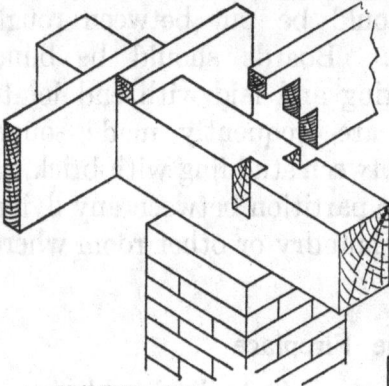

FIGURE 58

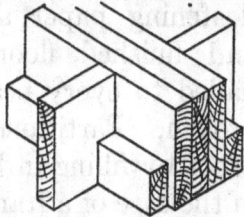

FIGURE 59

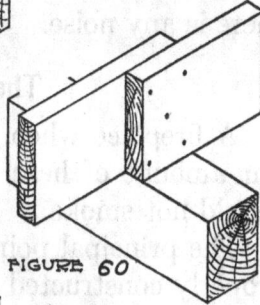

FIGURE 60

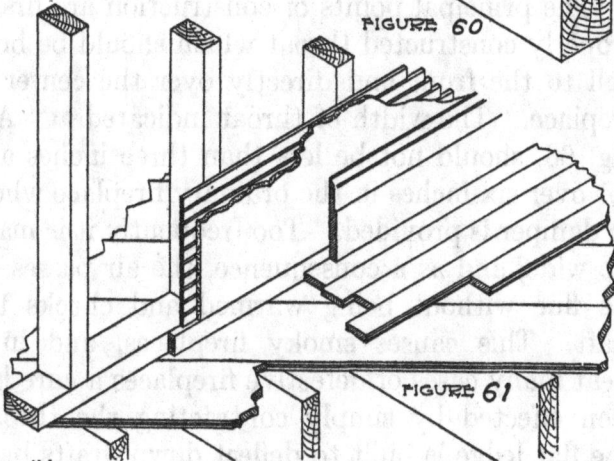

FIGURE 61

PLATE XI.

FRAMING.

floor angle to serve as a gauge for lath and plaster and as a nailing for base. The finished floor is shown extending under the base and is generally of seven-eighths-inch stuff, matched, and not over three inches wide. A layer of deafening paper should be put between rough and finished floors. Boards should be blind nailed to every bearing and laid with end joints broken. Partitions are frequently made sound proof by filling in between studding with brick, as in the case of a frame partition between any living room and a kitchen, laundry or other room where there is any noise.

The Fireplace

A fireplace when properly built should give a fair amount of heat with ordinary attention and should not smoke.

The principal points of construction are first a properly constructed throat which should be built well to the front and directly over the center of fireplace. The width of throat indicated at "A," Fig. 63, should not be less than three inches and not over six inches in the ordinary fireplace where no damper is provided. Too frequently it is made too wide, and as a consequence, the air passes up the flue without being warmed and checks the draft. This causes smoky fireplaces, and in a great many cases of defective fireplaces a cure has been effected by simply contracting the throat. The flat ledge is built to deflect down drafts back into the warm rising air.

Each fireplace should have a separate flue; a good size being eight inches by twelve inches, and chimneys should be carried well above highest

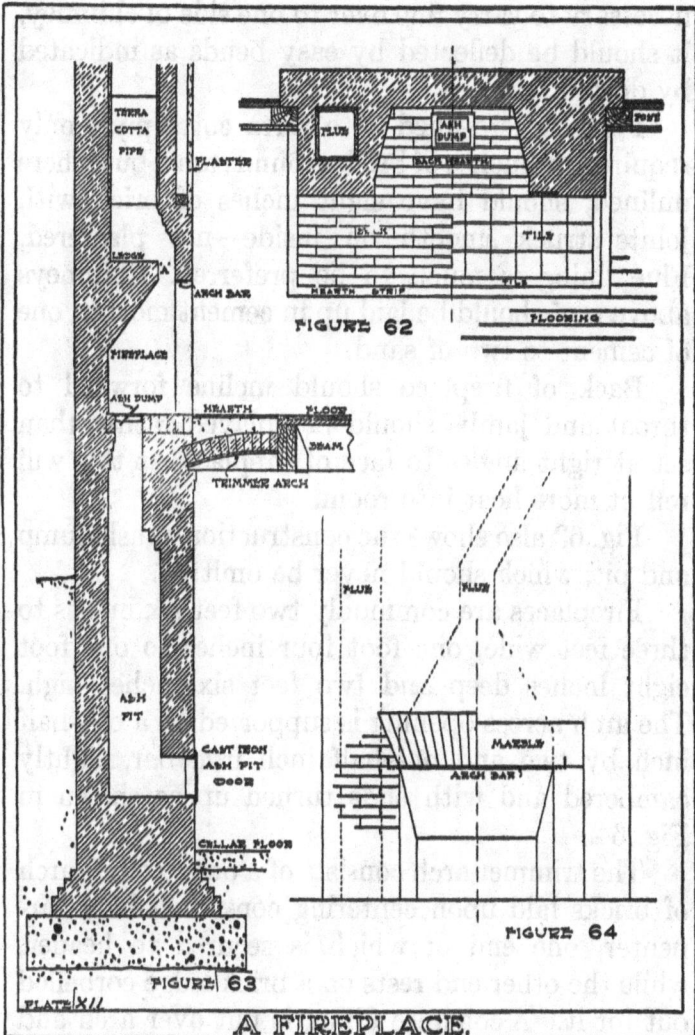

TERRA
COTTA
PIPE

PLASTER

FLUE

ASH
DUMP

BACK HEARTH

LEDGE

ARCH BAR

BRICK

TILE

FIREPLACE

HEARTH STRIP

TILE

FLOORING

FIGURE 62

ASH DUMP

HEARTH

FLOOR
BEAM

TRIMMER ARCH

FLUE

FLUE

ASH
PIT

MARBLE

ARCH BAR

CAST IRON
ASH PIT
DOOR

CELLAR FLOOR

FIGURE 64

FIGURE 63

PLATE XII

A FIREPLACE.

point of roof. The throat should extend entirely across the fireplace opening and should be gradually contracted to the flue directly over middle of fireplace, as shown by dotted lines in Fig. 64. If necessary to carry flue over to one side of chimney, it should be deflected by easy bends as indicated by dotted lines.

Flues, where lined with terra cotta pipe, only require four inches of brick around same, but where unlined, should have eight inches of brick with joints struck smooth on inside—not plastered. Flue lining is much to be preferred. Chimneys above roof should be laid up in cement mortar, one of cement to two of sand.

Back of fireplace should incline forward to throat and jambs should be splayed rather than set at right angles to face of fireplace, as this will reflect more heat into room.

Fig. 63 also shows the construction of ash dump and pit, which should never be omitted.

Fireplaces are commonly two feet six inches to three feet wide, one foot four inches to one foot eight inches deep and two feet six inches high. The arch across opening is supported on a one-half inch by two and one-half inch iron·bar, slightly cambered and with ends turned up as shown in Fig. 64.

The trimmer arch consists of a one rowlock arch of bricks laid upon centering constructed by carpenter, one end of which is secured to header, while the other end rests on a brick ledge corbelled out for it. A concrete filling is put over arch and brought to a level with rough flooring. On top of this tile hearth is set.

The back hearth and back and jambs of fireplace opening are usually of firebrick. The front hearth and facings are of brick, tile or marble.

Cellar Windows

Plate XIII illustrates the construction of a double hung sash window in a brick cellar wall. The illustration shows the construction so clearly that little explanation seems necessary.

Fig. 65 is a section through the head of the window. The cellar ceiling is shown plastered and the inside of the window is finished to correspond with the rooms on the upper floors. When the window occurs in an unfinished portion of the cellar, the finished woodwork, such as trim, stop bead, stool, and apron, is omitted, and a plain casing put on inside of frame. Space above window head, marked "A," should be filled up with mortar to make a draft-proof job.

Fig. 66 is a section through the meeting rails.

Fig. 67 is a section through the window sill. The space between brickwork and underside of wooden sill, marked "B," should also be filled up with mortar. The inside of brick foundation walls is shown furred with 1 by 2-inch strips and then lathed and plastered.

Fig. 68 is a section through the window jamb. Space marked "C" should be well slushed up with mortar.

Fig. 69 is an isometric view showing a little

more clearly the relation of the various members at the window head.

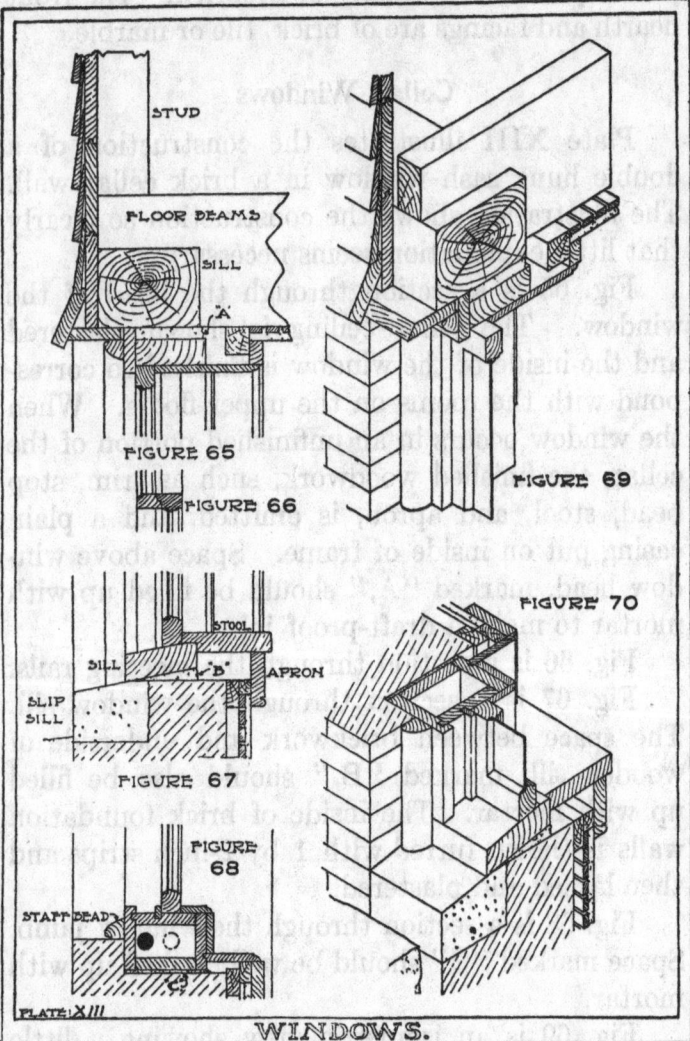

STUD

FLOOR BEAMS

SILL

A

FIGURE 65

FIGURE 66

FIGURE 69

STOOL

SILL

B

APRON

SLIP SILL

FIGURE 67

FIGURE 70

FIGURE 68

STAFF BEAD

PLATE XIII

WINDOWS.

Fig. 70 is an isometric view through the sill of window.

Windows in Frame Walls

Plate XIV illustrates the construction of a double hung sash window in a frame wall. It is what is known as a skeleton frame without a ground casing, and is the kind of window frame that is used in the cheapest grade of work.

Principal among its weak constructive features are the omission of grounds to nail trim to, the use of a single sill instead of a sill and sub-sill and the omission of a ground casing, without which a true pulley stile cannot be insured.

Fig. 71 is a section through the head.

Fig. 72 is a section through the jamb.

Fig. 73 is a section through the sill.

Fig. 74 is an isometric view of head.

Fig. 75 is an isometric view of jamb and sill.

Plates XV and XVI continue the illustration of double hung sash windows in frame walls. The first plate illustrates a skeleton frame with a ground casing.

Fig. 76 is a section through the window head and could be improved by the use of grounds nailed to studs to serve as a gauge for plastering and as a nailing for the trim. The tops of all windows on exterior are almost always exposed to the weather and, as indicated in this case, should be well flashed with tin or copper.

Fig. 77 shows a section through the meeting rails.

Fig. 78 is a section through the jamb of the window and shows the ground casing, marked "G C." When grounds and ground casings are

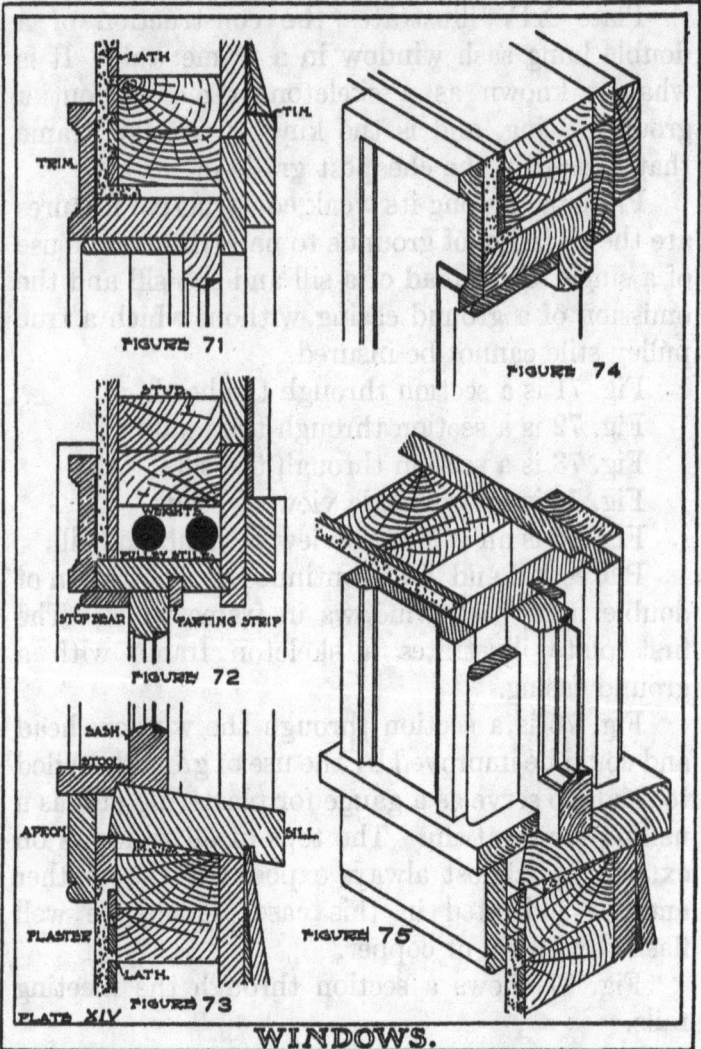

FIGURE 71

FIGURE 74

STUD.

WEIGHTS

PULLEY STILES

STOP BEAD   PARTING STRIP

FIGURE 72

FIGURE 75

SASH

STOOL

APRON   SILL.

JAMB

PLASTER

LATH.

FIGURE 73

PLATE XIV

WINDOWS.

omitted, the trim must always be wide enough
to get a nailing into the studs.   The outside
archi trave should always be at least one and
one-eighth inches thick or better, one and

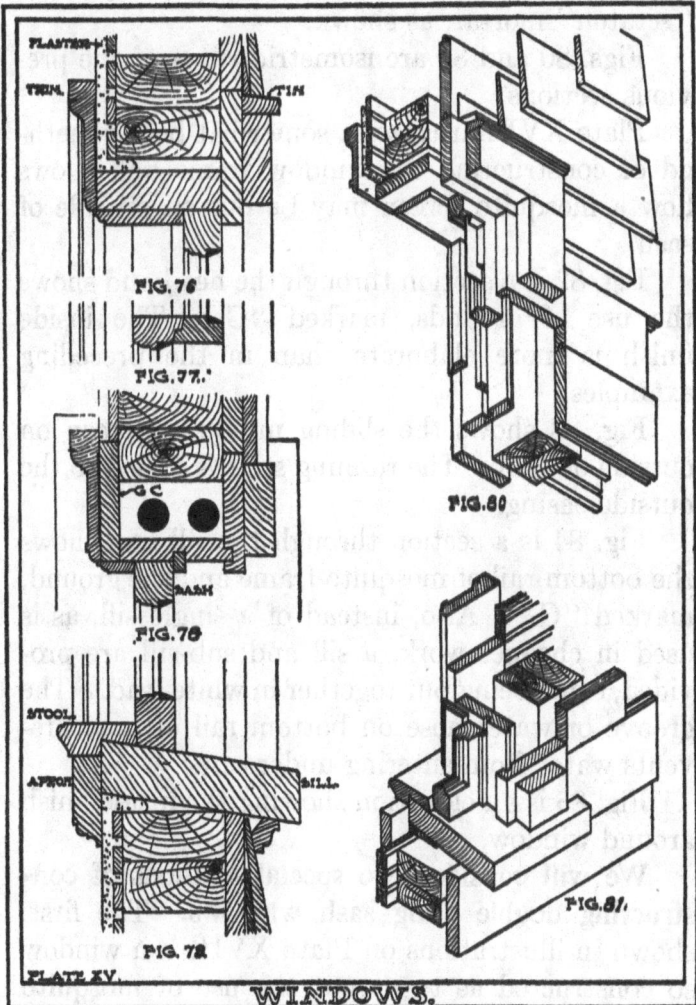

FIG. 76.

FIG. 77.

FIG. 79

FIG. 80

FIG. 81

FIG. 78

PLATE XV.

WINDOWS.

three-eighths inches, to receive clapboards or shingles.

Fig. 79 is a section through the sill of the window. The openings around sills and heads of windows should always be plastered up with "scratch" mortar, as shown.

Figs. 80 and 81 are isometric views of the previous sections.

Plate XVI illustrates a somewhat better method of constructing the window frame and shows how a mosquito screen may be put on outside of sash.

Fig. 82 is a section through the head and shows the use of grounds, marked "G." The inside finish is more elaborate than in the preceding examples.

Fig. 83 shows the sliding mosquito screen on outside of sash. The running strip is nailed to the outside casing.

Fig. 84 is a section through the sill and shows the bottom rail of mosquito frame and the ground, marked "G." Also, instead of a single sill, as is used in cheaper work, a sill and sub-sill are provided, same being put together in white lead. The groove or water nose on bottom rail of sash prevents water from entering under same.

Fig. 85 is an elevation showing the inside finish around window.

We will consider two special methods of constructing double hung sash windows. The first, shown in illustrations on Plate XVII, is a window so constructed as to permit the use of mosquito screen, and blinds with swivel slats outside of the sashes.   This is accomplished by putting the outside casing "D," over the sheathing boards "C," which makes a wider box for sash weights and

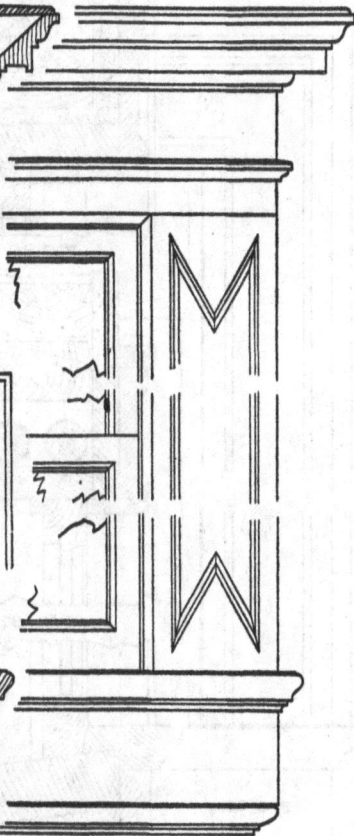

TIN

FIG 82

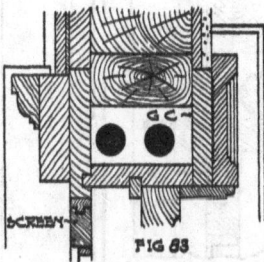

GC

SCREEN

FIG 83

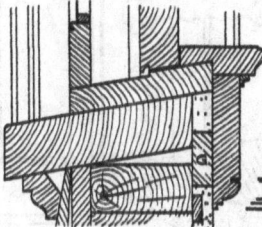

FIG 84   FIG 85

PLATE XVI

WINDOWS.

allows the piece "A" to be set for mosquito screen. The space between mosquito screen and blind "B" is required for blind fasteners.

Fig. 86 is a section through the window head.

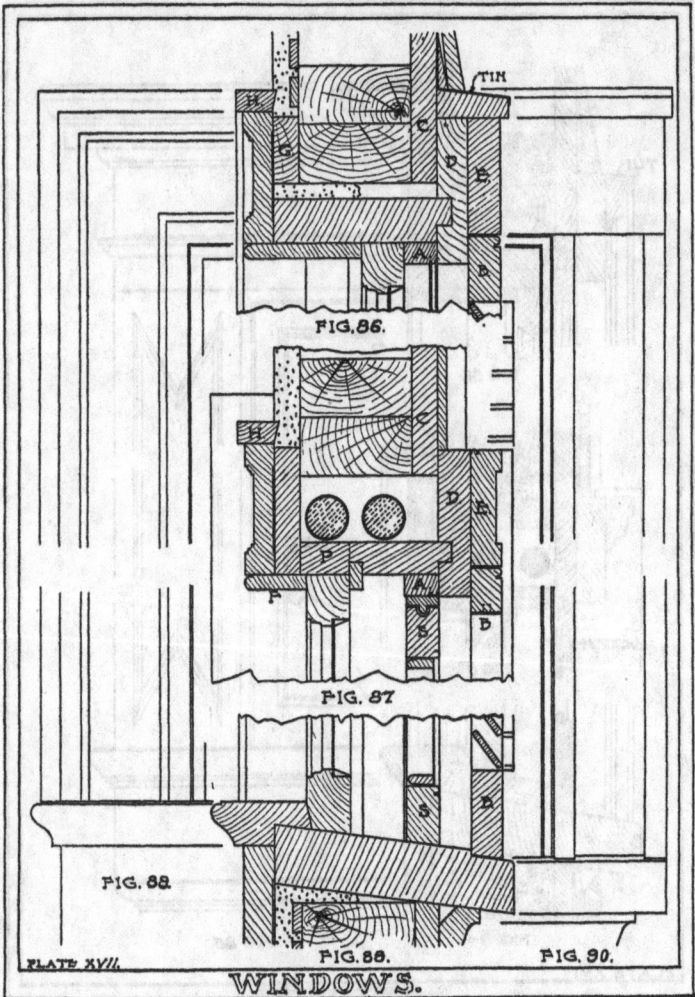

FIG. 86.

FIG. 87

FIG. 88.

FIG. 89.

FIG. 90.

PLATE XVII.

WINDOWS.

The ground "G" serves as a gauge for plastering and as a nailing for the trim and should always be used in the best work.

Fig. 87 is a section through the jamb of the window. The stop bead "F" should never be less than one and three-fourths inches wide so as to allow proper space for the window shades. At "P" a pocket is formed in pulley style for access to sash weights.

Fig. 88 is a section through the sill of the window. The sill should be grooved out three-eighths of an inch for pulley stile.

In all good work a back mould "H" should always be provided. This mould has a beveled edge which may be planed off to fit the unevenness of the plaster work.

Figs. 89 and 90 are respectively interior and exterior elevations of the window.

Plate XVIII illustrates a somewhat better method of construction than any previously shown. The adventage of having mosquito screen and blinds outside of sashes is secured in this case by using four by five-inch studs. The construction is known as a "box frame," the back casing "A" forming the box and insuring a rigid pulley style and consequently accurately fitting sashes. The window sill forms the bottom of the weight box, and is grooved out three-eighths of an inch for the pulley stile.

Fig. 91 is a section through the head of the window.

Fig. 92 is a section through the jamb of the

window. The use of the strip of wood "B," dividing the weight box, is an improvement used only in the best grade work.

Fig. 93 is a section through the sill of the win-

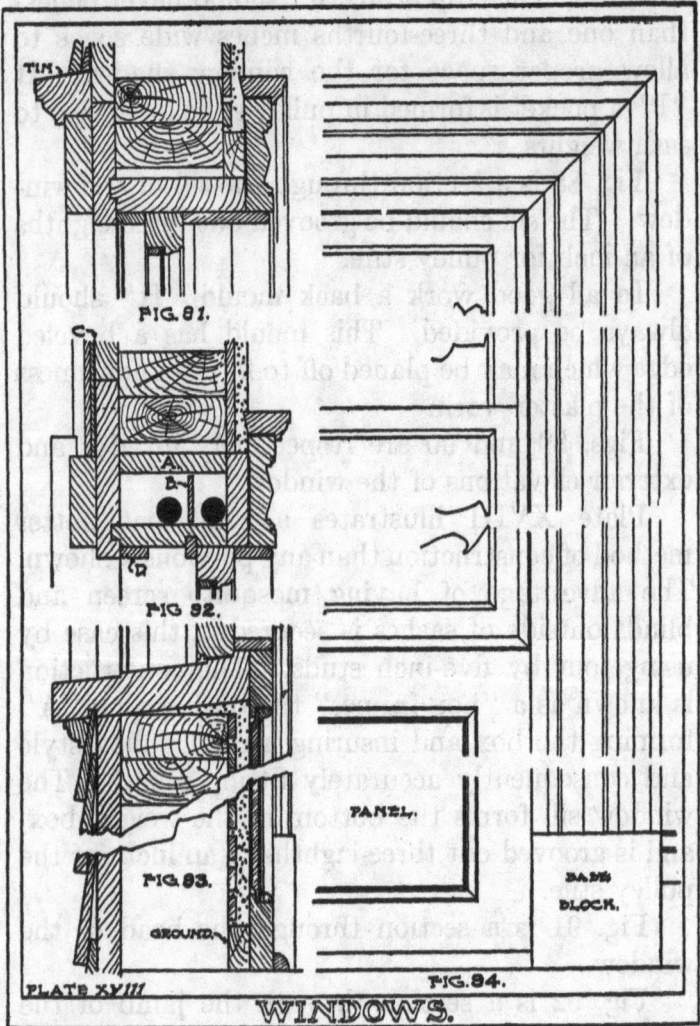

TIN.

FIG. 91.

A
B

C
D

FIG. 92.

PANEL.

SASH
BLOCK.

FIG. 93.

GROUND.

PLATE XVIII

FIG. 94.

WINDOWS.

dow, and shows the use of the moulded panel back under the window in place of the stool and apron finish.

Fig. 94 is an interior elevation of the window.

Plate XIX takes up the **construction of a double-hung sash frame in a double-plastered wall**, with mosquito screen and blinds outside of the sashes and ample space for window shades on the inside stop head. Also, the inside finish is of a somewhat better character than in preceding examples.

In locations exposed to severe cold weather and penetrating winds the double-plastered wall is particularly desirable. The walls are constructed of the usual two by four-inch studs, with one four by four-inch or two two by four-inch studs at all openings. The outside of the wall is sheathed with matched boards, laid horizontally or diagonally, preferably the latter way, and the sheathing paper is then put on; being well lapped at all corners and around all openings. On top of this the shingles, clapboards or other covering material is placed.

The inside of the wall is lathed and plastered two coats—scratch coat and brown coat. One-inch by two-inch grounds are nailed to studs as indicated at "G." A one-inch air space is then formed by means of one-inch by two-inch furring strips, marked "F," and the wall is again lathed and plastered, this time with three coats of plaster. All spaces around head, sill and jambs of window frame should be well filled up with scratch mortar, so as to be absolutely wind-proof.

In Fig. 96, where marked "P," two pockets are formed in the lower part of the pulley stile, for access to the sash weights. When the strip of wood dividing weight box is omitted, one pocket, usually on the inner half of the pulley stile, is

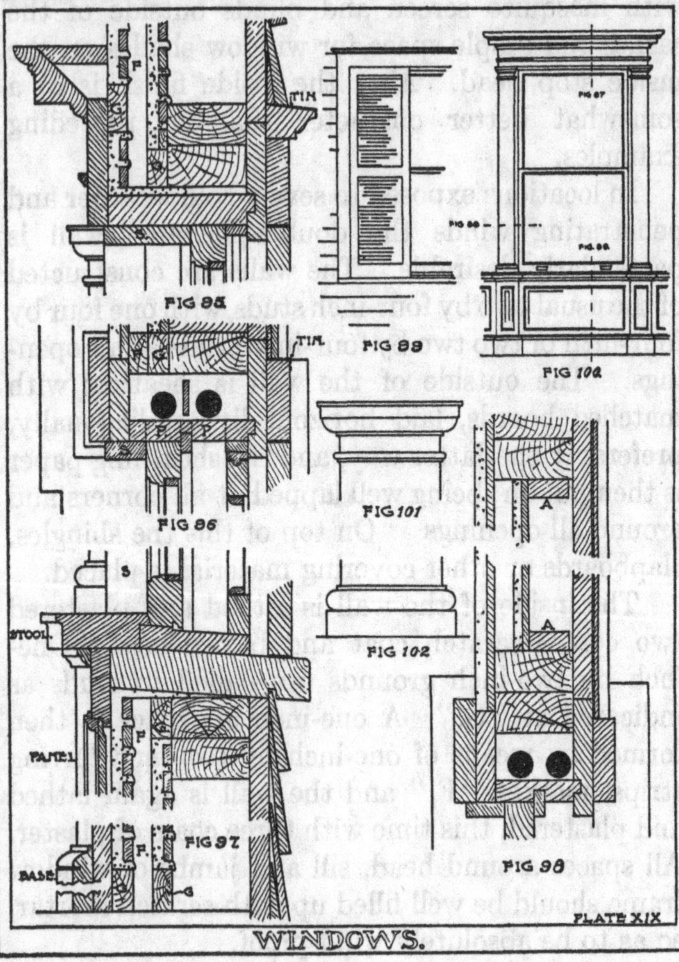

FIG 95

FIG 96

FIG 97

FIG 99

FIG 101

FIG 102

FIG 87

FIG 88

FIG 89

FIG 100

FIG 98

WINDOWS.

PLATE XIX

sufficient. The flashing shown at the outside architrave is used only in the best grade of work.

The interior finish consists of a pedestal base or wainscoting, the top member of which forms the window stool; pilaster jambs and an entablature head. With a finish of this kind the stop bead, marked "S," should always be made thick enough to take up the projecting mouldings of the cap and base, which butt against it. The panel under the window should be constructed so that it can be readily removed in case it cracks or is damaged, by taking off the panel moulding. The frieze of the entablature should be on a line with the face of the pilaster, and the face of the wainscoting should be kept on a line with the plinth of the pilaster base. Cap and base are shown in Fig. 101 and 102.

Fig. 99 is an exterior elevation and Fig. 100 an interior elevation of the window.

Another method of constructing double-plastered walls and a window frame for same is shown in Fig. 98. In this case one-inch by two-inch furring strips, marked "A," are nailed to the studs as a bearing for the lath and are plaecd so as to allow a one-inch air space between the inside plastering and the back plastering. The frame could be improved by placing the outside casing outside of the sheathing boards, thereby giving enough width to the pulley stile to permit of both screen and blinds outside of sashes.

In Plate XX we illustrate a storm-resisting window, with a double set of sashes, in a double-plastered wall, for use in a location exposed to very severe weather throughout the entire or greater part of the year.

Three by six-inch studs are used for the outside walls and the window has a box frame with a sepa-

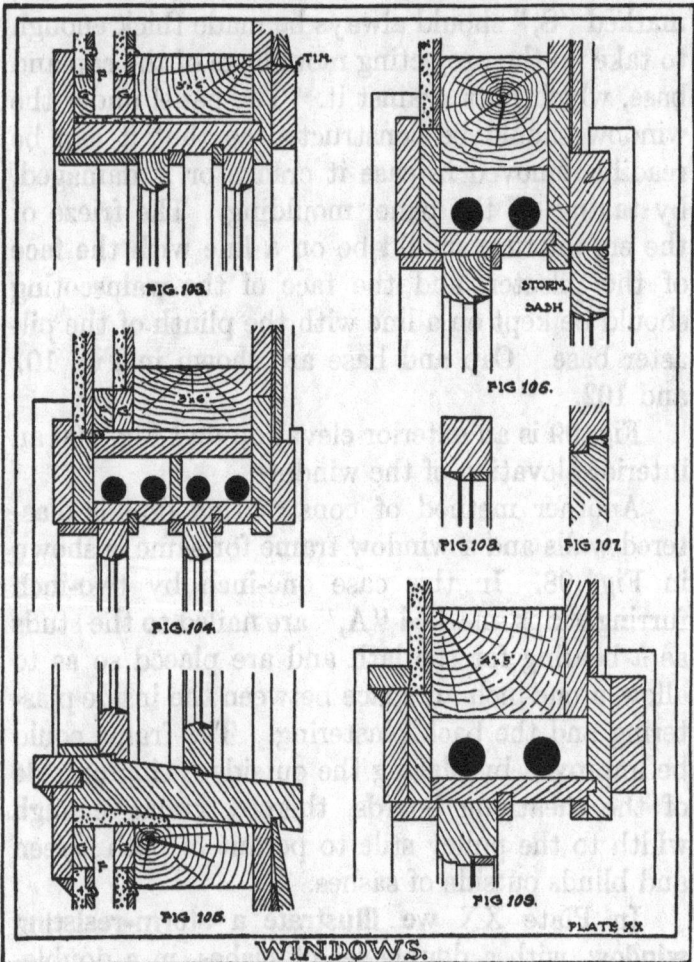

FIG. 103.

STORM SASH

FIG. 106.

FIG. 104.

FIG. 108. FIG. 107.

FIG. 105.

FIG. 109.

PLATE XX

WINDOWS.

rate weight box for each set of sashes. The sill should be gotten out of two-and-one-half inch stuff.

Fig. 106 shows how storm sash may be substituted for blinds when cold weather sets in. The thickness of storm sash and outside architrave should be at least one and one-half inches, or better, one and three-quarter inches. The joint at the meeting stiles of storm sash is shown in Fig. 107. Sash are usually rebated one-half inch.

Fig. 108 shows a double-glazed sash for exposed locations. The principal objection to their use is that dust will sooner or later get between the sheets of glass, which, with the sweating in cold weather, will soil and streak the glass where it is inaccessible for cleaning.

Fig 109 shows a window frame constructed for the use of heavy sashes the use of four-inch by five-inch studs and the placing of the outside casing over the sheathing boards gives the required width to the pulley stile.

We will conclude the study of double hung windows in frame walls; illustrating in Plate XXI a window with inside sliding blinds and in Plate XXII a double hung window in a brick veneer wall.

In Plate XXI we have shown the **inside blinds sliding in grooves on the window jambs,** and, when not in use, sliding down in a pocket (Fig. 112) behind a moulded panel back. The pocket is covered by a hinged stool "A," which raises up as indicated by the dotted lines when it is desired to use the blinds.

Rather than have the sliding grooves and blinds

project in the room, we have made the wall thicker by the use of three by six-inch studs, with the exception of that portion of wall under window

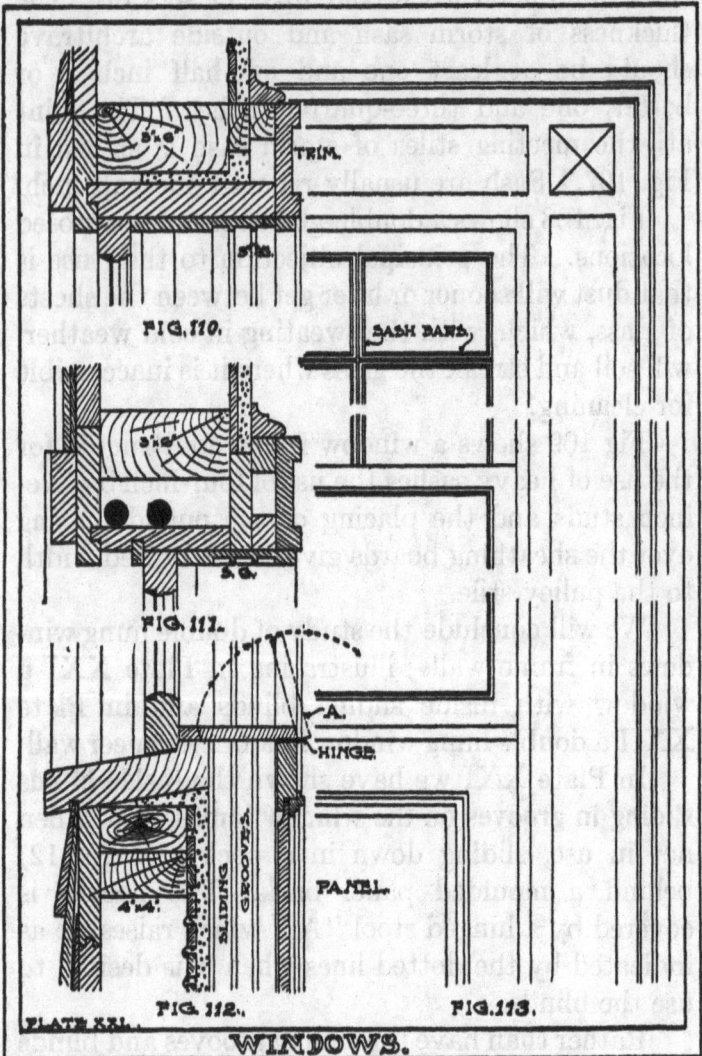

FIG. 110.

FIG. 111.

FIG. 112.

FIG. 113.

PLATE XXI.

WINDOWS.

frame where we have used two two by four-inch studs. When the wall is not made thick enough to take up the extra thickness of blind box and slides, the whole window is made to project in the room, causing a more or less unsightly appearance. The window frame itself is constructed in pretty much the usual manner, the exceptions being the extra width of window head and pulley stile caused by thickness of wall and the omission of the inside stop beads, in place of which we have the sliding grooves marked "S. G.

The plastering behind the blind box should not be omitted. The scratch and brown coats, however, are sufficient. All spaces around window frame should be well filled up with scratch mortar.

In Plate XXII the wall is constructed of two by four-inch studs, doubled at openings; plastered on the inside; sheathed diagonally on the outside with matched boards; then covered with waterproof sheathing paper; and then with four inches of brickwork. The veneer of brick should be tied to the frame wall every five courses, opposite every stud, with patent veneer ties, shown in Fig. 117. The tie is made of one-eighth-inch galvanized steel wire, and is far superior to the iron nails frequently used.

Brick veneer construction for exterior walls is largely used in many sections of the country, and seems constantly growing in favor.

The window frame is constructed in the usual manner, with a moulded staff bead to cover the joint of brickwork and window frame.

Fig. 114 is a section through the window head. Fig. 115 is a section through the jamb. Fig. 116 is a section through the sill, and shows the wood sill lapping over the stone sill. The stone sill is

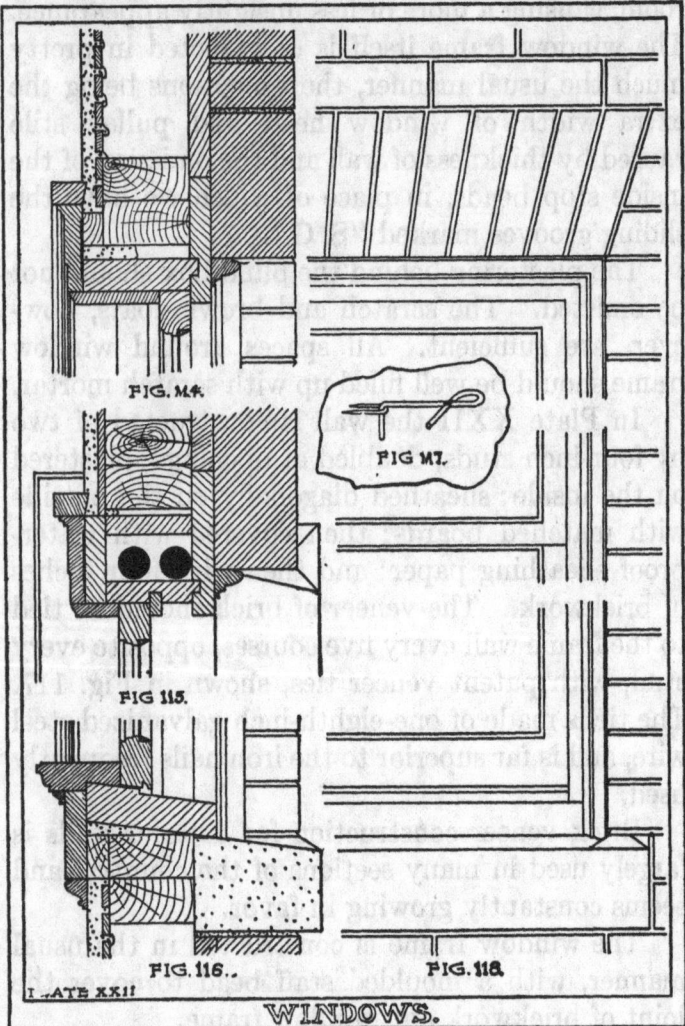

FIG. 114.

FIG. 117.

FIG. 115.

FIG. 116.

FIG. 118.

PLATE XXII.

WINDOWS.

beams for the support of the floor joists which bear directly over the opening. These iron beams are necessary when there is not sufficient space between the window head and the under side of the floor joists, to turn a brick relieving arch on top of a timber intel.

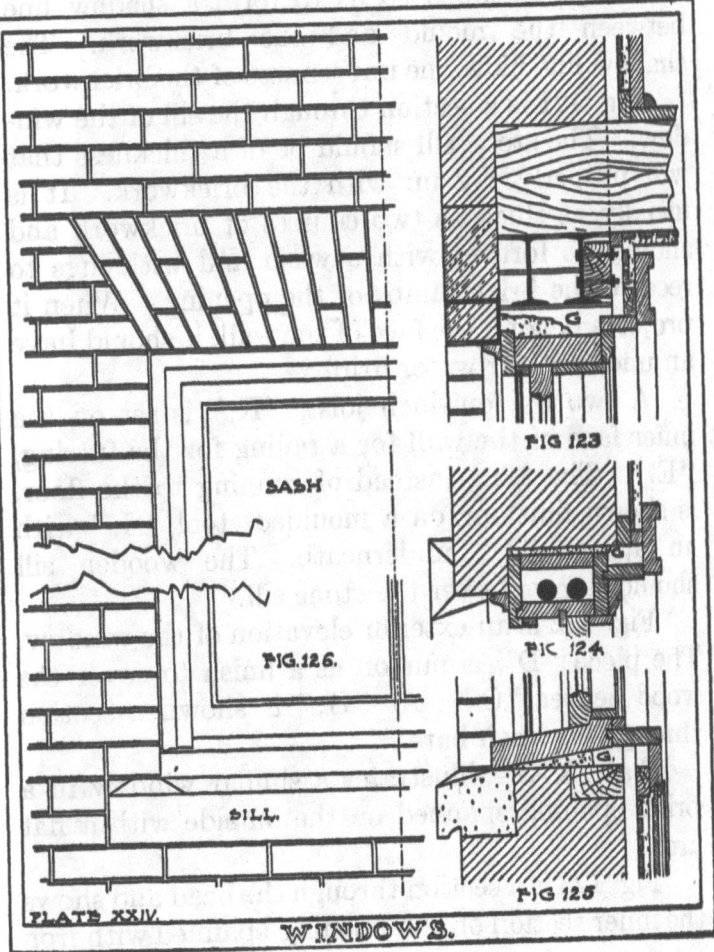

PLATE XXIV.   WINDOWS.

made the thickness of two courses of brick, four inches longer than the width of brick opening, and six inches deep. It is made with a wash and with lugs at each end, projects one inch beyond the face of the wall, and extends back to the studding, over the sheathing. Fig. 118 is an exterior elevation of the window.

Windows in Brick Walls

We will start a consideration of double hung sash frames in brick walls.

Plate XXIII illustrates a well constructed frame in a thirteen-inch wall; the window finishing with an arch on the outside and a square head on the inside.

Fig. 119 is a section through the window head at the center line. The opening is spanned on the outside by a segmental arch "B" of face brick, rubbed to the required shape and laid on a temporary wood center. This center should not be "stuck" until the mortar has thoroughly set.

The inner eight inches is spanned by a permanent wood center "C," usually constructed of two-inch spruce and made of sufficient length to give a bearing of four inches on the wall on each side of the window opening. On top of this center, a two-rowlock relieving arch "A," is turned.

The wall is furred on the inside as indicated at "E," and is then lathed and plastered; grounds being set as shown.

This furring of late is frequently omitted and the plaster applied on the brick wall itself, after it has been made impervious to water by a heavy coat of one of the many waterproof paints now on the market. This paint prevents the dampness, which penetrates the wall, from discoloring the plaster work.

Fig. 120 is a section through the jamb of the

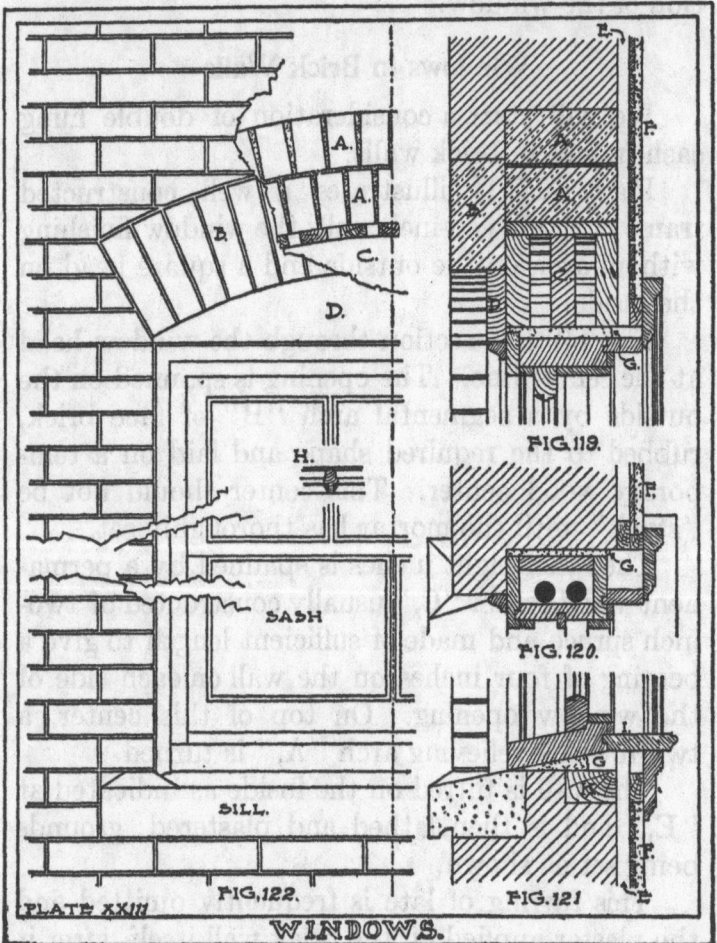

FIG. 119.

FIG. 120.

FIG. 121.

PLATE XXIII

FIG. 122.

WINDOWS.

A piece of two by four stuff is bolted to the web of the inner iron beam as shown, to secure a nailing for the furring.

A two and one-half-inch by a three and one-half-inch angle iron is shown under the face arch, and is provided so as to prevent any settlement of the arch. Without some support of this kind, flat arches are very likely to sag in the center, causing a very unsightly appearance.

Fig. 124 is a section through the jamb.

Fig. 125 is a section through the sill. It shows the trim running to the floor and the space under window finished with a panel back.

Fig. 126 is an exterior elevation of the window.

Continuing the consideration of window construction, we illustrate in Plate XXV, a double hung sash window in an eighteen-inch brick wall.

Fig. 127 is a section through the window head. The opening is spanned on the outside with an arch of stone and on the inside a timber lintel is provided, and a rowlock relieving arch turned on top of same. Relieving arches are usually constructed with one rowlock to each eighteen inches or fraction thereof in the width of the brick opening. The timber lintel is constructed with two or three centers of two-inch stuff, cut to the required curvature. On top of these centers are nailed narrow wood strips called lags.

Fig. 128 is a section through the window jamb. The calking shown in this and other sections is to keep out penetrating winds. This calking is commonly done by filling around all openings, as shown, with scratch mortar, but in the highest grade work is done by hand-calking all the spaces around frame with oakum.

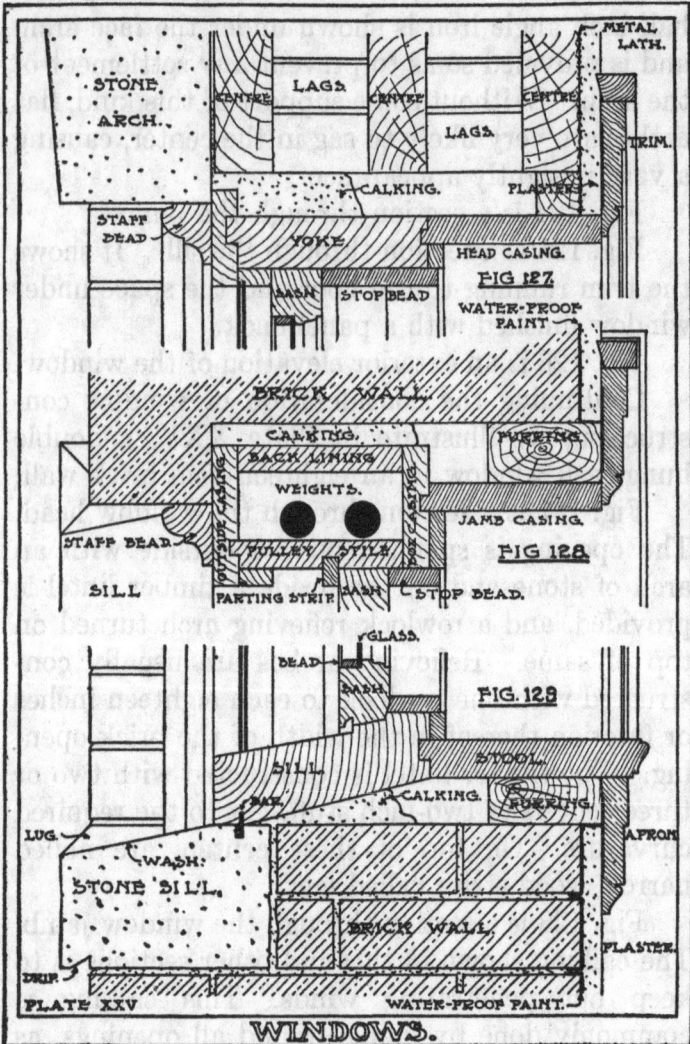

PLATE XXV

WINDOWS.

It will be noticed that furring and lathing of inside of walls is omitted, and the plaster applied directly to brickwork. When this is done, the wall should be thoroughly coated with a water-proof paint so as to make it absolutely impervious to moisture and dampness, which would discolor the finished plaster work. There are several of these paints now on the market, and, when properly applied, make a wall absolutely damp-resisting. Before the paint is applied, the mortar joints should be raked out enough to give a clinch for the plaster, as shown in Fig. 129.

This section is taken through the window sill, and its principal feature is the joint of the wood and stone sills. This joint is made water-tight by means of a galvanized iron bar or tongue which is let into a slot on the underside of the wood sill and lead-calked into a corresponding reglet in the stone sill. The stone sill is cut with wash and lugs on top, and with a water drip on the lower edge.

The sash are shown one and three-quarters inches thick and glazed with plate glass, bedded in putty and held in place with wood beads.

Metal lath is nailed to the timber lintel to afford a clinch for plaster. This lath will also be required over any recesses or pipe chases in masonry walls.

Plate XXVI illustrates another method of constructing a double hung window frame in a masonry wall. The opening is spanned on the outside by a moulded stone lintel and the inner eight inches of the wall is carried on an iron lintel

(Fig. 130). An iron lintel is usually provided when it is not convenient to turn a rowlock relieving arch over a timber lintel, as in the case of floor

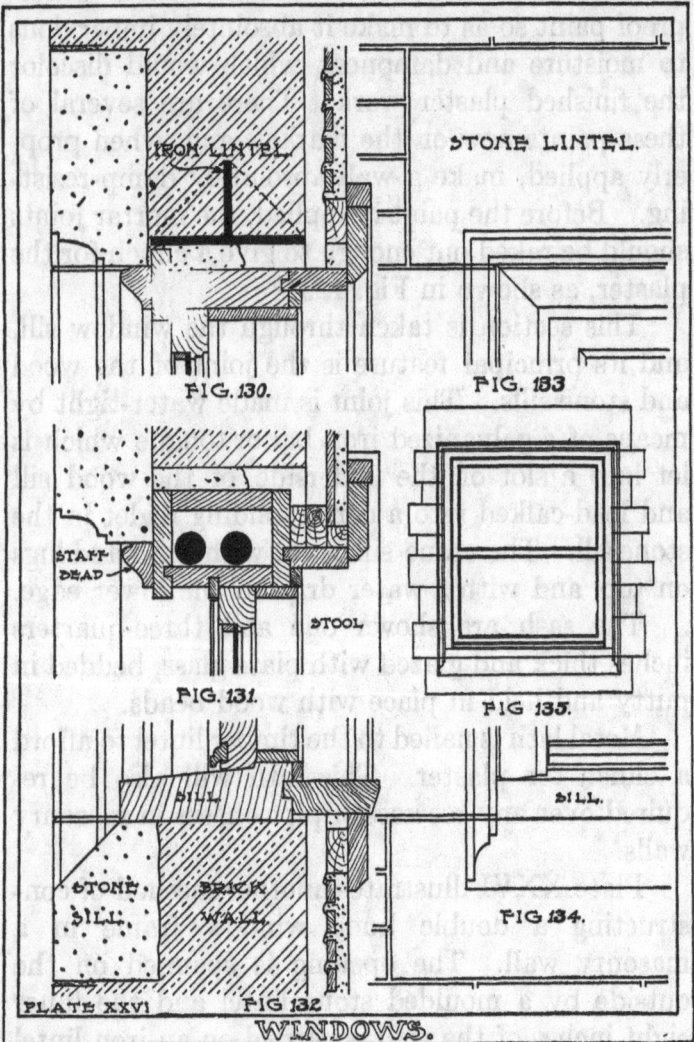

IRON LINTEL.

FIG. 130.

STONE LINTEL.

FIG. 133

STAFF
DEAD

STOOL

FIG. 131.

FIG 135.

STONE
SILL.

BRICK
WALL.

SILL.

SILL.

FIG 134.

PLATE XXVI   FIG 132

WINDOWS.

beams bearing over opening, so close to window head as to make it impossible to turn an arch underneath them. They are also used over wide openings in preference to arches.

Fig. 131 is a section through the jamb of window. Stone quoins of various widths and heights are provided for masonry jambs.

Fig. 132 is a sill section and shows a moulded drip on bottom rail of sash.

Figs. 133-135 are exterior elevations.

The use of inside blinds on double hung windows in masonry walls is considered. Plate XXVII is a double hung window frame in a thirteen-inch brick wall. A stone lintel is provided on the outside over the opening; in depth it should equal the height of four or five courses of brick; its thickness should be not less than the brick reveal, nd it should have a bearing of about four inches on each brick jamb so as to take up with the brick joints. The opening is spanned on the inside with a timber lintel or center made up of three pieces of two-inch stuff, cut to the required curve, with cross strips across top called lags. On top of the wood center or lintel a rowlock arch is turned. Rowlock arches are usually made with one rowlock for each eighteen inches or fraction thereof in the width of masonry opening. The rise is usually made one inch for every eighteen inches of width of opening.

The sill of the masonry opening is of stone, the thickness of two courses of brick, of a depth that will overhang the wall on the outside about one and one-half or two inches, and will extend under the wooden sill not less than two inches. The sill is cut with a wash and lugs and on the underside of projecting part has an undercut or water nose.

The wall is furred on the inside with two by four-inch stuff, placed sixteen inches on centers, so as to give the required extra width for blind boxes.

The caulking of interstices at head sill and jambs of frame is shown of "scratch" mortar, but in the best grade of work is usually of oakum hand caulked, so as to make the construction absolutely wind proof.

Fig. 136 is a section showing the usual construction at the window head. The wide soffit may be paneled if desired. A moulded staff bead should always be used to cover the joint at the intersection of the frame with the masonry, on the outside. The lath, plaster and grounds are applied to the furring in the usual manner. The trim is mitred and tongued or doweled together at the head, a feature employed only where expense is not an all important consideration.

Fig. 137 is a section through the jamb and shows the construction of the blind box and the method of folding blinds. It will be observed that a special hinge is used on blinds, which prevents them from catching or sticking in the box. All the woodwork of box, which would be exposed to view when the blinds are closed, should be made of the same material as, and should conform to, the finish of the balance of the room. All the blind stiles should be rebated as shown.

The blinds are all shown with rolling slats, but frequently the blind fold, which is exposed to view when blinds are folded back in the pocket, is made paneled instead of with slats, so as to give the effect of a paneled window jamb.

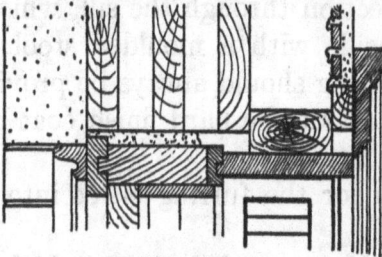

FIG. 136.

FIG. 137.

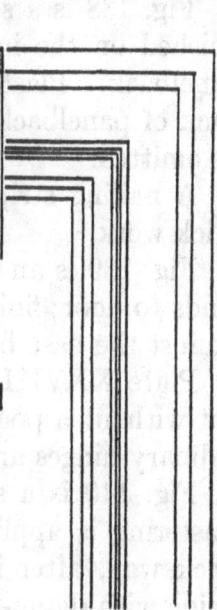

FIG. 139.

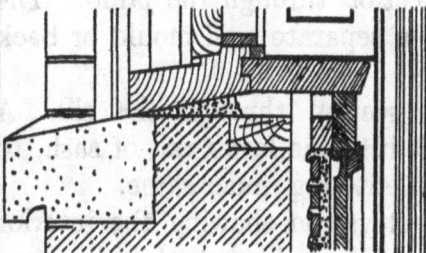

FIG. 138.

PLATE XXVII

WINDOWS.

The inside stop bead should be made at least one and three-fourth inches wide, or better two inches wide, so as to give plenty of space for shades between blinds and sashes.

The use of a strip dividing the weight box would be an improvement.

Fig. 138 is a section through the sill, which is finished on the inside with a moulded stool and panelback. Plastering should always be provided back of panelbacks, but the hard finish coat may be omitted.

A nailing strip for the furring is let into the brick work.

Fig. 139 is an elevation of the trim, which extends to floor; finishing on a base block or plinth, against the base butts.

Plate XXVIII shows a similar window frame, but without a pocket or box for the blinds. The ordinary hinges are used in this case.

Fig. 140 is a section through the head. The plastering is applied directly to the sixteen-inch brick wall, after it has first been coated on the inside with damp-resisting paint.

Fig. 141 is a section through the jamb. The trim is made with a separate wall mould or back band.

Fig. 142 is a section through the sill. A moulded drip is let into the lower rail of sash, to prevent water from entering under same.

Fig. 143 is an elevation showing the interior trim which finishes on a moulded stool and apron; both having returned ends.

A double-hung sash window in a brick wall, with inside folding blinds set in a projecting box.

The brick wall is thirteen inches thick and the opening is spanned on top by a stone lintel and has a stone sill cut with a wash on top, lugs at each end

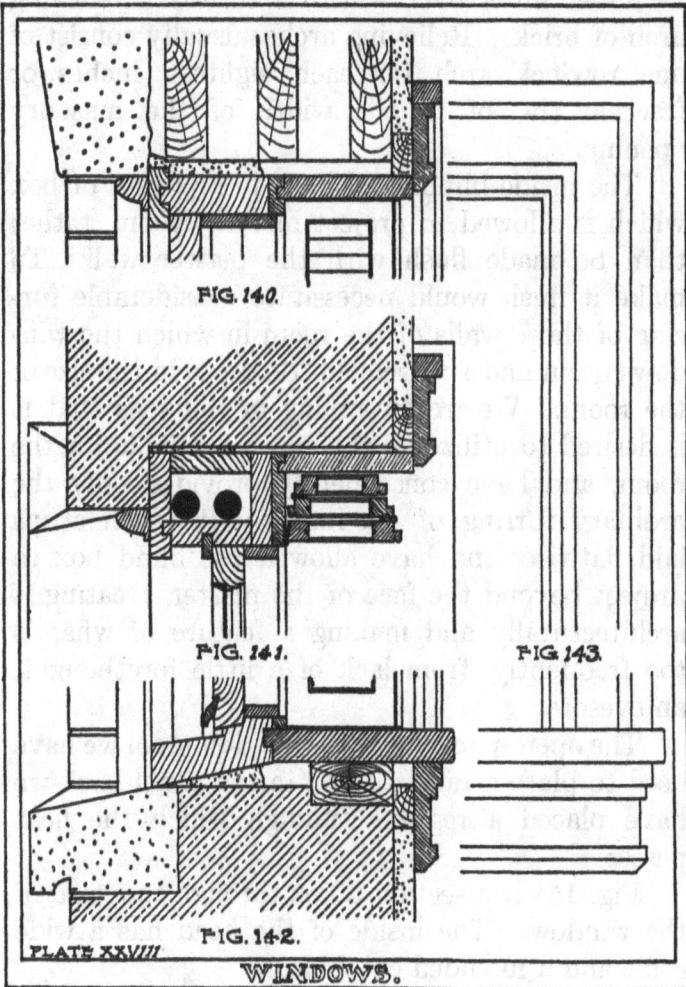

FIG. 140.

FIG. 141.

FIG. 143.

FIG. 142.

PLATE XXVIII

WINDOWS.

to receive the brick jambs, and drip or water nose on underside where it projects beyond the face of the wall.

The frame is set so as to give a four-inch reveal. The inner two-thirds of the opening is spanned by a timber lintel over which is turned a relieving arch of brick. Relieving arches usually consist of one rowlock arch for each eighteen inches or fraction thereof in the width of the masonry opening.

The inside blinds fold back in a pocket or box which is allowed to project into the room, rather than be made flush with the plaster wall. To make it flush would necessitate considerable furring of those walls of the room in which the windows occur and a consequent decrease in the size of the room. We are presuming in this case that it is desired to utilize all the space available for the room, and have consequently provided only the ordinary furring of one-inch by two-inch strips laid flatwise, and have allowed the blind box to project beyond the face of the plaster, treating it architecturally and making a feature of what is too frequently, from lack of a little forethought, an eyesore.

The open space under the window stool we have used to place a radiator and in the panel back we have placed a register through which the heat passes.

Fig. 144 is a section taken through the head of the window. The inside of the head has a wide soffit and a moulded entablature.

Fig. 145 is a section taken through the jamb of
the window. When the blinds are folded back in
the pocket, it has the appearance of a wide paneled
jamb, owing to the fact that the first fold of the

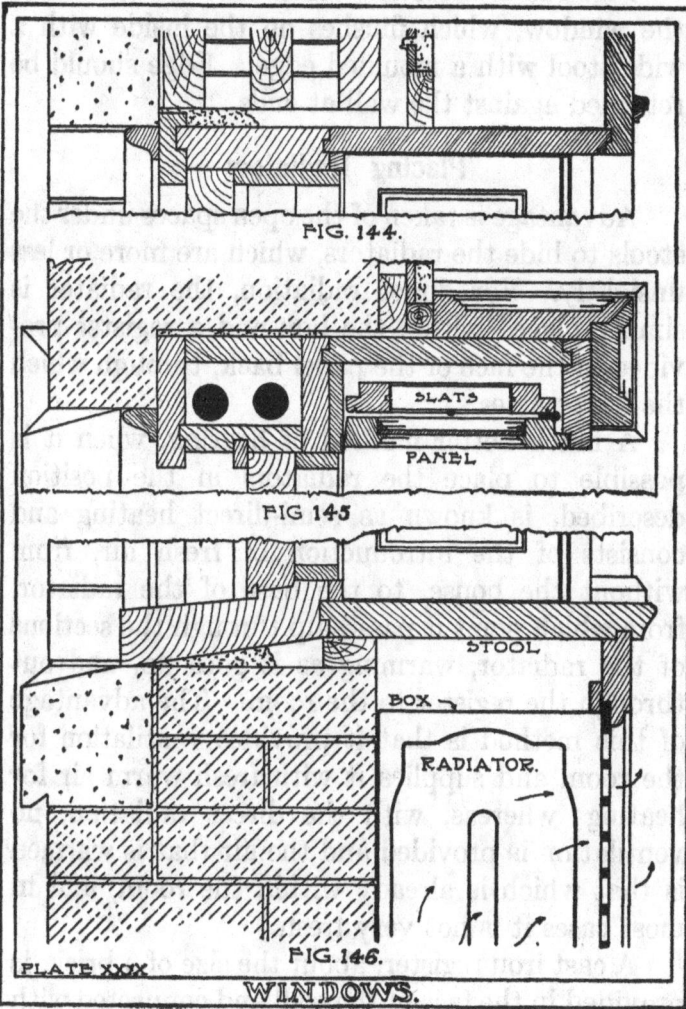

FIG. 144.

SLATS

PANEL

FIG 145

STOOL

BOX

RADIATOR.

FIG. 146.

PLATE XXX

WINDOWS.

blinds has a panel instead of rolling slats. The second fold, however, has slats. The exposed faces of the box are covered with paneled pilasters which finish with a small base on the top of stool.

Fig. 146 is a section taken through the sill of the window, which finishes on the inside with a wide stool with a moulded edge. Edge should be returned against the wall at ends.

Placing Radiators

Advantage is taken of the open spaces under the stools to hide the radiators, which are more or less unsightly. For direct radiation, the radiator is simply placed within the box and a register provided in the face of the panel back, through which the heat passes.

A much better method of heating, when it is possible to place the radiators in the position described, is known as semi-direct heating and consists of the introduction of fresh air, from without the house, to the base of the radiator, from which point it passes up through the sections of the radiator, warming as it goes up, and out through the register to the room. The advantage of this method is that it provides ventilation for the room and supplies it with fresh warm air for heating; whereas, with the direct radiation, no ventilation is provided and the air that is warmed is that which is already within the room, and in most cases it is not very fresh.

A cast iron register, about the size of a brick, is provided in the face of the wall and connected with the base of the radiator by means of a small galvanized iron duct. The fresh cold air from outside the house enters through this passage and is controlled by means of a damper within the duct.

In severely cold weather the damper may be closed entirely and the radiator used for direct radiation.

When radiators are boxed in as shown, they should be set within a galvanized iron box, or the stool, panel back and other surrounding woodwork should be lined with galvanized iron as a protection against fire.

Frequently the radiator is placed under the window stool, but not boxed in. In such a case the inside of the brick wall below the stool is furred and plastered two coats, scratch and brown, and a panel back provided immediately in front of plaster and behind the radiator. The stool, which is wide, is supported on two wooden brackets, one under each pilaster, and about the same width as pilaster. This, however, is only for direct radiation. The underside of the stool should be covered with galvanized iron. The radiator, which is exposed in the room, should be neatly finished.

A double hung sash window in a brick wall, with inside blinds folding in a slanting box.

The brick wall is thirteen inches thick and the opening is spanned on top by a stone lintel four inches deep, eight inches longer than the width of the opening, and four brick courses in height. Back of the lintel, the inner two-thirds of the wall is carried across the opening on a rowlock arch turned over a wood center. Rowlock arches are usually made with one rowlock for each eighteen inches or fraction thereof, in the width of the opening, and are segmental in form with a one-inch rise for each foot of span. Key bricks of the lower rowlocks should not be set until the arch is ready for the key bricks of the top rowlocks. The centers for the arches are usually set by the carpenter under the direction of the mason. Temporary centers for face arches, should not be struck until the mortar with which arches are laid, has thoroughly set and hardened.

The window frame is set so as to give a four-inch reveal. The joint between the brick jamb or lintel and the wood frame is covered with a moulded staff head nailed to the frame. This staff head should be moulded in such a manner as will give a channel or deep recess on the side adjoining the brickwork

The opening on the exterior has a stone sill two brick courses in height, eight inches longer than the width of the opening, and of a proper depth to extend under the wood sill at least two inches. The stone sill is cut with a wash and has raised lugs or stools at each end to receive the brick jambs or imposts. The underside of projecting portion is cut with a water nose or drip. The sill should be bedded in mortar at each end under the imposts but should not be bedded in the center, as any settlement of the building would be likely to crack it if bedded. When building is nearly completed, the open space or joint under the sill should be pointed up with mortar, worked well into the joint.

The inside blinds fold back in a pocket or box, set on a slant, giving the appearance of a deep, paneled, splayed jamb. The first fold of the blinds is paneled and the inside soffit has a panel to correspond with it. The blind box projects into the room, with the trim returned about box to plaster wall. The head is formed like an entablature, with a crown moulding, a facia and a mould planted on facia. The stool is deep, carried across opening, and returned about box to plaster wall. The stool is supported at each end, under the trim, on carved wood brackets. The joint between the plaster and the underside of the wood stool is covered with a small cored mould which breaks around the wood brackets and returns against wall. The blind box projects into the room, with the trim finish as the other woodwork of the room in which it occurs. This is also true of the inside stop head.

Fig. 147 is a section taken through the head of the window. The panel shown in the inside soffit should be arranged so as to be easily removable in case of splitting or warping. This may be readily done, if constructed as indicated, by removing the small coved wood moulds which hold it in position. These moulds should be fastened to the frame rather than to the panel. In this way the panel is set free so that it may contract or expand without affecting adjacent work.

Fig. 148 is a section taken through the jamb of

the window. The second and third folds of the blinds are provided with rolling slats. The first fold may also be provided with slats, but a panel makes a better appearance and does not catch

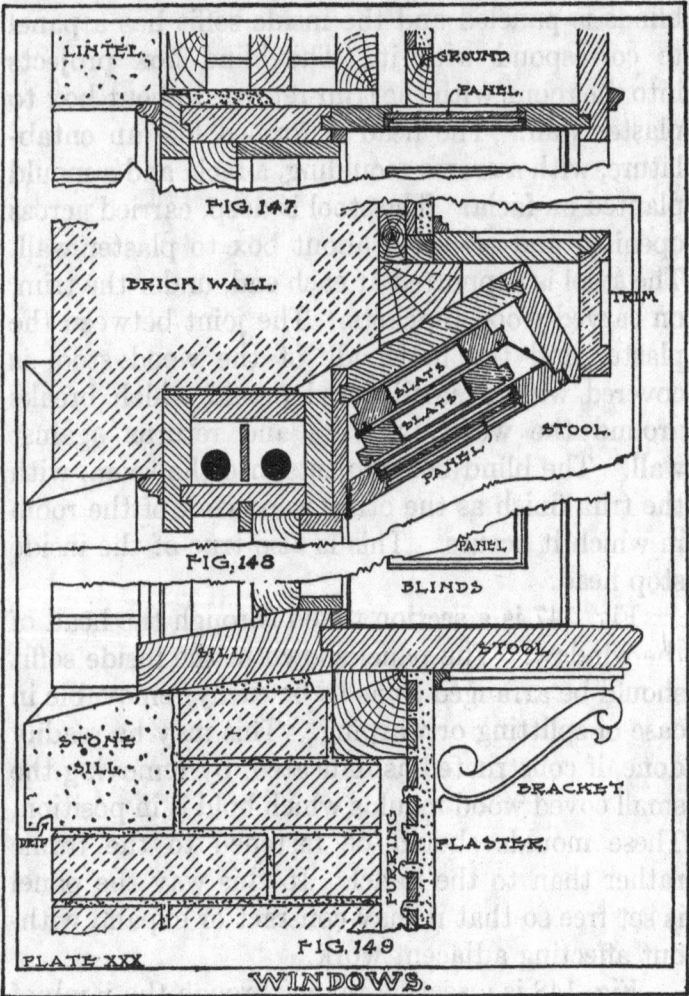

FIG. 147.

LINTEL.

SECOND PANEL.

BRICK WALL.

TRIM

SLATS

SLATS

STOOL.

PANEL

FIG. 148

PANEL

BLINDS

SILL

STOOL

STONE SILL.

BRACKET.

DRIP

FURRING

PLASTER.

FIG. 149

PLATE XXX

WINDOWS.

dust the way the slats would. It will be observed that a special hinge is used on the blinds, which prevents them from catching or sticking in the box. All the blind stiles should be rabeted as shown and should have a head on edge to lessen the prominence of the joint. The first fold is rabeted over the side of the box which acts as a blind stop. The use of the tongue dividing the weight box is a feature used only in the best grade work.

All interstices and spaces between the window frame and the masonry opening should be made absolutely wind proof. In ordinary work this caulking is done by caulking the spaces with oakum and pointing up with mortar.

Fig. 149 is a section taken through the sill of the window. The wood sill is ploughed for the inside stool. The space under the projecting stool is an excellent place for a radiator. If one is put there, the underside of the stool should be covered with asbestos paper and tinned.

Plate XXXI illustrates windows in both frame and brick walls with inside blinds projecting in the room and folding back flat against wall.

Fig. 150 illustrates the window in the frame wall. This wall is constructed of two-inch by five-inch studs set sixteen inches on centers. Studs are doubled about window and door openings and are doubled for the head and sill of each window and the head of each door. One row of herring-bone crossbridging is provided between the studs for each story.

The outside of the framework is covered with

one-inch by eight-inch matched or ship-lapped hemlock or North Carolina boards. These boards should be laid diagonally in preference to horizontally as they make a more rigid wall when so laid.

This sheathing is covered with a heavy tarred or rosin sized building paper, well lapped at all joints and well turned in at all corners and angles.

The exposed covering of the exterior of the wall consists of shingles laid five inches to the weather with broken joints and of random widths. Shingles over six inches wide should be split. Shingles should be well nailed with two galvanized nails to each one.

The inside of the wall is plastered on wooden lath and grounds are nailed to the studs to form a gauge for plastering and to give a nailing for the trim. The strip on which the studs are hung should be of the same material and finish as the adjoining trim. The inside stop head also serves the purpose of a blind stop.

The blinds are hung on a special hinge as shown, which permits them to be folded back against the wall, clear of the trim.

This manner of folding the blinds is not quite so good as the methods given in previous pages, as the blinds are rather unsightly and are great dust catchers. This method, however, is resorted to quite frequently owing to lack of space for a blind box.

The window frame is known as a box frame and the space between the box and the stud is either filled with "scratch" mortar or caulked with oakum, so as to make it wind proof.

Fig. 151 shows the same construction adapted to a masonry wall. The window frame is set so as

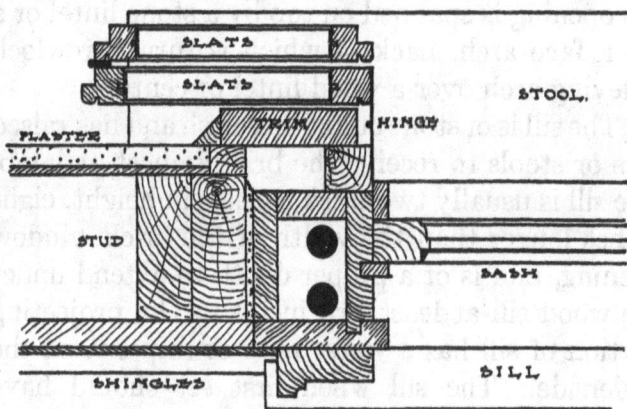

FIG. 150

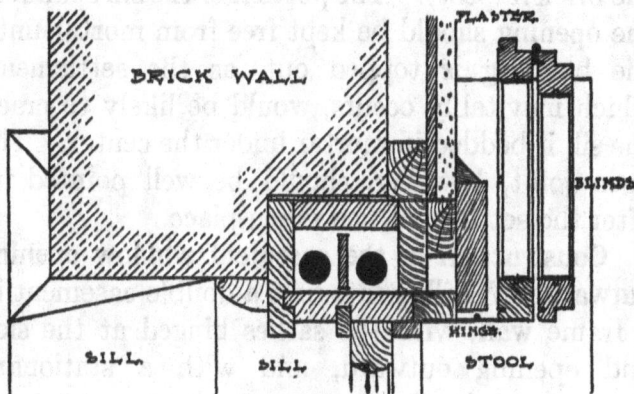

FIG. 151

PLATE XXXI

WINDOWS.

to give an eight-inch reveal. The masonry jamb is rebated for the frame and the joint of the frame with brickwork is covered with a moulded staff bead.

The construction of the heads and sills is about the same as in previous window frame illustrations.

The wall is of brick, thirteen inches thick and the opening is spanned on top by a stone lintel or a brick face arch, back of which is turned rowlock relieving arch over a wood lintel or center.

The sill is of stone cut with a wash and has raised lugs or stools to receive the brick impost or jamb. The sill is usually two brick corners in height, eight inches longer than the width of the brick window opening, and is of a proper depth to extend under the wood sill at least two inches. The projecting portion of sill has a water nose or drip cut on the underside. The sill when first set should have mortar placed only under the ends which receive the brick imposts. The portion of the sill spanning the opening should be kept free from mortar until the building is topped out, as the settlement, which inevitably occurs, would be likely to crack the sill if bedded in mortar under the center. The open joint, however, should be well pointed up after the settlement has taken place.

Construction of the casement window opening outward. We illustrate first a double casement in a frame wall, with the sashes hinged at the side and opening outward, and with a stationary transom overhead.

The wall is constructed of two by four-inch

studs, placed sixteen inches on centers and doubled for jambs, heads and sills of openings. It is covered on the outside with matched sheathing boards, heavy tarred felt and random with shingles. The inside is wood lathed and plastered three coats.

The frame is moulded and rebated as shown, and one and three-quarter inches thick. The sashes require heavier stiles and rails than the sashes for double hung windows do and in cases where it is desired that as little wood as possible shows, the sashes are made of cherry or other suitable hardwood and are reduced in size. The stiles and bottom rail of sashes, in addition to being rebated, have grooves cut in same as shown, so that any water which may beat its way in will run out through the grooves.

The sashes are shown glazed with double thick glass, bedded in putty, sprigged, and back-puttied. For large sashes or where plate glass is used wood beads or moldings are better than the putty for securing the glass in place.

The sill is rebated for the lower rail of sash and is ploughed for stool and shingles. All spaces about the frames are calked with oakum and pointed with "scratch" mortar.

The trim or architrave is molded and formed out of seven-eighths inch stuff, is blocked on the back and is provided with a back band and a neat wall mold. This wall mold being small may be bent to fit the slight unevenness of the plaster. The inside stop bead should be of the same material as the trim and similarly finished. In the better class of buildings the stop bead is secured in place by means of round head brass or bronze screws set in a sunken socket of brass or bronze, instead of being nailed in place. By this means the stop bead is easily removed without marring the wood or varnish or may be slightly shifted in or out, if the sash binds or rattles, by merely loosening the screw a little. Stop beads should never be less than one and three-quarter inches wide so that window shades may be set on same.

Fig. 152 is a section taken through the head of the window. The top of the outside casing is covered with tin turned down over the outer edge and neatly tacked, and extending up under the shingles or other exterior covering about six inches. A little wood fillet is placed over the cap as shown, to give the first course of shingles the required tilt.

Fig. 153 is a section taken through the jamb of the window. The block or furring strip which is shown nailed on to the side of the stud, is necessary as a nailing for the lath which cannot obtain a direct bearing and nailing on the stud, owing to the position of the ground to which the inside architrave is nailed.

Fig. 154 is a section taken through the sill of the window. As an additional precaution against rain, beating in under the lower rail of sash a molded water nose might be let into the rail on the outside or a groove might be cut on the underside of the rail near its outer edge.

Fig. 155 is an exterior elevation of the window and the dotted lines with the numbers indicate the cuts from which the sections illustrated are taken.

Fig. 156 is a horizontal section taken through

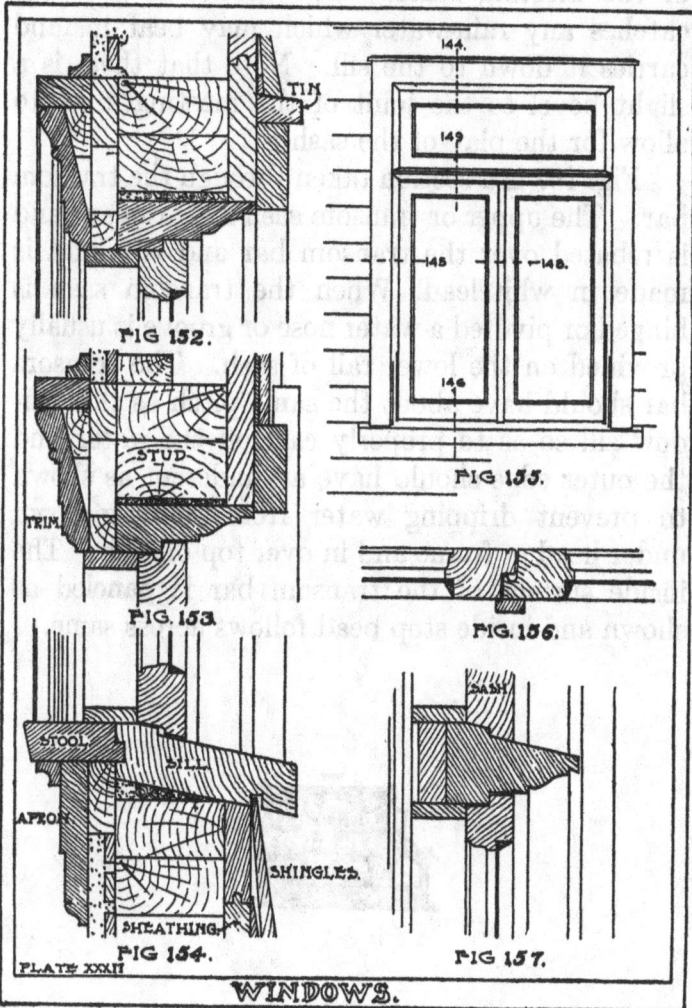

FIG 152.

FIG. 153.

STUD

TRIM

STOOL

APRON

SHINGLES.

SHEATHING.

FIG 154.

PLATE XXII

FIG. 155.

FIG.156.

FIG 157.

WINDOWS.

the meeting stiles of the sash. The standing leaf is rebated and the inner edge is beaded. The other leaf is rebated and grooved and has a flat band or astragal planted on the outside to cover the joint of the meeting stiles. The groove in the stile catches any rain-water which may beat in, and carries it down to the sill. Note that there is a slight bevel to the joint of the meeting stiles to allow for the play of the sashes.

Fig. 157 is a section taken through the transom bar. The upper or transom sash is stationary and is rebated over the transom bar and the joint is made in whitelead. When the transom sash is hinged or pivoted a water nose or groove is usually provided on the lower rail of sash. The transom bar should have about the same pitch as the window sill so as to properly cast off the water, and the outer edge should have an undercut as shown to prevent dripping water from running down under head of frame and in over top of sash. The inside surface of the transom bar is paneled as shown and inside stop bead follows across same.

Part II

COVERING OF ROOFS. Shingling over a hip-ridge — Method of shingling — Old-fashioned method shown — To cover a circular dome with horizontal boarding — To describe the covering of an ellipoidal dome — Simple iron bracket — Common shingling hatchet — Vertical section of circular dome

In slating or shingling a roof, great care should be taken at the hips, ridges and valleys. Where the roof is shingled, two or three courses should be left off at the ridge until the two sides are brought up, then the courses left off should be laid on together, and in such a manner as to have them lap over each other alternately. This can easily be done if the workman uses a little judgment in the matter; and a roof shingled in this manner will be perfectly rain-tight, without the ridge-boards or cresting. In valleys, the tin laid in should be sufficiently wide to run up the adjacent sides far enough to prevent back-flow from running over it. Ample space should be left in the gutter to permit the water to flow off freely. There is a general tendency to make these waterways too narrow, which is frequently the cause of the water backing up under the shingles, causing leakage and premature decay of the roof.

There are several methods of **shingling over a hip-ridge;** one is the old and well-tried method of shingling with the edges of the shingles so cut that the grain of the wood runs parallel with the line of the hip, as shown in Fig. 158. Here it will be seen that the shingles next to those on the hip have the grain running up and down at right angles with the eave. In Fig. 159 we show a front view of the same hip, which will give a better idea of what is meant by having the grain parallel with

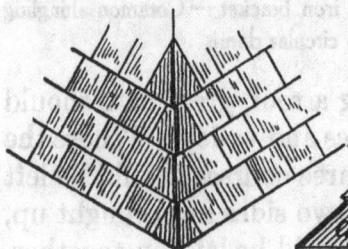 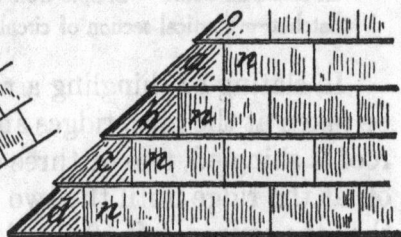

FIG. 158   FIG. 159

the line of hip. abcd show the cut of hip shingles and nnnn the common shingles.

The proper way to put in these shingles is to let the ends run over alternately and then dress them to the bevel of the opposite side of the roof; this is shown better at obd, where the edges of the shingles are shown that are laid on the other side of the roof, and are not seen in the drawing. Another method now frequently used is to cover the hip with strips of tin; this is used quite extensively and found to be one of the best methods in use.

Method of Shingling.

There is an unwritten law that all carpenters know how to shingle—much better than they really do. Our observation has been that shingling is one of the most important matters in house construction. In the early days a house consisted of practically nothing but a roof to keep the rain out. And even to-day, no matter how well the foundation and floors are made, or how plumb and beautiful the walls are, or how well decorated on the inside, or how nice the roof was framed by one skilled in roof framing, or how well it was sheathed, it is practically worthless unless it is well shingled. There is no place like the roof for a carpenter to show his mechanical skill, for any one that can saw to a line and drive a nail can lay a floor, put on siding and put up inside finish.

We wish to impress on the mind of the young mechanic that no matter what part of the trade he is working at, it is of some importance, and there is always an opportunity to learn something.

Some old mechanics that have had twice as much experience in house building say nail as high as possible. They say the higher in the shingle you nail the more air will get to all parts of the shingle and it will last longer, which is undoubtedly true. Another with possibly even more experience will say nail just as low as possible and have the nail covered with the next course, and gives for his reason that high winds and heavy rain and snow storms will not drive in.

Some mechanics think there is nothing like a six-inch shingle for a roof, and point to what a nice roof is made of slate all of the same size. It is all right for slate, but in the opinion of some all wrong for wood shingles, for if one cracks, and near the center as they generally do, it makes a crack three courses long, and should a few more crack there are bound to be bad leaks in the roof.

Fig. 160 shows three courses of shingles being laid on the roof to three chalk lines, and you will

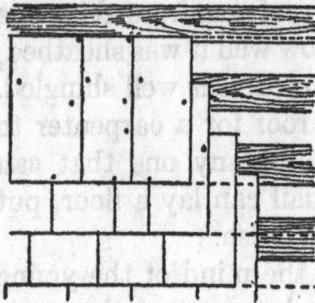   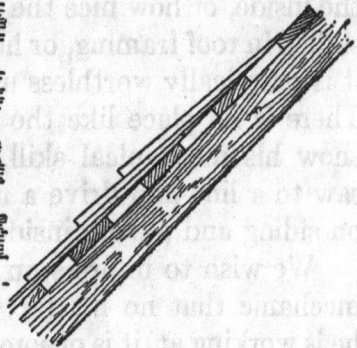

FIG. 160   FIG. 161

notice that not a single joint comes over another, not even from the first to the fourth courses, so if one should split right at another joint it only makes a crack less than two-thirds the length of a shingle. By driving the nails one a little above the other they are not as liable to crack as if they were right opposite each other.

Fig. 161 is an end view and shows that not only is there practically three courses on all parts of a

roof, but a good part of the fourth should be there
also.

Fig. 162 shows an old-fashioned way, and when
we got so high we had to get a new rest for our feet
we nailed brackets on the roof with tenpenny nails,
and put a plank on them. Fig. 163 shows a more
modern way and how it is done. It is now the
most common way, and does not make a leak, as
the tenpenny nails would, although sometimes a
shingle nail makes a hole. To avoid that, drive

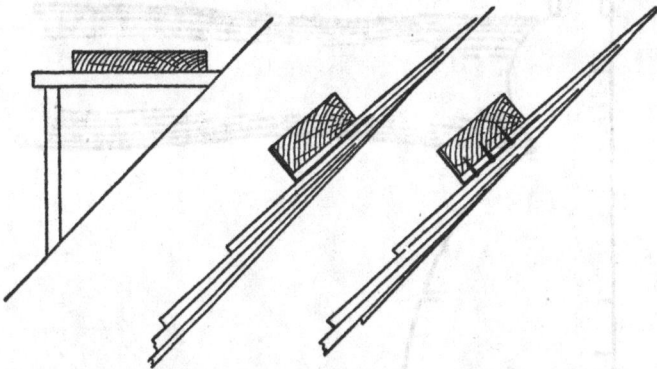

FIG. 162   FIG. 163   FIG. 164

the shingle nail below where the nails were driven
in the roof when you take it down, and it makes
the holes break joints.

Some carpenters to avoid the danger of leak
just mentioned, put the shingle on tip end up,
shingle it in, and then saw it off when the scaffold
is taken down. Others shingle over it and drive
it out with a hand ax.

Fig. 164 shows a very simple iron bracket,

which has many good points, but there is nothing quite as handy and cheap as the shingle.

Fig. 165 is a very common shingling hatchet. File marks at one and one-half and two inches (the projection that is generally on a barn or shed roof—as most houses nowadays have gutters) and other marks at four and one-half and five inches

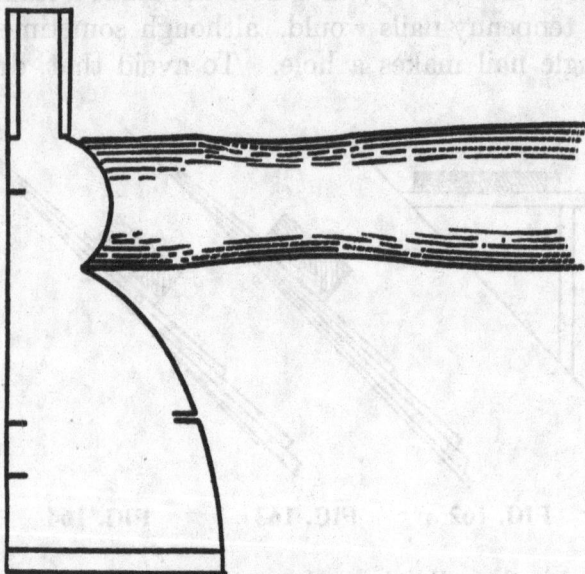

FIG. 165

(which are ordinary courses), and it makes a very convenient all-around rule, hammer and hatchet, and in the hands of a mechanic a very good roof can be laid even without the aid of chalk lines or straight edge.

To cover a circular dome with horizontal boarding, proceed as follows:

Let ABC (Fig. 166) be a vertical section through the axis of a circular dome, and let it be required to cover this dome horizontally. Bisect the base in the point D, and draw DBE perpen-

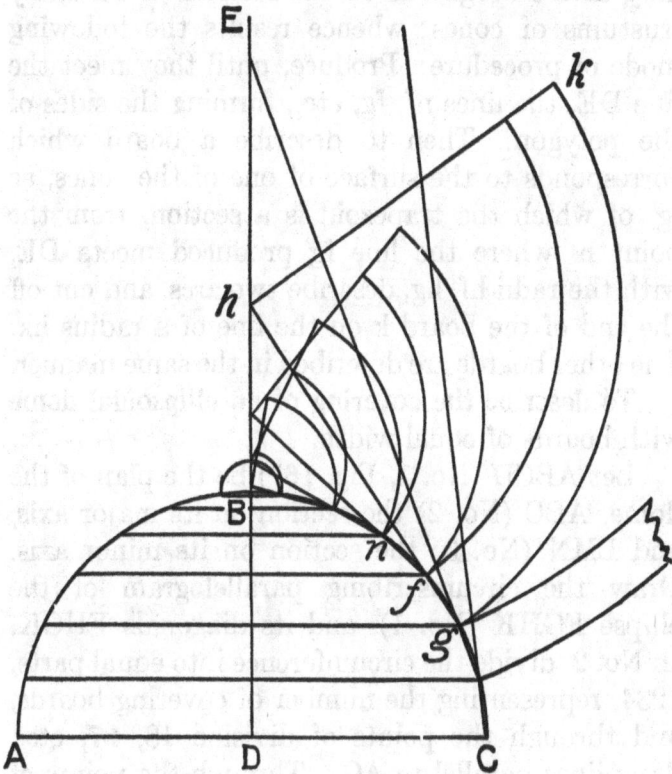

FIG. 166

dicular to AC, cutting the circumference in B. Now divide the arc BC into equal parts, so that each part will be rather less than the width of a board, and join the points of division by straight lines, which will form an inscribed polygon of so

many sides; and through these points draw lines
parallel to the base AC, meeting the opposite sides
of the circumference. The trapezoids formed by
the sides of the polygon and the horizontal lines,
may then be regarded as the sections of so many
frustums of cones; whence results the following
mode of procedure: Produce, until they meet the
line DE, the lines nf, fg, etc., forming the sides of
the polygon. Then to describe a board which
corresponds to the surface of one of the zones, as
fg, of which the trapezoid is a section, from the
point h where the line fg produced meets DE,
with the radii hf, hg, describe two arcs, and cut off
the end of the board k on the line of a radius hk.
The other boards are described in the same manner.

To describe the covering of an ellipsoidal dome
with boards of equal width.

Let ABCD (No. 1, Fig. 167) be the plan of the
dome, ABC (No. 2) the section on its major axis,
and LMN (No. 3) the section on its minor axis.
Draw the circumscribing parallelogram of the
ellipse FGHK (No. 1), and its diagonals FHGK.
In No. 2, divide the circumference into equal parts,
1234, representing the number of covering boards,
and through the points of division 18, 27, etc.,
draw lines parallel to AC. Through the points of
division draw 1p, 2t, 3x, etc., perpendicular to
AC, cutting the diagonals of the circumscribing
parallelogram of the ellipse (No. 1), and meeting
its major axis in ptx, etc. Complete the parallelograms, and their inscribed ellipses corresponding
to the lines of the covering, as in the figure. Pro-

duce the sides of the parallelograms to intersect
the circumference of the section of the transverse

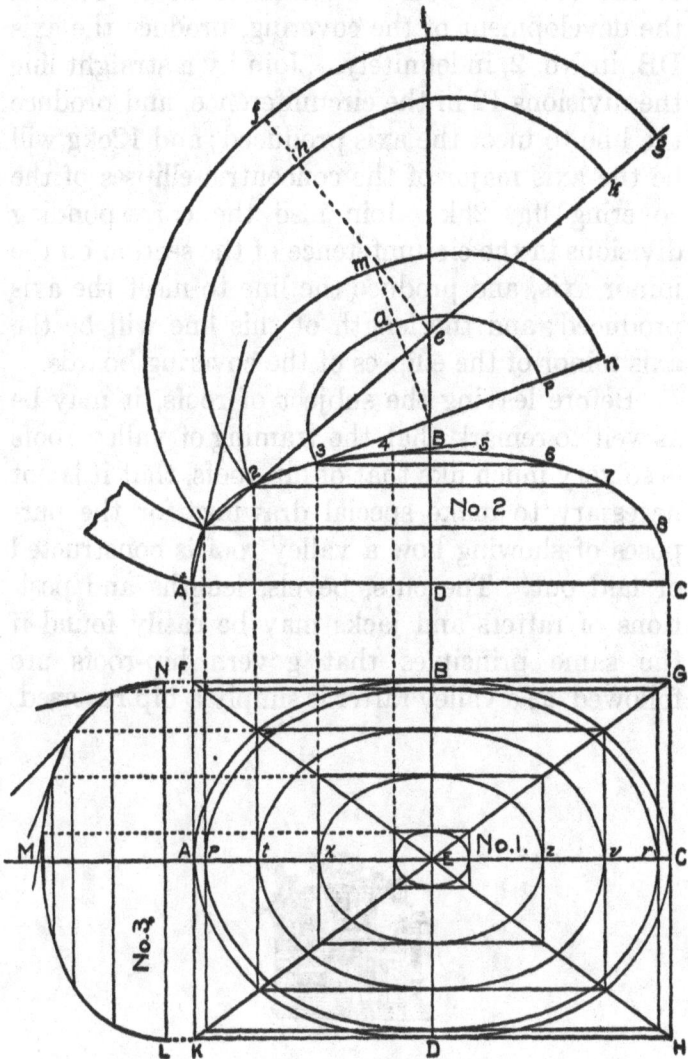

FIG. 167

axis of the ellipse in 1234, and lines drawn through these, parallel to LN, will give the representation of the covering boards in that section. To find the development of the covering, produce the axis DB, in No. 2, indefinitely. Join by a straight line the divisions 12 in the circumference, and produce the line to meet the axis produced; and 12ekg will be the axis major of the concentric ellipses of the covering lfg, 2hk. Join also the corresponding divisions in the circumference of the section on the minor axis, and produce the line to meet the axis produced; and the length of this line will be the axis minor of the ellipses of the covering boards.

Before leaving the subject of roofs, it may be as well to remark that the framing of valley roofs is so very much like that of hip-roofs, that it is not necessary to make special drawings for the purposes of showing how a valley roof is constructed or laid out. The cuts, bevels, lengths and positions of rafters and jacks may be easily found if the same principles that govern hip-roofs are followed, as a valley rafter is simply a hip reversed.

Part III

MOULDINGS. Roman mouldings — Ovals, or quarter round — Torus, or half round — Cavetto, or hollow — Cyma Recta — Cyma Reversa — Scotia — Fillet — Apophyge — Astragal — Bead — Mitering mouldings — Cutting of a spring moulding when the horizontal portion has to miter with a gable or raking moulding — Obtaining the section of a raking mould that will intersect a given horizontal moulding — Descriptive geometry — Lines for the cuts in a miter box — Finding proportions of a small moulding — To obtain correct cut of a veneer — To get the contour or outline for a raking moulding, Etc.

Mouldings are classed as Roman, Grecian, and Gothic.

The **Roman mouldings** are all formed of parts of circles, and can therefore be struck with compasses. The Grecian are principally composed of parts of curves known as the conic sections—such as the ellipse or hpyerbola. They are otherwise nearly similar to the Roman, which are therefore illustrated in this place as being the simpler and the more generally used.

Fig. 168. The moulding of which this is a section is called the **Ovolo, or quarter round.** The fillet, or straight edge projecting beyond the curved portion, is to be drawn first, and then the horizontal, which represents the depth or bottom line of the moulding. Now produce the bottom line of the fillet, and on it, from the point at which the curve is to start, mark off the width of the moulding.   This point, marked O in the cut, is the center from which the quadrant is to be struck.

Fig. 169 is called the **Torus, or half-round**. Having drawn the fillet, and the line representing the bottom of the moulding, draw a line at right angles to these. Bisect the width of the curved part, and the bisecting point will be the center.

Fig. 170 is the **Cavetto, or hollow.** This is a quarter-round, the curve turning inward. It is thus precisely the reverse of the ovolo.

Fig. 171 is a section of the moulding called the **Cyma Recta.** The exact form of this moulding is

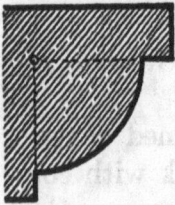

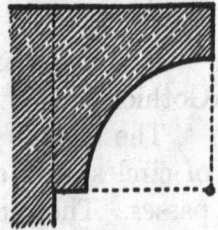

FIG. 168   **FIG. 169**   **FIG. 170**

to a certain extent a matter of taste, since the curve may be made more or less full, as shown in the three examples, Figs. 171, 172, and 173. To describe Fig. 171, draw a perpendicular across the depth of the moulding, and bisect it. From the bisecting point as a center point describe a quadrant; through the center draw a horizontal line, and from the point where the quadrant already drawn touches this line mark off the radius; then from this point as a center describe the second quadrant, which will complete the form. In this and subsequent curves of combined arcs the greatest care is necessary, so that the one may glide smoothly into the other without showing any break or thickening at the joining. To describe

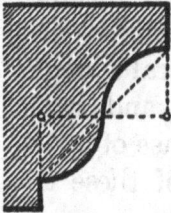 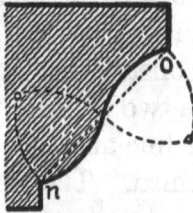 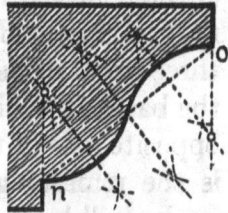

FIG. 171   FIG. 172   FIG. 173

the Cyma Recta shown in Fig. 172, which is the form most generally used, let n and o be the points to be united by the moulding. Draw the line n o, and bisect it; with half n o as a base describe an equilateral triangle on the opposite sides of the line; then the apices of the triangle will be the centers from which the curves are to be struck.

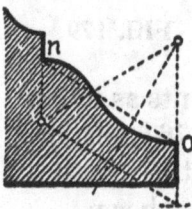 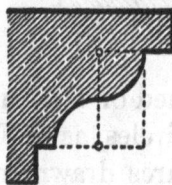 

FIG. 174   FIG. 175   FIG. 176

To describe Fig. 173, or others the curves of which are required to be more flat than in the last figure, draw the line n o as before, and bisect it. Bisect these two divisions again, and the centers will be on these bisecting lines, according to the form required; for, of course, the longer the radius the flatter the curve will be.

If it is required that the curve should be more full at the lower than at the upper part, it may be effected in the following manner, which is shown in Fig. 174: Having drawn n o, divide it into three equal parts; construct an equilateral triangle, the base of which is two of these thirds, and on the opposite side of the line another, the base of which is the remaining third. The apices of these triangles will be the centers for the curves.

Fig. 175 is the **Cyma Reversa**. In this moulding the curve bulges outward at its upper part, its fulness being regulated by the taste of the designer.

FIG. 177   FIG. 178   FIG. 179

Thus it may be formed of two quadrants as in Fig. 175; or of two semicircles, as in Fig. 176; or it may consist of the two arcs drawn from the apices of triangles as in the cyma recta already shown.

Fig. 177 is the **Scotia**. This is the hollow moulding, sometimes consisting of a semicircle only: viz., the reverse of the torus. In other instances as in Fig. 177, it is composed of two quadrants; and in others it is drawn from three centers, as in Fig. 178. To draw this, divide the depth of the moulding into three equal parts, and with one-third describe the quadrant r u; produce the horizontal r u, and from r set off i, equal to half u r. At n erect a perpendicular, and mark on it n k, equal to i u; draw i k, and bisect it; produce the bisecting line until it cuts n k in s. Draw s i and produce it. From i, with radius i u, draw the next portion of the curve, meeting s i produced; then complete the curve by an arc drawn from s with radius s n.

A **fillet** is the small flat edging used to separate two larger mouldings, to strengthen their edges,

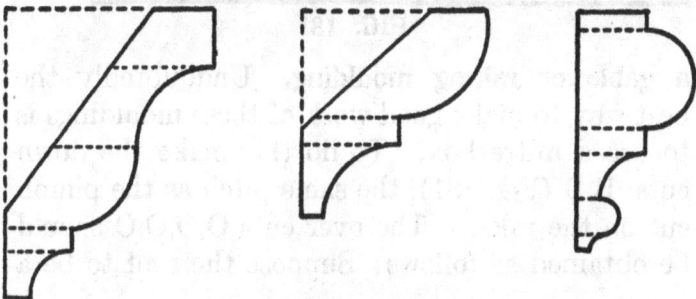

FIG. 180

or to form a cap or crowning to a moulding. The fillet is one of the smallest members used in cornices, architraves, bases, and pedestals. When placed against the flat surface of a pedestal, it is usually joined to it by a small quarter-round hollow called the **Apophyge** (Fig. 179).

The **torus,** when worked very small, is called the **Astragal** (Fig. 180); but when worked so as not to project, as on the edge of boards to be joined, it is called a **bead.**

Mitering Mouldings.

One of the most troublesome things the carpenter meets with is the **cutting of a spring moulding when the horizontal portion has to mitre with**

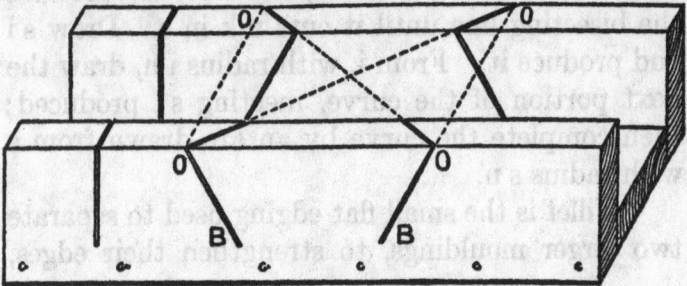

FIG. 181

a gable or raking moulding. Undoubtedly the best way to make good work of these mouldings is to use a mitre box. To do this make the down cuts, B,B (Fig. 181), the same pitch as the plumb cut on the rake. The over cuts O,O,O,O should be obtained as follows: Suppose the roof to be a

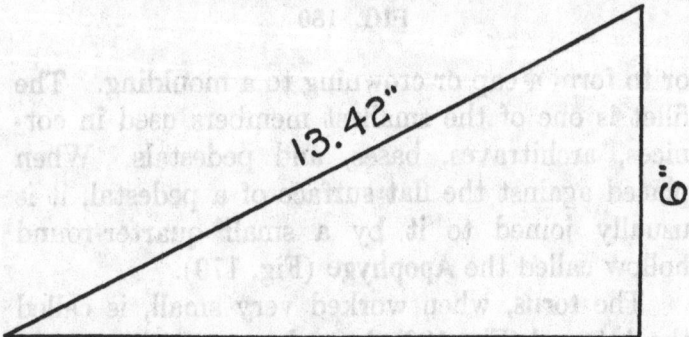

FIG. 182

quarter pitch—though the rule works for any pitch when followed as here laid down—we set up one foot of the rafter, as at Fig. 182, raising it up 6 inches, which gives it an inclination of quarter pitch; then the diagonal will be nearly 13½ inches. Now draw a right-angled triangle whose two sides forming the right angle, measure respectively 12 and 13½ inches, as shown in Fig. 183.

The lines A and B show the top of the mitre box with the lines marked on. The side marked

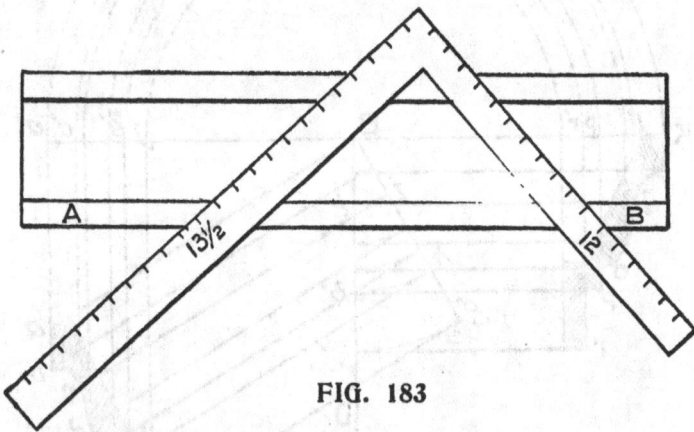

FIG. 183

13½ inches is the side to mark from; this must be borne in mind, and it must be remembered that this bevel must be used for both cuts, the 12-inch side not being used at all.

Another excellent method for **obtaining the section of a raking mould that will intersect a given horizontal moulding**, is given below, also a manner of finding the cuts for a miter box for same. The principles on which the method is based being, first, that similar points on the rake and horizontal parts of a cornice are equally distant from vertical planes represented by the walls of a building; and, second, that such similar points are equally distant from the plane of the roof. Representing the wall

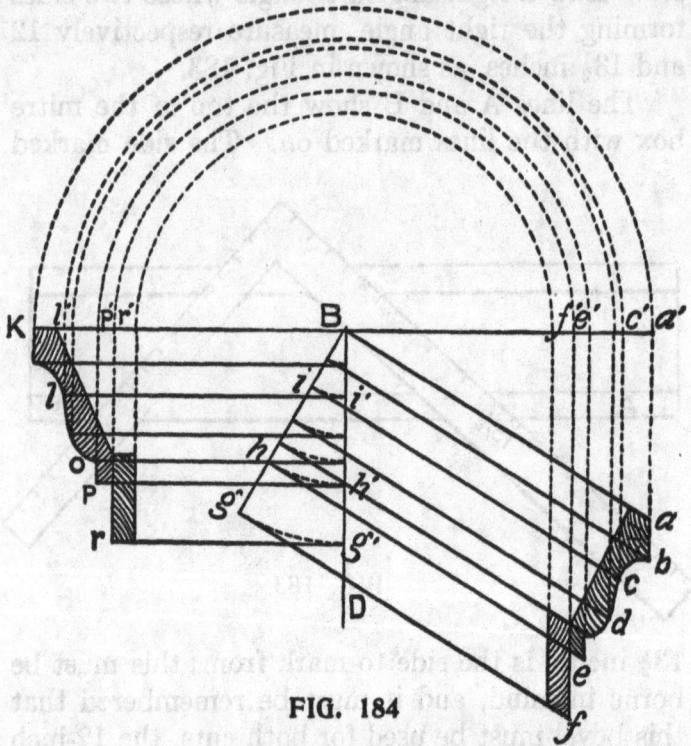

FIG. 184

faces of a building by the line DB (Fig 184), and a section of the horizontal cornice by DBabcdef—Bab being the angle of the roof pitch, draw line aa', cc', ff', parallel to DB and intersecting the line ka', which is drawn at right angles to DB through the point B; then, with B as a center, describe the arcs a'k, c'l', etc., intersecting the same line ka' on the opposite side of DB. This gives the point k at the same distance from DB as the points a and a', and the line ll' at the same distance as cc'. The rest of the same group of parallel lines are found to be similarly situated with respect to DB.

From **Descriptive Geometry** we have the principle, that if we have given two intersecting lines contained in a plane, we know the position of that

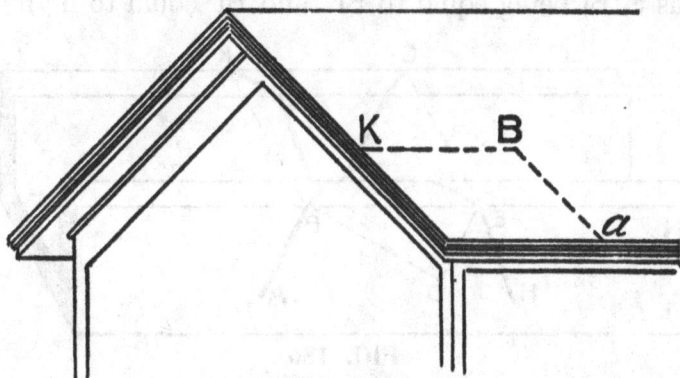

FIG. 185

plane, hence we may represent the plane of a roof by the line Ba and Bk (Figs. 184 and 185); and since it will be most convenient to measure the distances required in a direction perpendicular to that plane, in following out the principle draw lines from the points cef, etc., parallel to Ba and intersecting the line Bg, which is made prpendicular to Ba. This gives us on Bg the perpendicular distance of the points cef, etc., from the line Ba. From the intersection of these lines with Bg, and with B as a center, describe arcs intersecting the line DB at i'h'g', etc.; from these intersections with DB draw lines l'l, h'p, g'r, etc., parallel to Bk, until they intersect the first group of lines drawn perpendicular to Bk, and the intersection of each set of two lines drawn from the same point on the horizontal section will give the similar point of the rake section. Taking the point l, for example, we have, as before proved, l at the same distance from DB as c, and i being at the same distance from Ba as c, Bi being equal to Bi', and Bi' equal to ll', ll'

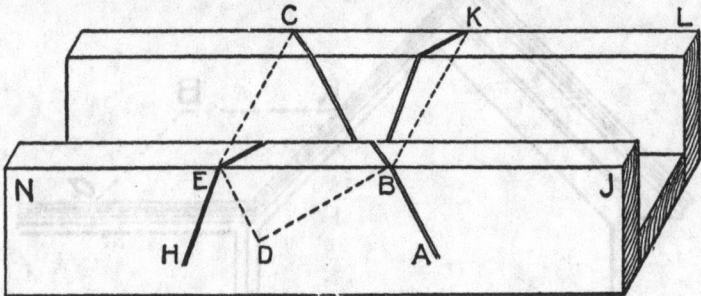

FIG. 186

is equal to Bi'; and consequently l is the same distance from Ba, which is in accordance with principle already shown. The intersection of each set of lines being found and marked by a point, the contour of the moulding may be sketched in, and the rake moulding, of which the section is thus found, will intersect the given horizontal moulding, if proper care has been taken in executing the diagram.

Fig. 186 shows **how to find the miter cut for the rake moulding,** the cut for the horizontal one being the same as for any ordinary moulding. Take an ordinary plain miter-box, NJL, and draw the line AB, making the angle ABJ equal to the pitch angle of the roof. Draw BD perpendicular to AB, and extend lines from B and E square across the box to K and C; join BC and EK. ABC will be the miter cut for two of the rake angles; HEK will be the cut for the other two angles, the angle HEN being equal to the angle ABJ. In mitering, being horizontal and rake moulding, that part of the moulding which is vertical when in its place on the cornice, must be placed against the side of the box.

Lines for the cuts in a miter box, for joining spring mouldings, may be obtained as follows: If we take BS (Fig. 187), the moulding showing the spring or lean of the member, and DE the miter required, then proceed as follows: With A as a center, and the radius AG, describe the semicircle FHGC; then drop perpendiculars from the line FC, at the points F, A, H, G, and C, cutting the miter line as shown at the line ID. Draw IE parallel to FC, then from I draw IS, which will be the bevel for the side of the box, and the bevel CR will be the line across the top of the box. The miter line, as shown here, is for an octagon, but the system is applicable to any figure from a triangle or rectangle to a polygon with any number of sides.

Fig. 188 shows the manner of finding proportions of a small moulding which is required to miter with a larger one, or vice versa. Let AB be

the width of the larger moulding, and AD the width
of the smaller one; construct with these dimensions
the parallelogram ABCD, and draw its diagonal
AC. Let AB be the section of the moulding which
we wish to reduce to member with a moulding the
width of AD. Draw any number of parallel lines

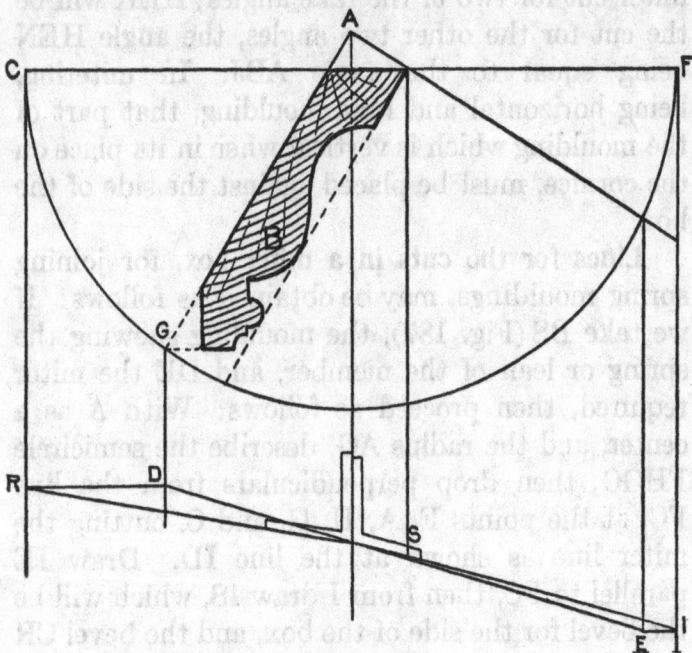

FIG. 187

to BC cutting the line AC, from which points draw
lines parallel to DC and beyond the line AD.
From the latter set off the thickness of the moulding on the corresponding lines as 1–2 will give the
contour of the mould for the lesser width or vice
versa.

To Obtain Correct Shape of a Veneer. Fig. 189 exhibits a method of obtaining the correct shape of a veneer for covering the splayed head of a

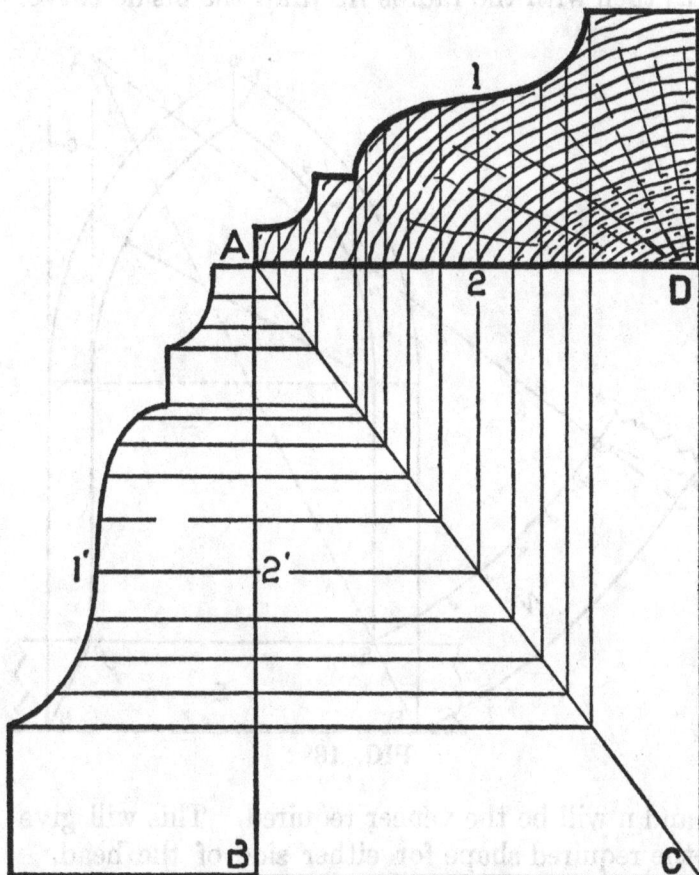

FIG. 188

gothic jamb. E shows the horizontal sill, ef the splay, FA the line of the inside of jamb, o the difference between front and back edges of jamb,

BA the line of splay. At the point of junction of the lines BA, FA, set one point of the compasses, and with the radius AB draw the outside curve of n; then with the radius AS draw the inside curve,

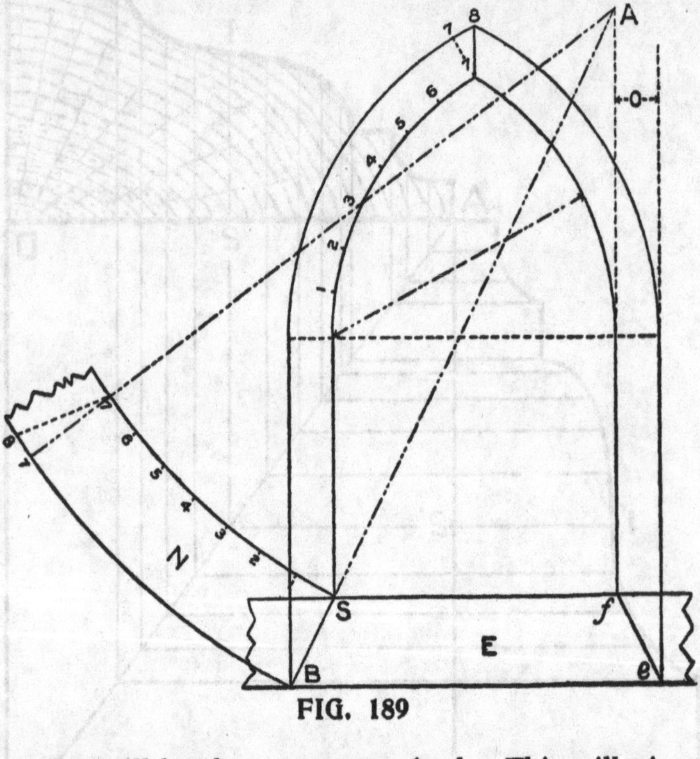

FIG. 189

and n will be the veneer required. This will give the required shape for either side of the head.

To get the contour or outline for a raking moulding, proceed as shown in Fig. 190. The horizontal moulding is divided into any number of parts, equal or unequal, as shown at abcdefg. The line AC shows the rake or inclination. Draw lines parallel with AC, from a to D, b to v, etc. Drop a line AB, perpendicular to AC, at any convenient point on the rake, make the distance AC equal to ho; then drop the lines pgrst, and where these lines cut the lines abcdefg, these points of contact will be in the curve line of the rake moulding, as at Dvwxyz. From these points trace the curve, which will be the proper shape for the moulding.

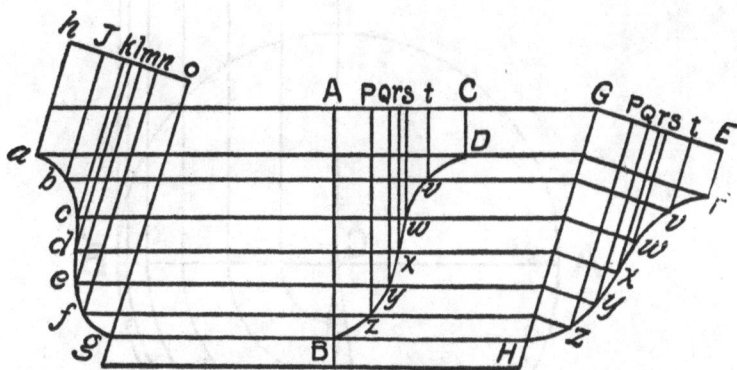

FIG. 190

The divisions and lines shown at GHEF give the proper shape for the moulding at the top return.

To Cover a Circular Dome with Vertical Boarding. The upper part of Fig. 191 represents the elevation of the dome, and the lower part represents the plan and the shape of the board in the stretch-out. Divide the diameter of the dome into spaces equal the width of the boards to be used. Divide one side of the elevation into any number of equal parts, as 1, 2, 3, etc., and from these points draw parallel lines down to the

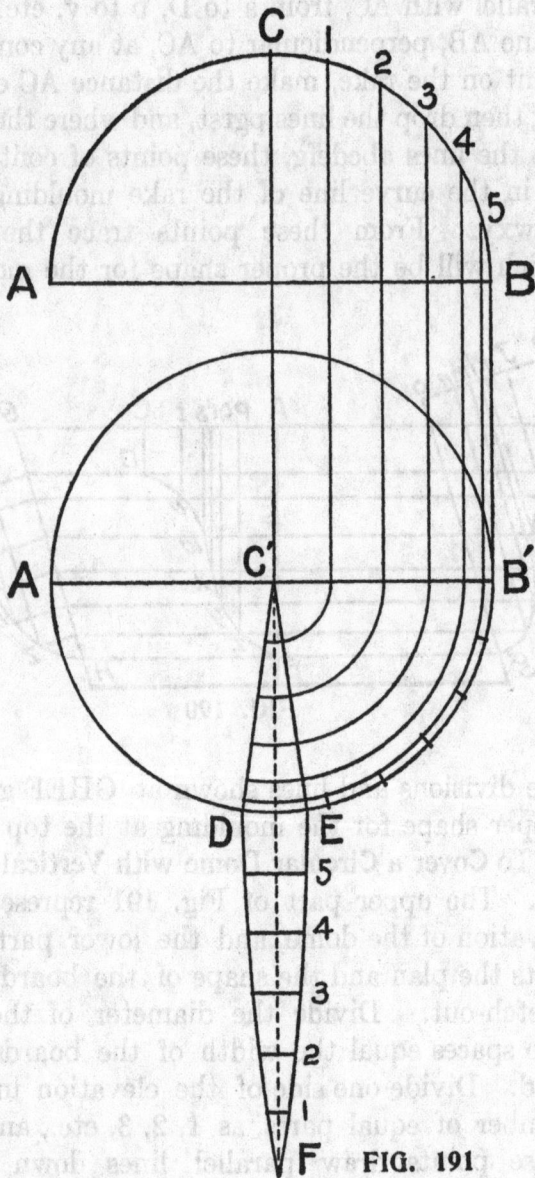

FIG. 191

diameter of the plan as at C″B″. From C″ as a center draw lines C″D″ and C″E which represents the board in its respective place on the dome. From C″ as center and with the points 1, 2, 3, etc., on the diameter swing around to C″D and C″E as shown; and on the line C″F lay off the distance 5, 4, 3, etc., which will be the correct length or stretch-put of the board. The length of the cross lines between C″D and C″E governs the length of the cross lines of the board. Care should be taken to not have the boards too wide, as the narrower the boards the better they will fit the dome.

Sometimes the workman will find it necessary **to miter in circles between two lines of mouldings,** and to do so the circular moulding must be made with a diameter large enough to have the solid

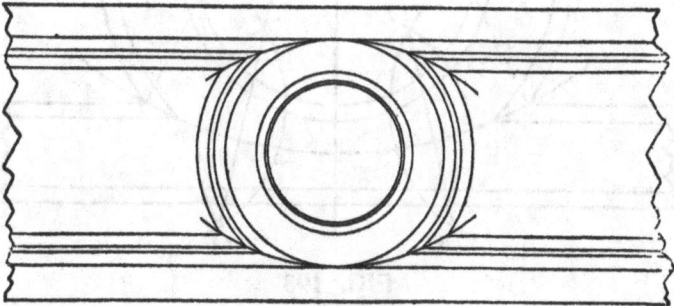

FIG. 192

wood adjoin the solid wood of the running mouldings, as shown at Fig. 192. It will be seen that the points of juncture of the various members of the mouldings do not run in a straight line, thus making the miter-joint a little curved, which ad-

mits of the mouldings working together accurately
without requiring to be pared.

Figs. 193, 194, 195, and 196 show how various
joints are made by the junction of circular mouldings and straight mouldings, and mouldings with
more or less curvature.

A spring moulding is one that is made of thin
stuff, and is leaned over to make the proper pro-

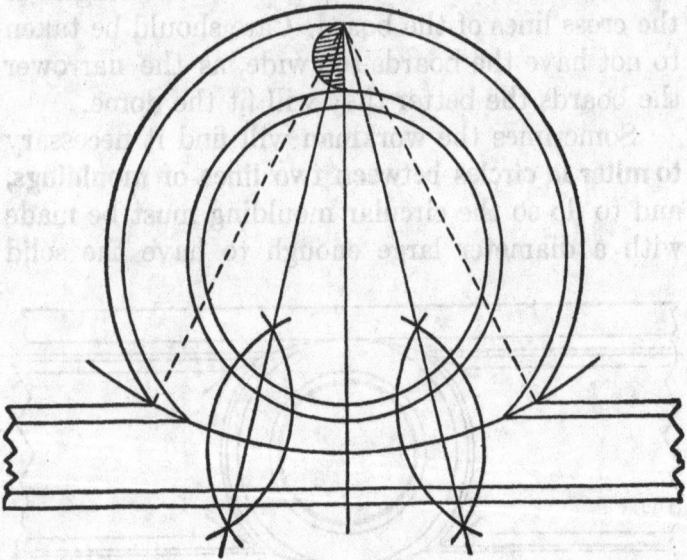

FIG. 193

jection, as shown at Fig. 197. A shows the spring
moulding; B the space left vacant by the leaning
of the moulding. These mouldings are difficult to
miter, more particularly so when the joint is made
with a raking moulding that springs also. Some
of the methods given for obtaining the cuts for

raking mouldings may be used for cutting these
mouldings when the work is straight, but when
circular the application of other methods is sometimes necessary. Many times the workman will

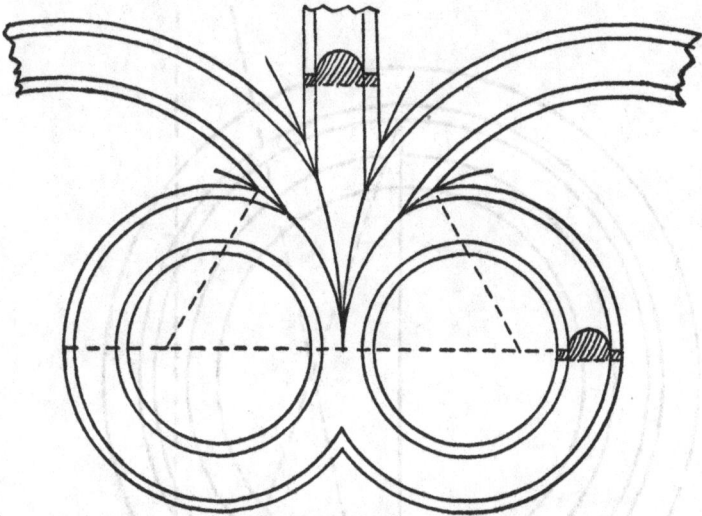

FIG. 194

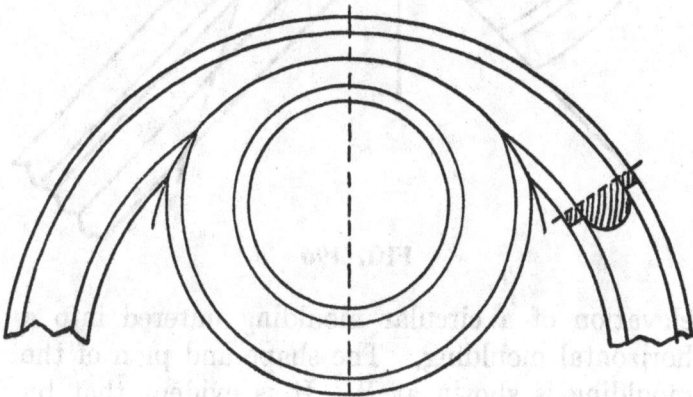

FIG. 195

come across very knotty operations of this kind to
work out, and the following diagrams will then
prove exceedingly useful: Fig. 198 exhibits an

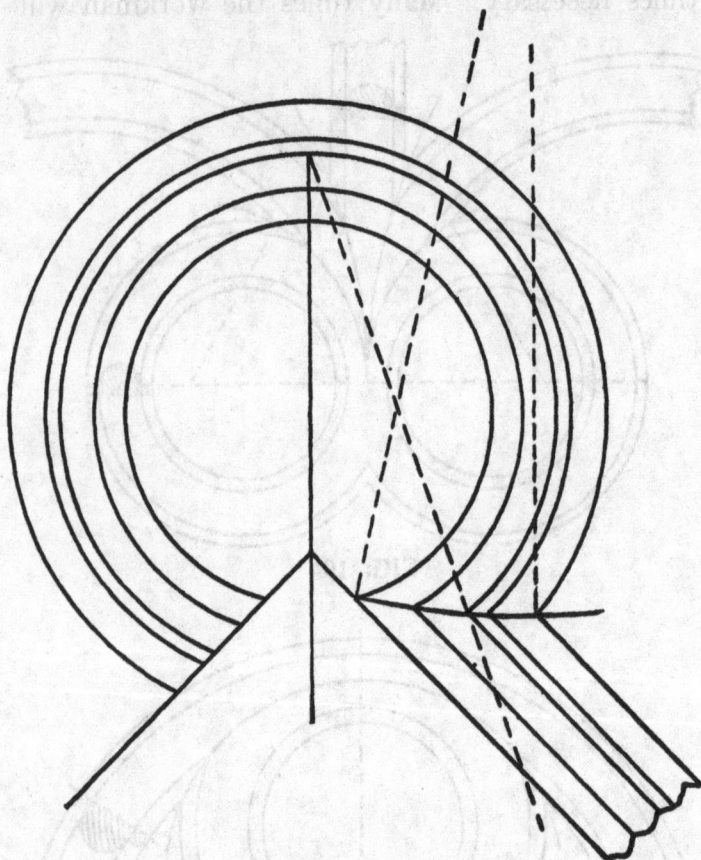

FIG. 196

elevation of a circular moulding mitered into a
horizontal moulding. The shape and plan of the
moulding is shown at B. It is evident that by
producing the line AD to intersect the center line

of the arc at C, the central point will be obtained, from which the circular piece required for the moulding may be described. EA and FD give

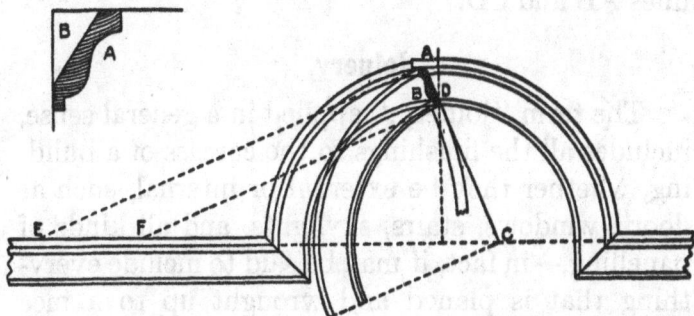

FIG. 197   FIG. 198

the radius for the curves of both edges when the stuff is in position, as shown in the elevation.

Fig. 199 shows the application of the same rules to a **circular elevation of a different form standing over a straight plan.** The back lines of

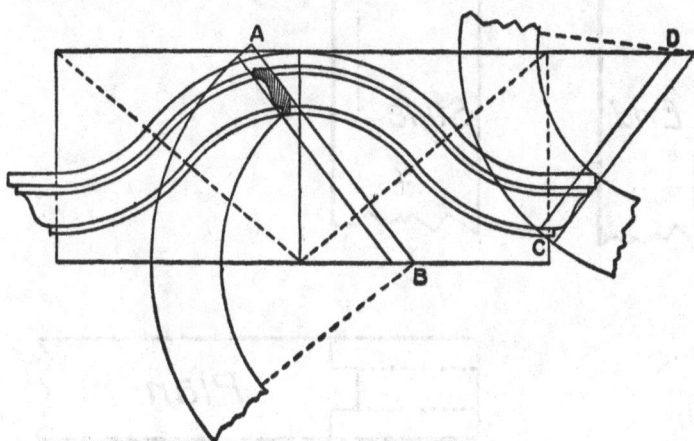

FIG. 199

the moulding are produced until they bisect a horizontal line drawn through the center, from which the circular cornice was struck, as shown by the lines AB and CD.

Joinery.

The term "joinery," applied in a general sense, includes all the finishings to the carcass of a building, whether they be external or internal, such as doors, windows, stairs, skylights, and all kinds of panelling,—in fact, it may be said to include everything that is planed and wrought up to a nice smooth and often ornamental face, and framed

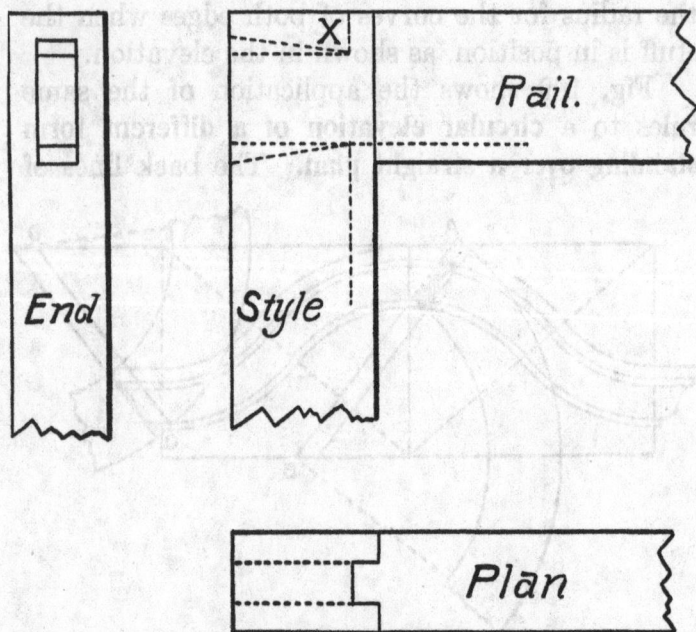

FIG. 200

together in such workmanlike and precise manner
that it is difficult to discern the joints unless
helped by the different grain and color of the stuff,
as the wood is called.

It is intended to devote most of this chapter to
a description of the various joints, mouldings, and
terms which are used in this particular branch of
building; but the principal and most common
joints, etc., will be pointed out to the student when
they are met in the different kinds of joiner's
work.

Framed Joints.—The mortise and tenon is the
chief and most common of all the joints used in
joinery framing. All rails of doors, for instance,

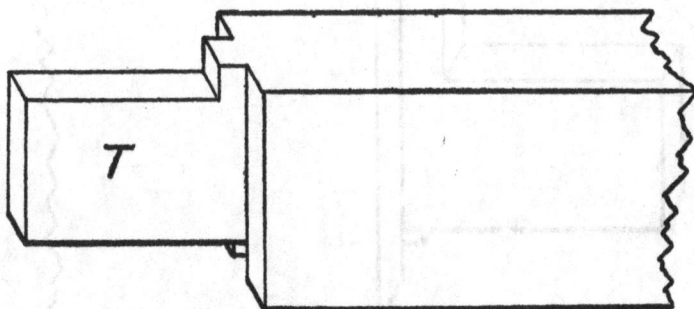

FIG. 201

are tenoned through the styles and wedged up
tight, as Fig. 200, the tenon usually being one-
third of the thickness, and with a haunch left on as
at X, to fill up the panel groove on the styles and
strengthen the joint. This is the origin of the
term haunched tenon. a sketch of which, as it is

on the rail before being wedged up into the styles (as shown in Fig. 200), is given in Fig. 201.

On wide rails, such as lock or middle or bottom rails, this tenon is in two parts in its depth, as in Fig. 202; and where the doors are more than two inches thick, or where the provision has to be

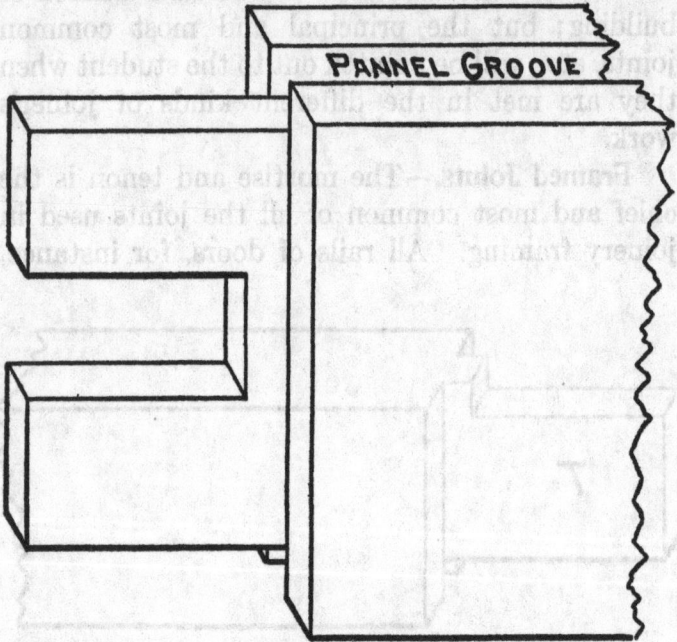

PANNEL GROOVE

FIG. 202

made for a mortise lock, which, when let into the center of the style, at lock rail height, completely cuts away a central single tenon, the tenons are double in thickness, and called double tenons, as Fig. 203.

A **bare-faced tenon** is a tenon with only one

shoulder, S, used chiefly in framed ledges and braced doors, where the rails are not so thick as the styles by the thickness of the boarding nailed on the rails between the styles. Fig. 204 illustrates a bare-faced tenon, one view being given of each side.

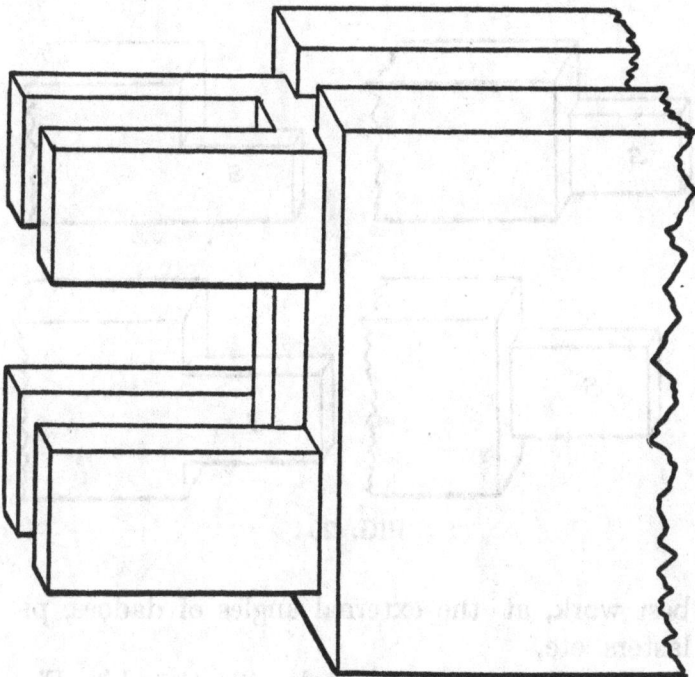

FIG. 203

A **housed tenon** is one let into a mortise with the section of the stuff let into the mortise to the depth of half an inch, as Fig. 205.

Angle joints.—Tongued angles are used for internal angles of skirtings, grounds, casings, etc., as Fig. 206.

Mitered angles are made by simply cutting half a right angle alternately off the two pieces, to be joined by nails at an external angle, as Fig. 207.

Mitered and tongued angles, a combination of the last two, are as Fig. 208, and are only used in

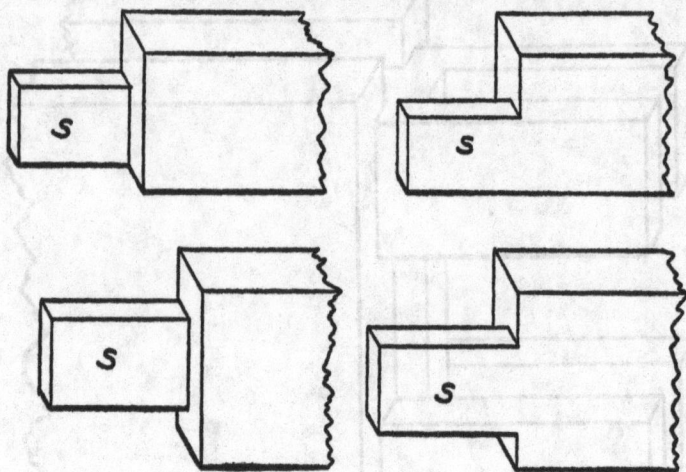

FIG. 204

best work, at the external angles of dadoes, pilasters, etc.

Beaded and tongued angles, illustrated by Fig. 209, are angles or joints of an ornamental as well as necessary character.

Return-beaded angles are suitable for all external miters where wear and tear would soon fetch the arrises off in soft woods. Fig. 210 shows one on wood framing, casings, etc., and Fig. 211 one as fixed to angles to plastering on wood, brick or stone walls, to which they are fixed by plugs, as will be explained hereafter, this being called a staff bead.

Where the two pieces of framing to be joined are of different widths or thicknesses the "mitered angle" is made as Fig. 212.   Another form of very good construction, is as Fig. 213.

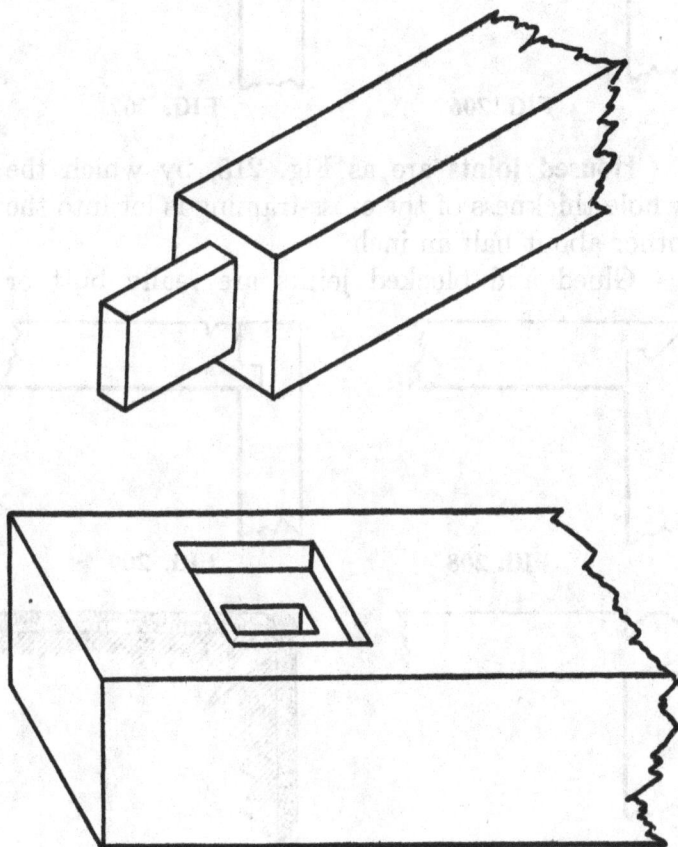

FIG. 205

Keyed mitered joints are not often used; but it is as well that the carpenter should know what they are. Fig. 214 is a view of the angle of one, XX being hardwood-slips let into the miters.

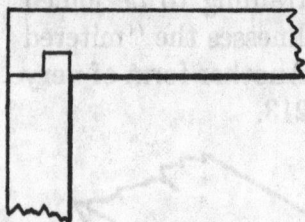

FIG. 206

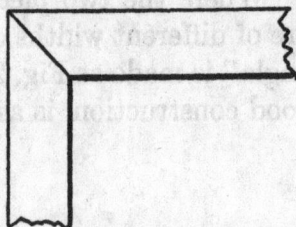

FIG. 207

Housed joints are as Fig. 215, by which the whole thickness of the cross-framing is let into the other about half an inch.

Glued and blocked joints are really butt or

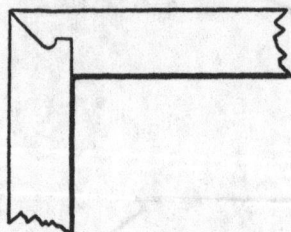

FIG. 208

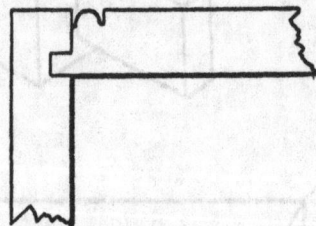

FIG. 209

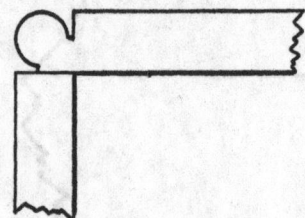

FIG. 210

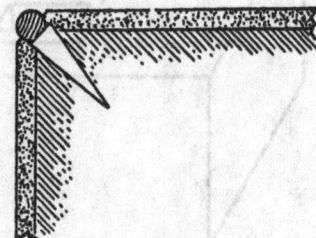

FIG. 211

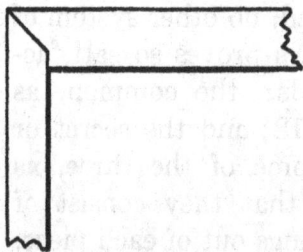

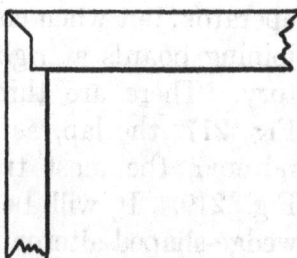

FIG. 212   FIG. 213

lapped joints secured by blocks glued to each piece of wood in the internal angle (Fig. 216).

Dovetails.—To make a good dovetail joint requires considerable skill and care on the part of the

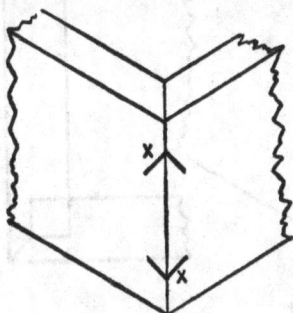

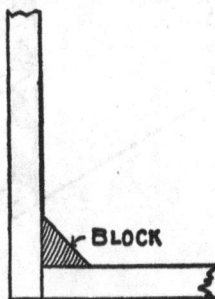

FIG. 214   FIG. 215   FIG. 216

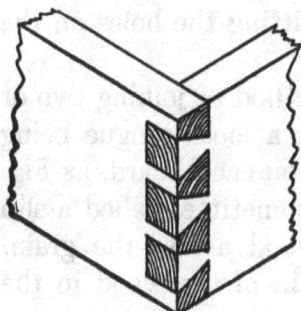

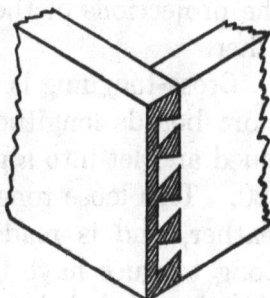

FIG. 217   FIG. 218

operator; but when completed, no other system of joining boards at right angles proves so satisfactory. There are three kinds: the common, as Fig. 217; the lap, as Fig. 218; and the secret or mitered, the most troublesome of the three, as Fig. 219. It will be seen that they consist of wedge-shaped alternate cuttings out of each piece,

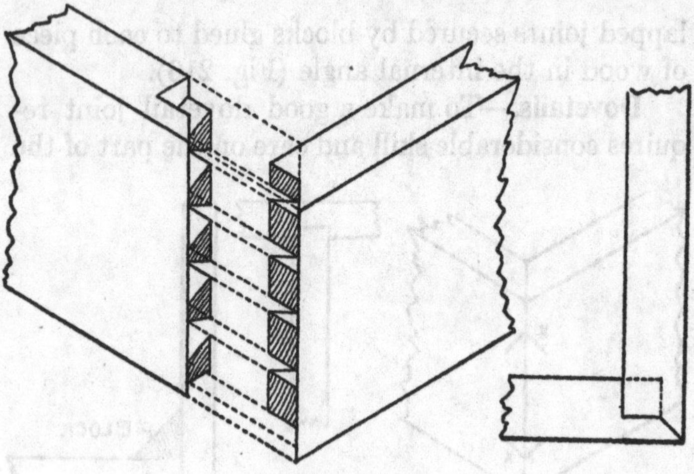

FIG. 219

the projections of the one fitting the holes on the other.

Cross-tonguing is the method of joining two or more boards longitudinally, a loose tongue being glued and let into a groove on each board, as Fig. 220. This loose tongue is sometimes called a slip feather, and is made of wood across the grain. Long tongues have the grain of the wood in the direction of their length.

Clamping is the method by which the ends of several boards are fastened together, as shown on the left hand of Fig. 221, while the right hand side of the same figure illustrates miter-clamping, by which the cross-grain end of the ordinary clamping is obscured.

Keying (Fig. 222) is a means of securing several boards together by a flush key, let in at the back

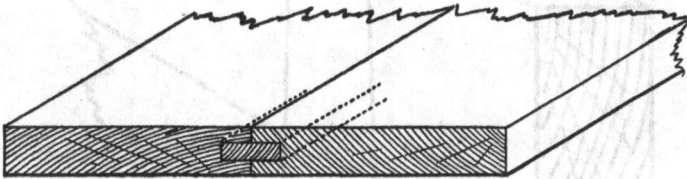

FIG. 220

in lieu of a projecting ledge, where the latter would be inconvenient on account of a level face being required on each side.   This is often used for wide door casings.

Keyed joints are also used to connect circular with straight or two pieces of circular wood, such as door frames, etc.   A shallow mortise is cut out

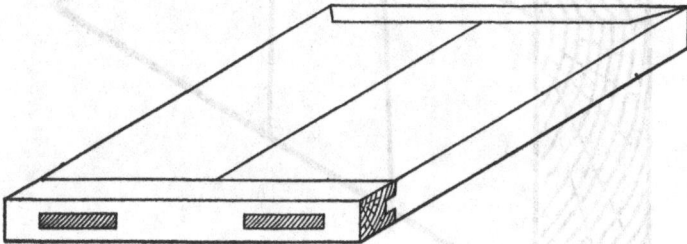

FIG. 221

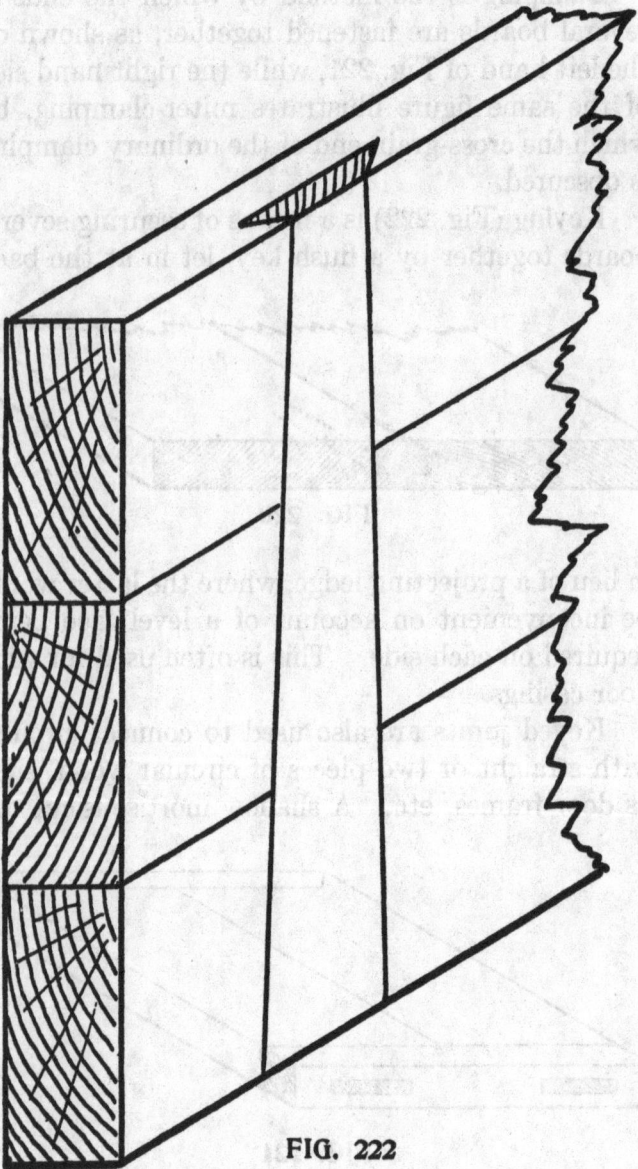

FIG. 222

of each part, and a hardwood key, (in the form of the letter I) connects the two together.

Scribing is the cutting, out of the face of one moulding, a hole of the contour of another to form a joint. It is chiefly used in joints of sashbars, internal angles of moulded skirtings, etc., and

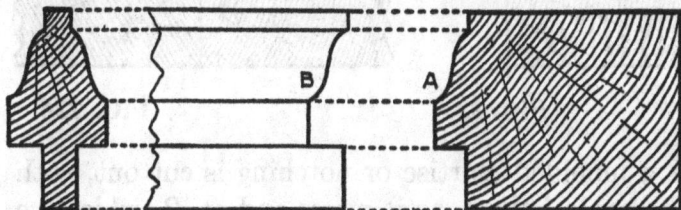

FIG. 223

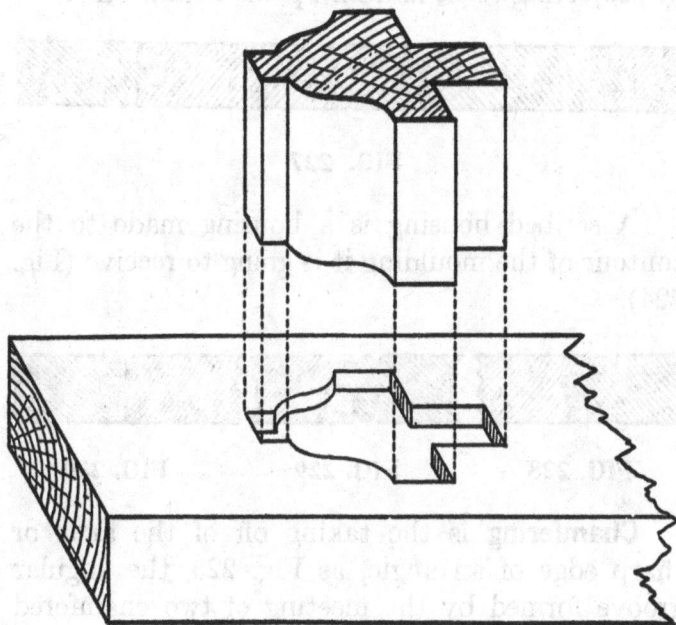

FIG. 224

really is a moulded mortise cut into another moulding to receive a moulded tenon of the same section as the mortise, but in a converse form. For instance, in Fig. 223, it will be noticed that on

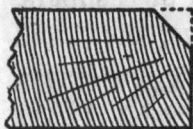 

FIG. 225   FIG. 226

A a moulded mortise or notching is cut out, with the ovolo hollow, as it were; and on B, which we will call the tenon, the cutting has the ovolo convex or projecting to fit and fill up the hollow on A.

FIG. 227

A scribed housing is a housing made to the contour of the moulding it is going to receive (Fig. 224).

FIG. 228   FIG. 229   FIG. 230

Chamfering is the taking off of the arris or sharp edge of an angle, as Fig. 225, the angular groove formed by the meeting of two chamfered angles (Fig. 226) being called a V-joint.

Match Boarding.—This is an arrangement of boards matched and put together with grooved and tongued joints, and their edges "shot" or planed truly, so that a fine joint can be made. Match boarding, otherwise called cleating, is of several kinds, the chief of which are:

Plain matched boards (Fig. 227).

Beaded-one-side match boarding (Fig. 228).

Beaded-both-sides match boarding (Fig. 229).

V-jointed match boarding (Fig. 230).

Part IV

JOINTS, STRAPS AND OTHER FASTENINGS. Lengthening ties — Scarfing — Simple method of scarfing — Building beams — When one piece is perpendicular to another — When the pieces to be joined are not at right angles to one another — Resistance at the joint — Effect of shrinkage and expansion — Maxims to follow

The joints in the framing of timber having to resist the strains to which the pieces are exposed, should be formed in such a manner that the bearing parts may have the greatest possible amount of effective surface. For should that part of the joint which receives the strain be narrow and thin, it will either indent itself into the pieces to which it is joined or become crippled by the strain; producing in either case a change in the form of the framing.

The effect of the shrinkage and expansion of timber should also be considered in the construction of joints. On account of the shrinkage of timber, dovetail joints should seldom be used in carpentry, as the smallest degree of shrinking allows the joint to draw out of its place; they can only be used with success when the shrinkages of the parts counteract each other; a case which seldom happens in carpentry, though very common in joinery and cabinet-making.

Joints should also be formed so that the contraction or expansion may not have a tendency to split any part of the framing. The force of contraction or expansion is capable of producing astonishing effects where the pieces are confined, which may sometimes be observed where framing has been wedged too tightly together in improper directions. The powerful effect of expanding timber is well know to quarrymen, as they sometimes use its force to break up large stones.

In forming joints the object to be attained should always be kept in view, as that which is excellent for one purpose may be the worst possible for another. With this consideration the subject will be treated under separate headings as follows:

Lengthening Ties.—The simplest and perhaps the best method of lengthening a beam is to abut the ends together, and place a piece on each side;

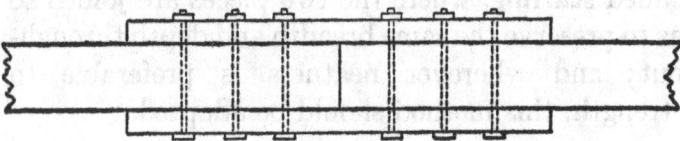

FIG. 231

these, when firmly bolted together, form a strong and simple connection. Such a method of lengthening a tie is shown by Fig. 231, and is what ship-carpenters call fishing a beam. It is obvious, however, that the strength in this case depends on the bolts, and the lateral adhesion and friction produced by screwing the parts tightly together.

The dependence on the bolts may be lessened by indenting the parts together, as shown by the upper side of Fig. 232; or by putting keys in the joint, as shown by the lower side of the same figure; but the strength of the beam will be decreased in proportion to the depth of the indents.

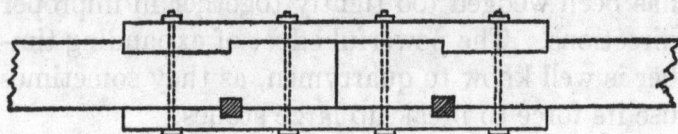

FIG. 232

The only reasons for not depending wholly on bolts are, that should the parts shrink ever so little, the bolts lose a great part of their effect; and the smallness of the bolts renders them liable to press into the timber, and thus to suffer the joint to yield.

The most usual method of joining beams is that called **scarfing,** where the two pieces are joined so as to preserve the same breadth and depth throughout; and wherever neatness is preferable to strength, this method should be adopted.

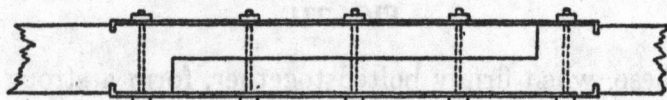

FIG. 233

Fig. 233 is the most **simple method of scrafing;** it depends wholly on the bolts, and in this and like cases it is best to put a continued plate of iron on each side to receive the heads of the bolts. The ends of the plates may be bent and let into the beams.

Fig. 234 is another very common method, but not so good a combination, as the bolts do not press the surfaces in a perpendicular direction; and an oblique pressure, such as would be likely

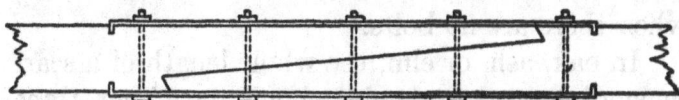

FIG. 234

to take place in this example must have some tendency to separate the joint, and it has no advantage in other respects.

Fig. 235 is a joint where bolts would not be absolutely necessary, but it is clear that the strength would not be quite so great as half that of an entire piece; the key, or double wedge in the center of the joint, should only be driven so as to bring the parts to their proper bearing, as it would be better to omit it altogether than to drive it so as

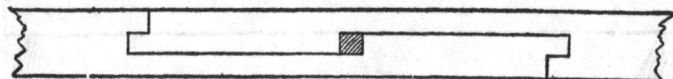

FIG. 235

to produce any considerable strain on the joint. It is not necessary that there should be a key, except when bolts are to be added, and then it is desirable to bring the joints to a bearing before the bolts are put in. The addition of bolts and straps, however, makes this an excellent scarf.

The following are some maxims that will be sufficiently accurate for practical purposes:

In oak, ash, or elm, the whole length of the scarf should be six times the depth or thickness of the beam, when there are no bolts.

In fir the whole length of the scarf should be about twelve times the thickness of the beam, when there are no bolts.

In oak, ash, or elm, the whole length of a scarf depending on bolts only, should be about three times the breadth of the beam; and for fir beams it should be six times the breadth.

When both bolts and indents are combined, the whole length of the scarf for oak and hard wood may be twice the depth; and for fir or soft woods, four times the depth.

Building Beams.—If two plain pieces of timber were laid upon one another, and supported at the ends, the pressure of a weight applied in the middle

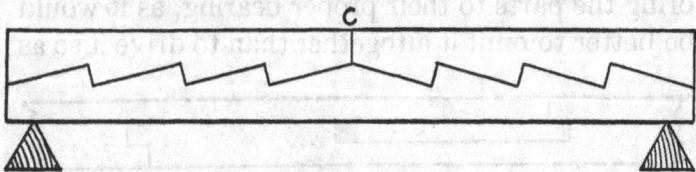

FIG. 236

would cause them to bend, and the surfaces in contact would slide against one another, the upper piece sliding towards each end upon the lower one. This sliding is effectually prevented by indenting the surfaces, as shown in Fig. 236, when the pieces are bolted together; but if the same indents be reversed, as in Fig. 237, they produce scarcely any effect, and nearly the whole strain is upon the bolts.

Wherever the principal strain on the beam may happen to be, to that point, as at C, Fig. 236, the indents should direct their square abutments; that

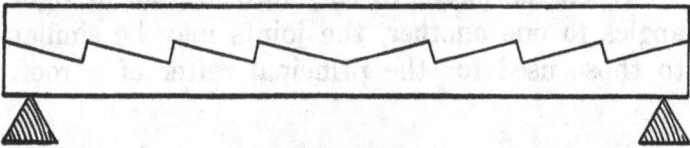

FIG. 237

is, toward the straining force. When the beam is uniformly loaded, the greatest strain is at the middle.

In drawings we frequently see all the indents put the same way, and sometimes as in Fig. 237, otherwise the preceding remarks would have appeared to have been unnecessary.

We will now proceed to describe some of the joints of most common occurrence, and endeavor to point out improvements that might be made in some of them.

When one piece is perpendicular to another, as, for example, a post upon a sill, the usual, as well as the most easy method, is to make the joint square, with a stub or short tenon of about one-fourth of the thickness of the framing, to retain it in its place.

But if the joint be not very accurately cut, the whole load will bear upon the projecting parts; consequently, the center of pressure will seldom coincide with the axis of the post, and its power of resistance will be much lessened.

If, instead of cutting the joint square, it were cut to form an angle, as shown by Fig. 238, then a very little care in cutting the joint would make the center of pressure coincide with the axis.

When the pieces to be joined are not at right angles to one another, the joints may be similar to those used for the principal rafter of a roof.

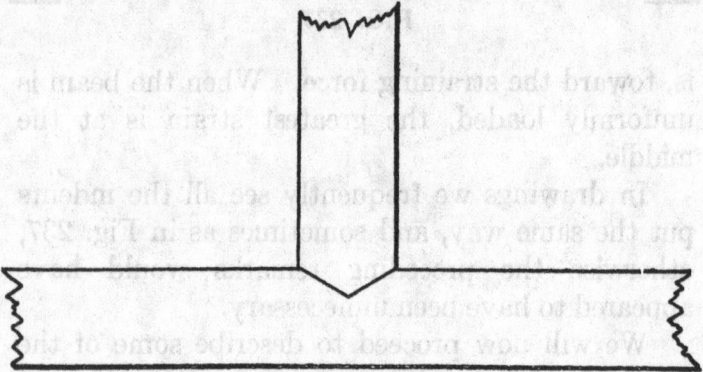

FIG. 238

It is necessary to state that the direction of the strains, as well as their magnitude, remains sensibly the same, whatever may be the form of the abutting joints, except so far as the form of the joint alters the points of bearing; which may in some cases cause the pressure to act with a leverage nearly equal to half the depth of the beam. The strength of the joint itself depends upon its form, as it may be so made that there will be a tendency to slide, which it would be well to avoid, without having recourse to straps.

The **resistance at the joint** is always most effectual when the abutment is perpendicular to the strain, but where the angle formed by the inner sides of the pieces is very acute, this kind of abutment cannot be obtained, at least not without wounding the tie too much.

Part V

STAIR BUILDING. To cut a pitch-board — Manner of applying the board — Section of stairs in position — Putting risers
and treads together — Housed or closed string — Manner of
finishing wall string — Sectional elevation through the steps —
Cut or open string at the foot of the stairs — Square of the
newel

In laying out stairs, it is first necesary to determine the height from the top of the floor which
the stairs start from, to the floor which they
are to land; also the run or distance of their horizontal stretch. This is found by dividing the
height into the number of risers desired in the
stairway. This usually results in fractions of an
inch to each rise and complicates the work.
However, the fractions may be avoided by the use
of the story pole by spacing it off with a pair of
compasses into the number of risers desired in the
stairs. This being done the rest is comparatively
easy because the run or horizontal stretch is not
usually limited as to exact space as in the case of
the rise of the stairs, and as there is always one less
tread than there are risers its length is determined
by the width given the treads. Thus if there are 14
risers in the stairs there would be 13 treads. If

the treads be 9 inches wide the run would be 9 times 13 or 117 inches, which is 9 feet and 9 inches as shown in Fig. 239. From this it will be seen

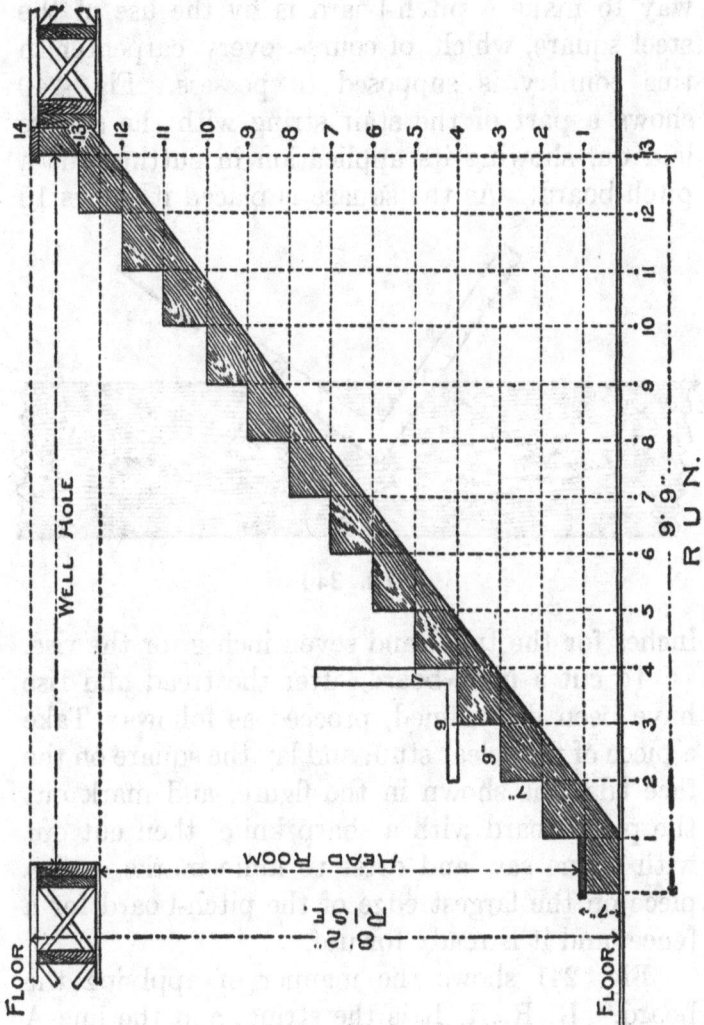

FIG. 239

that the run and rise of the individual step is taken on the steel square as shown in the illustration and determines the shape of the pitchboard. The way to make a pitch-board is by the use of the steel square, which, of course, every carpenter in this country is supposed to possess. Fig. 240 shows a part of the stair string with the square laid on, showing its application in cutting out a pitch-board. As the square is placed it shows 10

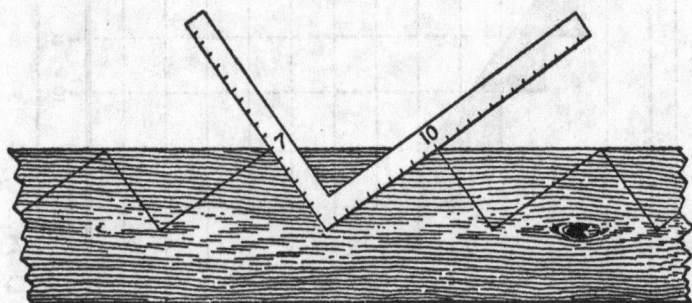

FIG. 240

inches for the tread and seven inches for the rise.

To cut a pitch-board, after the tread and rise have been determined, proceed as follows: Take a piece of thin clear stuff, and lay the square on the face edge, as shown in the figure, and mark out the pitch-board with a sharp knife; then cut out with a fine saw and dress to knife marks, nail a piece on the largest edge of the pitch-board for a fence, and it is ready for use.

Fig. 241 shows the manner of applying the board. R, R, R, R is the string, and the line A shows the jointed or straight edge of the string.

The pitch-board, p, is shown in position, the line
8⅓ represents the step or tread, and the line 7¾
shows the line of the riser. These two lines are of
course at right angles. This string shows five
complete cuts for treads, and six complete cuts for
risers. The bottom of the string at W is cut off
at the line of the floor on which it is supposed to
rest. The line C is the line of the first riser. This
riser is narrower than any of the other risers,
because the thickness of the first tread is always
taken off it; thus, if the tread is 1½ inches thick, the

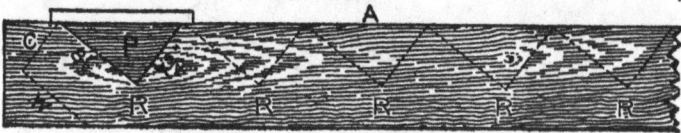

FIG. 241

riser in this case would only require to be 6¼
inches wide, as 6¼ and 1½ inches together make 7¾
inches. Another thing to be considered is the
string, which must be cut so that the line at W
will be only 6¼ inches from the line at 8⅓, and it
must be parallel with it. The first riser and tread
having been satisfactorily dealt with, the rest may
be easily marked off by sliding the pitch-board
along the line A until the line 8⅓ on the pitch-
board strikes the line 7¾ on the string, when another
tread and another riser are marked off in the same
manner.

Fig. 242 shows a portion of the stairs in position.
S, S shows the strings, which in this case are cut

square; that is, the part of the string to which the riser is joined is cut square across, and the "butt" or end wood of the riser is seen. In this case, also, the end of the tread is cut square off and flush with the string and riser. Usually in a stair of this kind the ends of the treads are rounded off similar to the front of the tread, and the ends project over the strings the same distance that the front edge projects over the riser. If a moulding or cove is

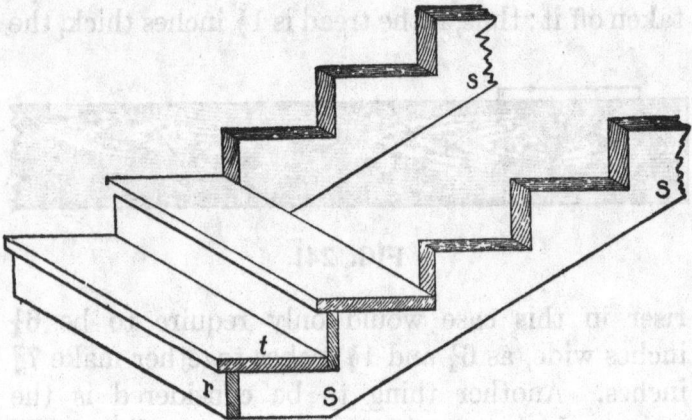

FIG. 242

used under the nosing in front, it should be carried round on the string to the back edge of the tread, and cut off square. The riser is shown at r, and it will be noticed that it runs down behind the tread on the back edge, and is either nailed or screwed to the tread.

Fig. 243 shows the customary way of putting risers and treads together. T, T shows the treads; R,R the risers; S,S the string; O,O the cove moulding under the nosing X,X. B,B shows the blocks that hold the tread and risers together. These blocks should be from four to six inches long, and made of dry wood. Their section may be from one to two inches square. On a tread three feet long, three of these blocks should be used at about

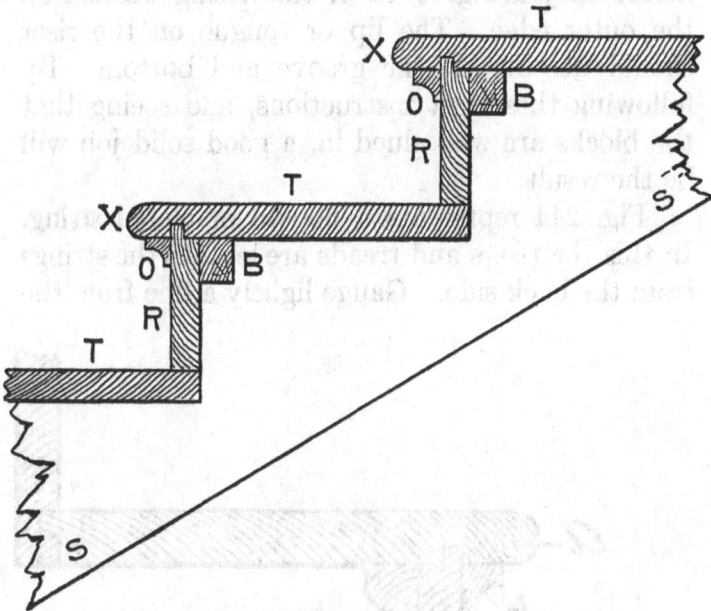

FIG. 243

equal distances apart, putting the two outside ones about six inches from the strings. They are glued in the angle. It will be noticed that the riser has a lip on the upper edge which enters into a groove in the tread. This lip is generally about ⅜ inches long, and may be ⅜ or ½ an inch in thickness. Care must be taken in getting out the risers, that they are not made too narrow, as allowance must be made for the lip. If the riser is a little too wide, it will do no harm, as the overwidth may hang down below the tread; but it must be made the exact width where it rests on the string. The treads must be made the exact width required before they are grooved or the nosing worked on the outer edge. The lip or tongue on the riser should fit snug in the groove and bottom. By following these last instructions, and seeing that the blocks are well glued in, a good solid job will be the result.

Fig. 244 represents a **housed or closed string.** In this the risers and treads are let into the strings from the back side. Gauge lightly a line from the

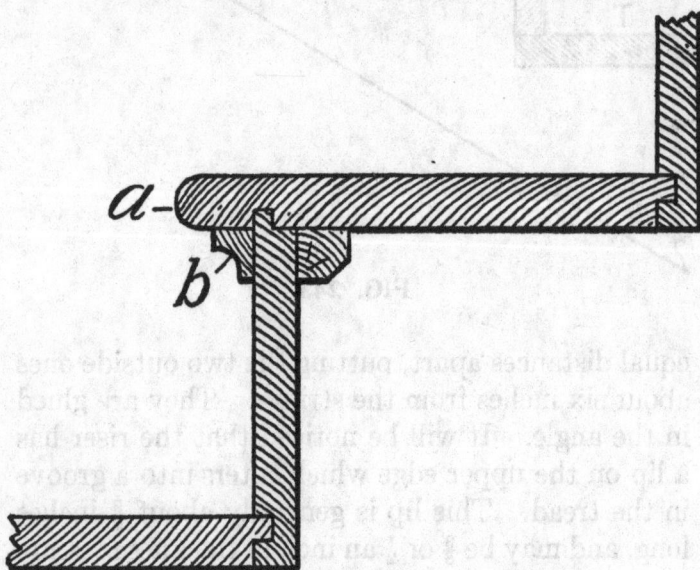

FIG. 244

upper edge of the string, the distance intended to stand above the treads as shown in the dotted line. On this line apply the pitch-board as explained on previous pages. In laying out housed strings it is as well to take the fence off the pitch-board, as it can be handled much better without it, as the long side will have to be kept close to the gauge line, to insure good work. The top lines for treads and the face lines for risers, are the lines that define the step, and cannot be changed; but the back line of the riser and the lower line of the tread should be made to run so that the housing or groove will be wider at the under side of the string than at the junction of the riser and tread at the nosing, where the grooves will be the same width as the riser and tread are in thickness separately. The nosing projects over the riser, as will be seen, and to mark this portion out it is usual to make a template or pattern for the purpose. Indeed it is best to make a template to lay out the whole housing of the tread, and in shape as the shaded part is shown in the illustration.

The reason the grooves are left wider at the back edge of tread is so that the wedge can be driven between the tread and the lower edge of the groove, to force the top side of the tread close to the upper edge of the groove, thus making a tight joint and insuring strength and rigidity to the whole structure. The risers are also wedged into place as will be shown later on. After the treads and risers are laid out on the string, a sharp pointed knife blade should be used to mark the lines for the face of the riser and top of tread, then a fine tenon saw should be used to saw down to the exact depth. This will not be difficult to perform when the hole forming the nosing recess has been bored to the proper depth. A gauge line should be made on the back edge of the string to indicate the depth of the housing. Care should be taken in removing the

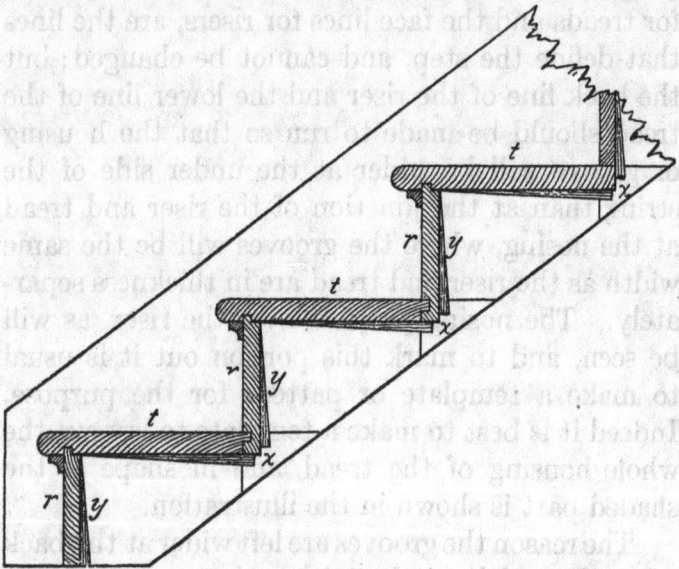

FIG. 245

wood from the grooves that too much is not taken or the grooves made too deep. A gauge for trying the depth may be made out of a piece of hard wood, say about four inches long and three inches wide, by about one-half inch in thickness. Make a tenon on the center of one end, about three-quarters of an inch in width, and cut the shoulders back sufficiently far enough to admit the tenon being long enough to touch the bottom of the groove or housing, when the shoulders rest on the face of the string.

In Fig. 245 we show a **sectional elevation through the steps.** The treads, t,t,t, and the risers, r,r, are shown in position. These are secured as will be seen by means of the wedges, x,x, and y,y, which are well covered with glue before they are inserted and driven home. Stairs made after this manner are strong and perfectly solid under foot.

Fig. 246 gives two views of a portion of a better class stair, a stair with cut and mitered string, or open string stair. In referring to the plan, WS shows the wall string; RS the rough string placed there to give the structure strength; and OS the outer or cut string. At a,a, the ends of the risers are shown, and it will be noticed that they are mitered against the vertical or riser line of the string, thus preventing the end wood of the riser from being seen. The other end of the riser is in the housing in the wall string. The outer end of the tread is also mitered at the nosing and a piece of stuff made or worked like the nosing is mitered against, or returned at the end of the tread. The end of this returned piece is again returned on itself back to the string, as shown in the upper portion of the cut, at n. The moulding, which is a five-eighths cove in this case, is also returned round the string and into itself.

The mortises shown at the black points, B,B,B, etc., are for the balusters. It is always the proper

thing to saw the ends of the tread ready for the
balusters before they are attached to the string,
then when the time arrives to put up the rail, the
back end of the mortise may be cut out, when the

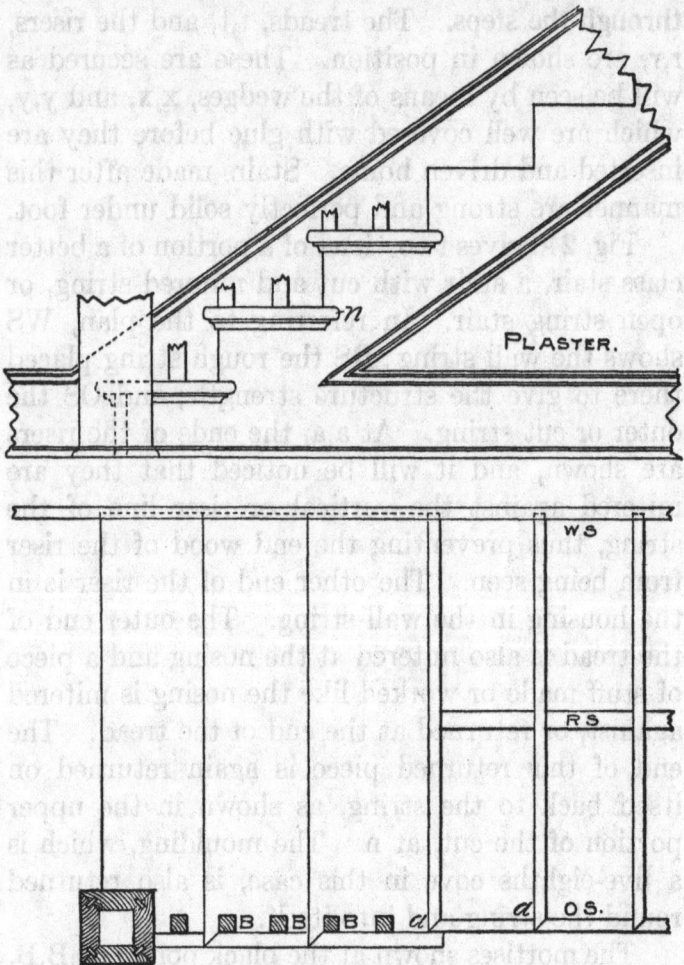

FIG. 246

tread will be ready to receive the baluster. The mortise is dovetailed, and, of course, the tenon in the baluster must be made to suit. The tread is finished on the bench, and the return nosing is fitted to it and tacked on so that it may be taken off to insert the balusters when the rail is being put into position.

Fig. 247 shows the manner in which the wall string is finished at the foot of the stairs. S shows

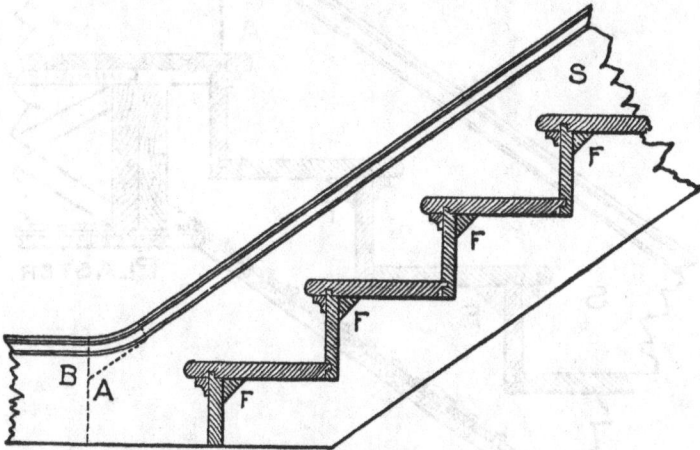

FIG. 247

the string with a moulding wrought on the upper edge. This moulding may be simple ogee, or may consist of a number of members, or may be only a bead, or the edge of the string may be left quite plain; this will be regulated in a great measure by the style of finish in the hall, or wherever the stairs are placed. B shows a portion of the baseboard, the top edge of which has the same finish as the

top edge of the string.   B and A together show the
junction of the string and the base.   The dotted
line shows when a piece of stuff has been glued on
to the string to make it wide enough at the junction
to get the ease-off or curve.   F,F show the blocks
glued in the angle of the steps to make them firm
and solid.

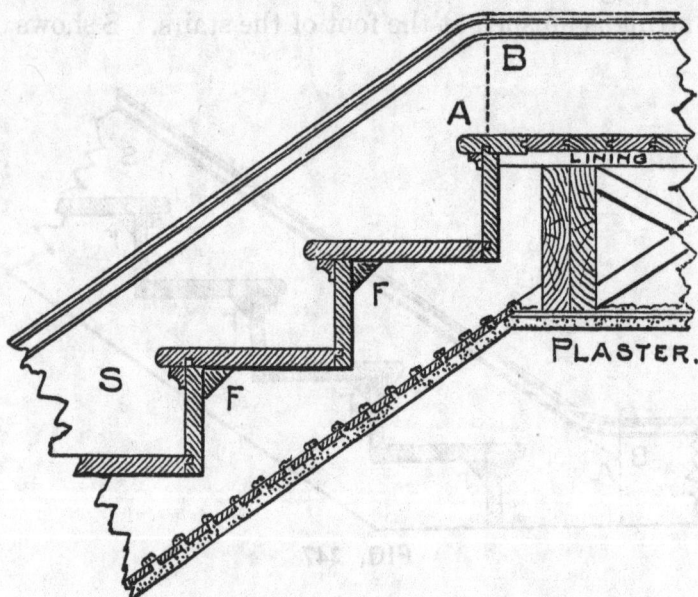

FIG. 248

Fig. 248 shows the manner in which the wall
string S is finished at the top of the stairs.   It will
be noticed that the moulding is worked round the
ease-off at A to suit the width of the base at B.
The string is cut over the floor horizontally and
vertically or plumb against the joists.   The plaster

line under the stairs and on the ceiling is also
shown.

Fig. 249 shows the **cut or open string at the
foot of the stairs,** and the manner of dealing with
it at its juncture with the newel post K. The
point of the string should be mortised into the
newel two, three or four inches, as shown by the
dotted lines, and the mortise made in the newel

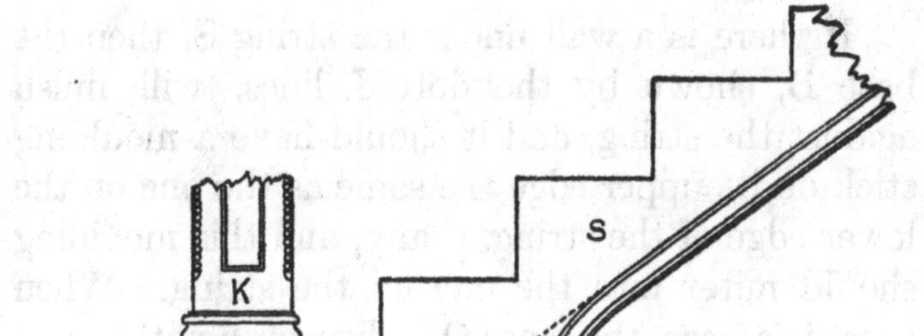

FIG. 249

should be made near the center, so that the center
of the baluster will be directly opposite the central
line of the newel post. The proper way to manage
this is to measure the central line of the baluster
on the tread, and then make this line correspond
with the central line of the newel post. By a careful attendance to this matter, much trouble will
be avoided where the turned cap is used to receive
the lower part of the rail. The lower riser, in a

stair of this kind, will be something shorter than the ones that follow it, as it must be cut between the newel and the wall string. A portion of the tread, as well as the riser, will also butt against the newel, as shown at W. If there is no spandril or wall under the open string it may run down to the floor, as shown at O. The piece O is glued on to the string, and the moulding is worked on the curve.

If there is a wall under the string S, then the base B, shown by the dotted lines, will finish against the string, and it should have a moulding stick on its upper edge the same as the one on the lower edge of the string, if any, and this moulding should miter into the one on the string. When there is a base, the piece O is dispensed with.

The square of the newel should run down by the side of the joist, as shown, and be firmly secured to it by iron knees or other suitable devices. If the joist run the other way, try and get the newel post against it, if possible, either by furring out the joist or cutting a portion off the thickness of the newel. The solidity of a stair, and the firmness of the rail, depend very much on the rigidity of the newel post.

Part IV

Safe Load for a Truss.

Question: Will you state the rule for getting the safe load to be carried by a truss? What the depth should be according to the length. I have a floor thirty feet wide and forty-two long and I want to put in a truss below. I want to truss up a ceiling thirty by forty-two feet, the one thirty feet to go through first and the forty-two to connect them ten feet from the end; and also the forty-two foot one. Now above these a hall which will be used for gatherings of all kinds. How deep will these trusses have to be to be safe? Steel beams will be safer than wood even though the latter be supported by truss rods with turn buckles, will they not? Would the thirty-foot truss have to be double? It would be safer, would it not? Please give me the formula for computing the strength of truss, how deep according to the length.

Answer: As we understand the question, he wishes to support a floor thirty by forty-two feet by means of trussed girders or steel beams, for if the room above is to be used as a dance hall, the trusses could not project above the floor. The information given is not sufficiently definite to enable one to say exactly what is required. In

154   PRACTICAL CARPENTRY

general we would say that the best way to support
such a floor would be by means of girders extending
across the building the narrow way, three girders

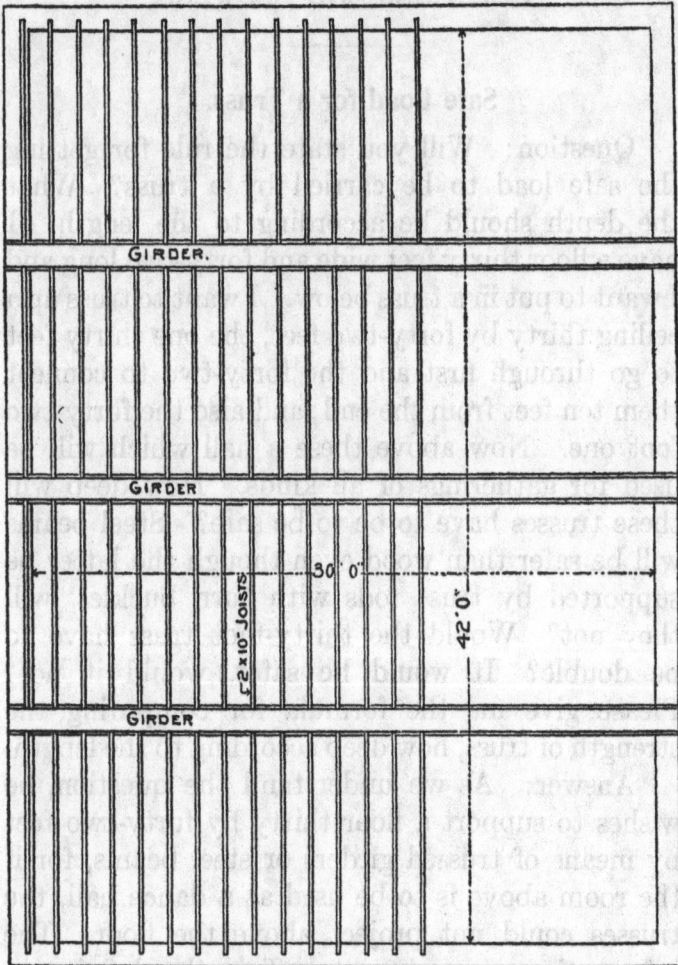

FIG. I. PLAN OF HALL FLOOR
SHOWING ARRANGEMENT OF JOISTS AND GIRDER

being used in the length of the building, so that they would be about ten and one-half feet on centers. These girders would support two by ten-inch joists placed parallel with the sides. Each girder would be required to support a floor area 10½ times 30 feet, which equals 315 square feet. If the floor has a plastered ceiling below, and double flooring, the dead weight, including that of the girder, will be about twenty-five pounds per square foot. For a dancing floor the live load should be taken at 100 pounds per square foot at least, which would make a total of 125 pounds per square oot. As each girder supports 315 square feet of floor, it must be capable of supporting 125 times 315, which equals 39 375 pounds.

To support this load with a thirty-foot span will require a twenty-inch, sixty-five pound steel beam, or a trussed girder such as is shown in Fig. 2.

Owing to the fact that the strength of a truss depends as much upon the inclination and arrangement of the individual members as upon their size, it is impossible to give simple rules for figuring the strength of trusses.

For a trussed girder similar to that shown in Fig. 2 it may be said that the depth of the girder, measured from the top of the beam to the center of the rod or rods, should, for economy, be about one-tenth of the span, and if the depth is less than one-twelfth of the span, the rods will have to be excessively large. As a rule carpenters, in building trusses of this kind, generally make the beam large enough, but almost always use rods that are too light, and do not drop the rods sufficiently, so that the trusses sag greatly and often have to be reinforced.

When the struts are placed at the third points of the span, the strain in the rods may be very closely computed by the following rules:

When the depth H equals three inches plus one-tenth of the span, the tension in the rods is found by multiplying the load on the girder by one and sixteen hundredths, and when H equals

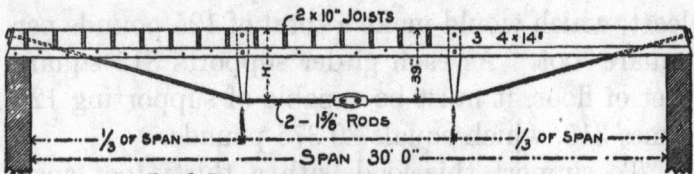

FIG. 2 DETAIL OF TRUSSED GIRDER.
DISTANCE CENTER TO CENTER OF GIRDERS, 10' 6"
SAFE LOAD FOR GIRDER, 40,000 LBS.

three inches plus one-twelfth of the span the tension in the rods is found by multiplying the load on the girder by one and three-eighths. Knowing the tension in the rods, the necessary size of the rods or rod may be found from a table giving the strength of rods.

The depth of the girder shown in Fig. 2 is three inches plus one-tenth of the span (3 inches plus 3 feet equals 39 inches), consequently the tension in the rods will be 1.16 times the load, or 1.16 times 39,375, which equals 45,675 pounds. This strain will require two 1⅝-inch rods. If we reduce the depth, H, to 3 inches plus 1–12 of the span or 3 inches plus 2½ feet, which equals 33 inches, the tension in the rods would be $1\frac{3}{8}$ times 39,375, which equals 54,140 pounds, which would require two $1\frac{3}{4}$-inch rods.

Design of a Truss.

Question: I have to put up a building for a paper mill; the building is eighty-four feet wide and two hundred and forty feet long. The trusses will carry three-inch white pine plank, six-ply paper with tar gravel; an allowance is to be made for snow. There will also be a center load of fifteen thousand pounds which will come on these trusses. Please give a sketch of what you would recommend for the place, giving diameter of timbers and size of rods. Also give a rule for figuring this kind of truss, giving the height and angle of struts and the spacing and length of the same.

Answer: The accompanying illustration shows the proper design of a truss to meet the conditions stated by our correspondent. The stress in each member in pounds is given by the number in parenthesis.

These stresses are those that would be produced by the weight of the truss and roof, a snow load of 36 pounds per square foot, and a load of 15,000 pounds suspended from the center of the truss all applied at the same time. The weight of the truss purlins, planking and gravel roofing was estimated at 24 pounds per square foot.

An allowance for snow of 36 pounds per square foot may seem large, but in some portions of the North the snowfall is very heavy.

The stresses were computed by the graphic method, and would require too much space to explain in these pages.

In regard to the height of trusses of this kind, the height at the center measured between center lines of the chords should not be less than one-tenth of the span, and it is not economical to make the height greater than one-sixth of the span. In this case the height at the center is about one-ninth of the span. The braces should have an inclination of about 45 degrees, and where purlins are

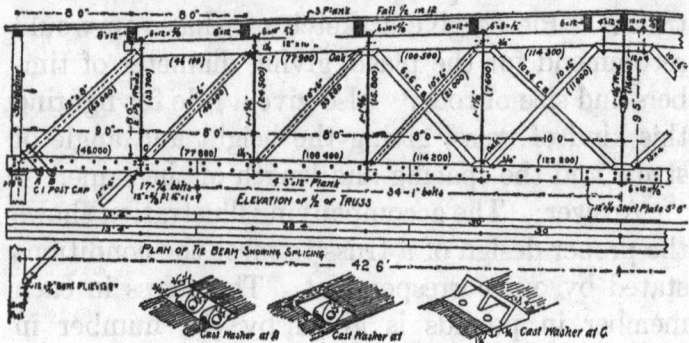

used the braces and purlins should be arranged so that the purlin will come over the top of the brace, or as near to it as practicable.

The purlins should be notched on to the truss one-half an inch, so as to hold the truss laterally, and the ends of the purlins should be tied together either by pieces of boards or iron straps.

In fitting the braces they should be located so that their center lines will intersect the center lines of the chords at the same point as the center line through the rods. This is shown in the engraving by dotted lines. For the braces in the center panels it was not practicable to do this, as it is desirable to bring the top of the brace near the six by ten purlin, so as to avoid a cross strain in the top chord.

The braces marked CB (counter braces) are inserted to provide for any unequal loading of the truss, as might be caused by more snow on one side of the truss than on the other. Under a uniform load, there will be no stress whatever in these braces.

They should not be fitted in place until the truss is in position and had an opportunity to settle to its bearings.

The sizes given for the rods, are for plain rods, not upset. Upset rods are about as expensive for this class of work as plain rods, and upset rods require larger holes in the chords. The factor of safety in the rods under full load is about three and one-half. In the timber from four to eight. The method shown for building up the tie-beam seems to the writer the most economical and as satisfactory as any that can be employed. In this case four three-inch planks are used, lapping each other fifteen feet. Two planks have sufficient strength to carry the entire stress, and the stress must be transferred to the other two planks by the bolts. It will require the full number of bolts shown to transfer the stress. The end of the tie beam is shown projecting four inches outside the post. If the outside wall is frame, a panel can be placed over end of truss. All truss timber should be of Georgia pine.

Size of Timbers in Truss Roof.

Question: Will you please tell me the proper size to have the timbers in a truss roof having a span of thirty-two feet, and what is the proper

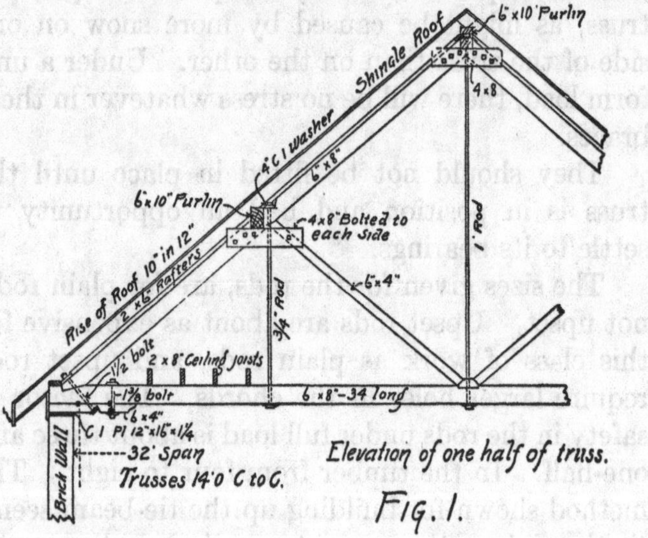

Elevation of one half of truss.

FIG. 1.

pitch to have the roof to look well and be substantial and self-supporting?

Answer: For either a slate or shingle roof, a pitch or rise of ten inches in twelve is both economical and pleasing in appearance. A steeper pitch is not objectionable except that it increases the length of rafters and consequently the cost. For the southern states, the dimensions given in Fig. 1 are about as small as should be used for trusses spaced fourteen feet on centers, and supporting a plastered ceiling. If there is no ceiling, and nothing to be supported but the roof, the truss rafters and tie beam may be made six by six, and the rods reduced to five-eighths and seven-eighths inches. The tie beam should be in one piece thirty-four feet long. Fig. 2 shows a detail for

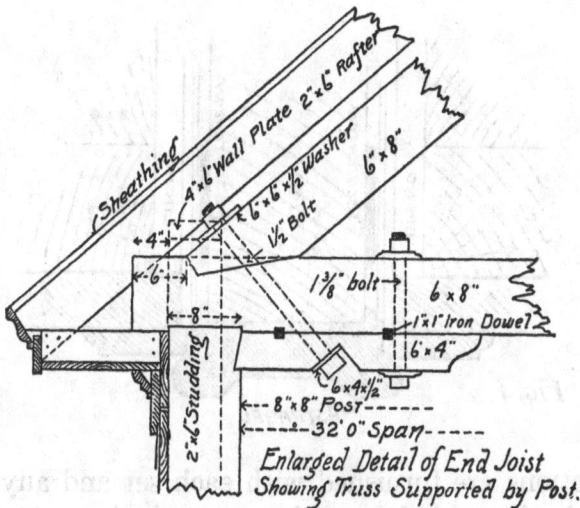

Enlarged Detail of End Joist
Showing Truss Supported by Post.

FIG. 2.

support on posts. If supported by a brick wall, the construction will be slightly modified as shown in Fig. 1. If the building is only one or two stories in height a twelve-inch wall will probably answer, but we recommend that it be reinforced under the trusses by a four by twenty-one-inch pilaster.

Hanging Sliding Parlor Doors.

Question: Please give the best method for hanging sliding parlor doors.

Answer: There are now many patented hangers on the market, each possessing more or less merit. Those with the steel track have largely superseded the wooden of a few years ago, and the hanging of same is an easy job to what it used to be. Full

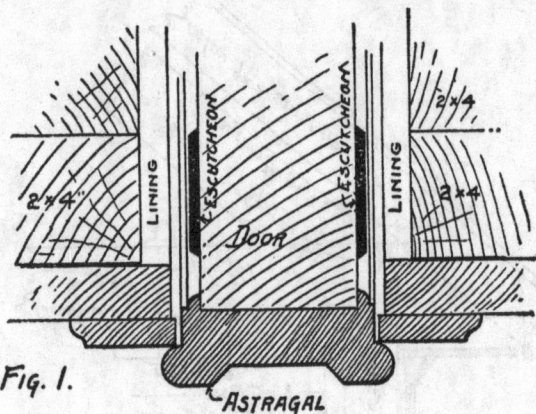

FIG. 1.

directions are furnished with each set and any of the leading kinds can be secured through the up-to-date hardware dealer. Any average workman should be able to put up the work. The main thing is to see that the partition rests on substantial bearings to prevent settlement, as this will necessarily throw the track out of level and affect the free working of the doors. Be sure to set the studding plumb and properly spaced for the pocket. Never set the studding flatwise with the door. Never allow a hot air pipe to run up beside the sliding door, when it is possible to place it in some other partition. Always double the studding at the jambs and be sure to make proper calculations for the opening so that when the finished work is in place, the full face of the door will show when closed. Be sure to have the woodwork over the opening perfectly rigid. Two well seasoned joists spiked together and set up edgewise make a good truss, or lintel, and an excellent

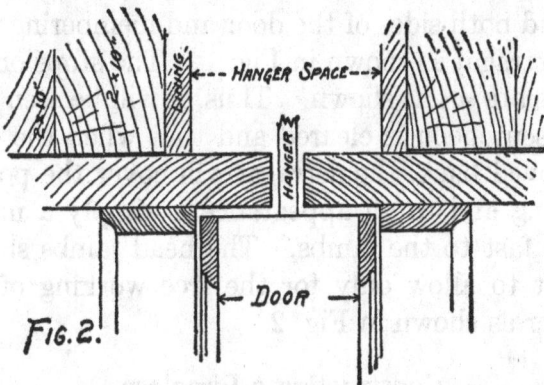

FIG. 2.

surface on which to secure the track. The short studs can rest on this lintel and can be retrussed by cutting in cross braces, or truss shaped braces can be put in above the hanger. In this, the workmen should take into consideration the load that is to be carried above and build accordingly. In good work, the pockets should be lined with tongue and grooved boards, which may be done with thin stuff, but whether this is done or not, be sure to have the pocket openings cut off at the back end so that there will be no connection with other openings in adjoining partitions and outer walls. This should be done for several reasons. First, to heat the house, because these openings will create a draft, then again if a fire gets started in a partition, these openings furnish an excellent draft to fan on the flames.

Another point we might call attention to and that is, the unsightly notching out of the stops to allow the raised escutcheon to pass into the pocket. This can be avoided by running a stop around both sides of the door and membering with the astragal as shown in Fig. 1. The stops on the jambs are set as shown. Thus, it will be seen that the escutcheon is cleared and that when the door is shoved back, the astragal will cover the pocket opening and to all appearance is simply a mould made fast to the jambs. The head jambs should be set to allow only for the free working of the hanger as shown in Fig. 2.

Constructing a Fireplace.

Question: Will you kindly illustrate how to construct a fire-place?

Answer: In the illustration herewith two flues are shown, one to extend to the basement floor, and is for use of stoves in adjoining rooms. While we have shown four openings to this flue there are not supposed to be more than two stoves in use at the same time, otherwise the draft is liable to be overtaxed, as the flues have designed calls for only one brick square opening, or from 64 to 70 inches, according to the size of the brick.

When thimbles are put in to make connection with adjoining rooms, the brick work should be corbled out to the full thickness of the wood partition, and a long thimble used to extend through the brick work, being careful not to let the thimble protrude

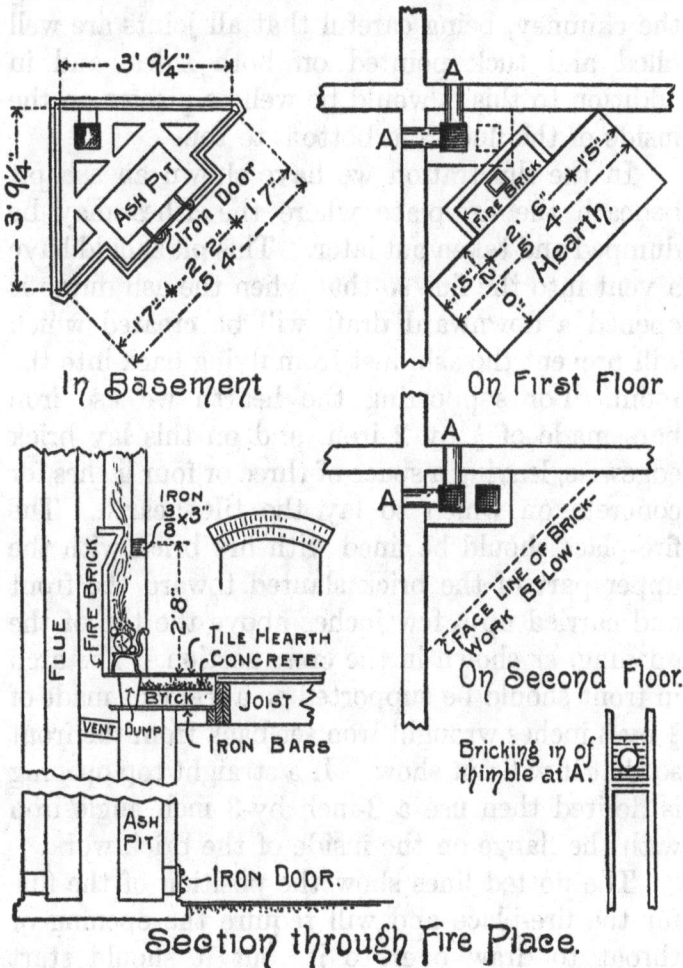

In Basement

On First Floor

On Second Floor.

Bricking in of thimble at A.

Section through Fire Place.

nto the flue space. At sketch A another way of widening the brick work at the thimble is shown, which is simply to cut in a cross piece between the studding, and on this build the extra brick work with all joints well filled with mortar. In all cases the thimbles should be set at the time of building the chimney, being careful that all joints are well filled and tuck pointed on both sides, and in addition to this it would be well to plaster on the inside of the flue from bottom to top.

In the illustration we have shown an ash pit beneath the fire-place where the ashes may be dumped and taken out later. This pit should have a vent into the flue so that when the ash dump is opened a downward draft will be created which will prevent the ash dust from flying back into the room. For supporting the hearth we use iron bars made of $\frac{1}{2}$ by 2 iron, and on this lay brick edgewise, leaving a space of three or four inches for concrete on which to lay the tile hearth. The fire-place should be lined with fire brick with the upper part of the brick slanted toward the front and carried up a few inches above the top of the opening, as shown in the cross section. The arch in front should be supported on a segment made of $\frac{3}{8}$ by 3 inches wrought iron set back from the front so that it will not show. If a straight top opening is desired then use a 3-inch by-3 inch angle iron with the flange on the inside of the brick work.

The dotted lines show the position of the flue for the fire-place and will require the opening or throat to draw over to it, but it should start straight from the fire place and gradually draw over to its position as shown. The face of the brick work should carry up to the ceiling of the first story and this gives ample space to make the proper bend in the flue. The flues should be independent from other openings. Cast iron hoods with damper attachment are quite often used to form the top of open fire-places and are set in place at the time of building the chimney. The top should be capped with Portland cement or with a 3 or 4-inch flat stone with openings cut to fit the flue openings.

How to Finish a Store Front.

Question: I would like to have you give illustrations and describe one or two ways of finishing off store fronts where rectangular cast iron columns are used. This refers especially to wood work in connection with same. I have never seen anything relating to the above subject.

Answer: Fifteen or twenty years ago it was quite the common way to use columns and cast lintels to support the upper work of store fronts, and we dare say most every contractor has oftentimes had a proposition of this kind to work out. The columns were, as a rule, selected from stock patterns—something that the local foundrymen happened to have on hand, to save time and cost of a new pattern, you know; consequently, the builder did not always know just what kind of a column, or how he was going to fit the woodwork to it, till he was ready to put it in. In the ac-

companying illustration are shown some of the
ways we have used in our own work.

In Fig. 1 is shown a plain box column, with
several openings in the back. These openings were
usually oval shape and large enough to insert the
hand. In that case, the woodwork can be made
secure by having a few holes drilled and using
stove bolts as shown.

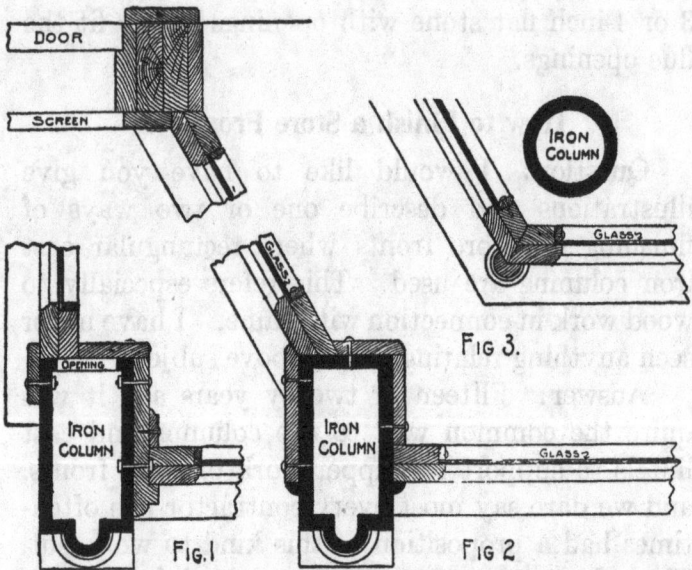

In this, the back is cast solid, and with quarter
rounds cast at the sides, which serves as a stop
for the face of the sash, and the woodwork back
of the sash is made secure by means of lug screws.
Now, since the casting is usually more or less

rough and not suitable to take on a finish to correspond with the interior woodwork, it is a good idea to get out the boards that form the stops wide enough to cover all of the iron work that would show from the inside. However, this kind of construction is fast giving way to the more metropolitan style of large plate glass windows with the least obstruction to the view. Architects and builders are now being quite frequently called upon to prepare plans for remodeling such fronts as shown in Figs. 1 and 2, the columns usually being removed and the masonry work above being carried on steel beams. This generally requires reinforcement at the ends, or pilasters, because the weight, instead of being distributed along the front, as is the case in the use of columns, falls on the piers or pilasters. The frame work can then be built in entirely of wood or light iron work, and future changes can be made without shoring up of the front. This construction is usually for a 25-foot front. If it be a 50-foot front, then two plain, round columns should be used at either side of the angle, at the splay of entrance, as shown in Fig. 3. These columns are independent of the woodwork and should have a firm footing below and independent of the watertable. The latter can then be set at any time in front of the column and the woodwork fitted to it. There being no extra weight on the watertable at the columns, will always keep in alignment. The columns in a front of this kind are not objectionable, as they can be used for decorative purposes, and, if given a finishing coat of aluminum paint, they rather add to than detract from the appearance of the building. There are now on the market a number of patented devices for sash bars and angle columns, each possessing more or less merit, but it is not our purpose to talk about these at this time, but more to show what may be done with wood to give a clear, open front.

Use of Metal Lath.

Question: Will you please give me some information on how to construct the walls of a frame house with 2 by 4-inch studding and shiplap for outside plastering? Should common or metal lath

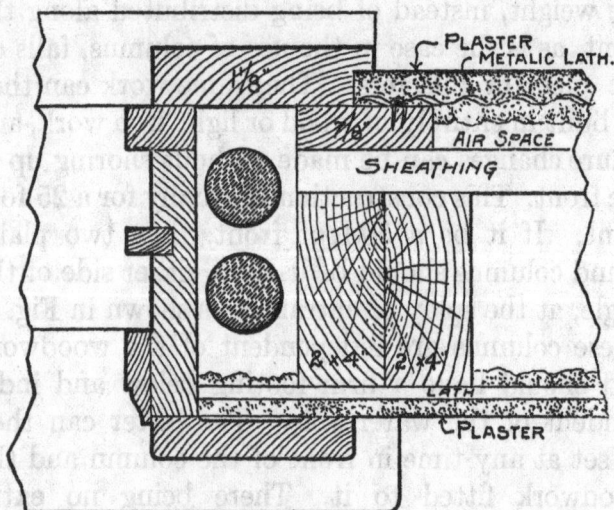

be used and should window and door frames be made different from those for a frame house with shiplap and siding?

Answer: We would prefer using expanded metal lath secured to strips as shown in the accompanying illustration. This gives a better clinch for the mortar than if the lath was stapled direct to the sheathing, besides it creates an air space and also a wider jamb at the windows, which is essential where large plate glass is used necessitating heavier sash than for the common double strength glass.   It is a good idea to plow or groove out the corner of the frame so that the mortar will extend under the edge of the frame.   The flashing cf the caps can be put on in the usual way and plastered over. Of course, it would be much easier as far as the plastering is concerned to set the frames after the plastering is done, but it would not make as tight a job, especially to prevent leakage at the top.   The frame work should be very substantial, otherwise a settlement or vibration will crack the plastering.

Proper Stair Finish.

Question: I am putting in a flight of stairs from a room that is painted white. Would like your advice as to finish of stairs. Would oak throughout look well with the white? Please give your opinion on the subject.

Answer: We should not advise the use of natural oak, but if oak finish is to be used for the stairs, a pleasing and effective combinatiowounld be to darken the oak with ammonia, or to use a sixteenth century effect, obtained by means of sulphuric acid diluted with an equal quantity of water. Forest green oak is also very pleasing in combination with white woodwork. Several manufacturers make green stairs suitable for this purpose, or the oak may be stained by dissolving verdigris in soft water or vinegar. Before using the stain on the actual work, try it on a small piece of the wood first. Mahogany hand rails and newels, with white spindles of a colonial design, always make an effective finish for a staircase that lead from a room or hall finished in white.

Kerfing a Riser.

Question: Will you please tell me how to get the distance between the depth of kerf, to kerf a riser for the first step of a stairway; also how much must be kerfed to bend the riser the required slope?

Answer: In Fig. 1, the distance AD the outside of the riser is longer than BC the inside. When kerfing, enough must be taken from the inside to take up this difference in the lengths. There must be no binding or it is apt to strain the fibers in the face of the wood, and if too much is taken out in kerfing the curved part of the riser will be weak. When the distance between the kerfs is two or three inches a V shaped cut must be made, but it is better however to have the kerfs closer or they will show on the face. No rule can be given that will cover all cases as the spacing of the kerfing depends upon the kind of wood and also upon the thickness of the stuff. The workman must depend largely upon his own judgment

for the different cases as they come up.   A fairly
good ratio between the radius of the curve and the
distance between the kerfs is 1 to 8, that is the
spacing of the kerfing about one-eighth of the
raidus, not over that and less if possible.   In Fig.
2, A is the distance between the kerfs and B should
equal C.   If it is desired to do the kerfing with one

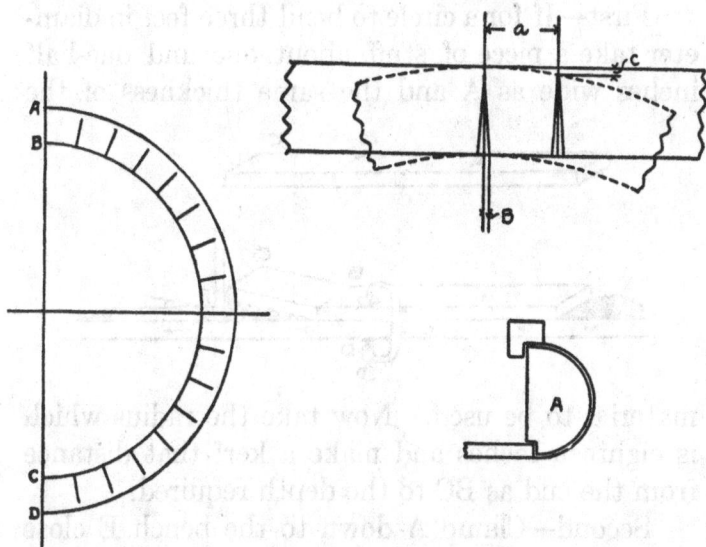

saw cut then by taking C equal to about half the
width of the cut of the saw, the distance between
the kerfs is approximately determined.   When the
radius is small it is advisable to do away with the
kerfing in the manner shown in Fig. 3.   A is a
solid block or one built up and cut to fit the curve.
The riser, where the curve begins, is cut down to
about three-sixteenths of an inch and glued to the
block A which is fastened to the floor.

Kerfing a Riser.

Saw kerfing is the simplest thing the mill man has to deal with, by the method shown with two simple illustrations, a piece of wood may be bent to any radius no matter how thick or thin the material may be, or how thick or thin the saw may be.

First—If for a circle to bend three feet in diameter take a piece of stuff about one and one-half inches wide as A and the same thickness of the

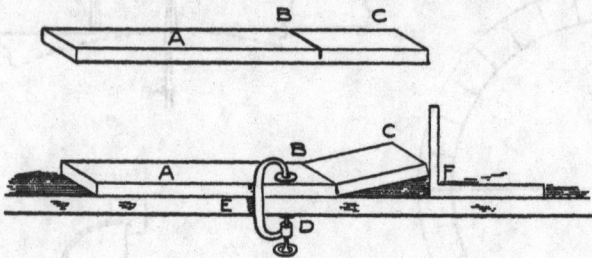

material to be used.   Now take the radius which is eighteen inches and make a kerf that distance from the end as BC to the depth required.

Second—Clamp A down to the bench E close to the kerf and raise the radius end till the cut comes together tight, and take the height with the steel square F from top of bench to underside of piece.   This will give the space between each kerf to bend the riser or anything that has a radius.

Stair Construction.

Question: In finishing the base in column partitions is it proper to extend the base board and cap mould around the base of the pedestal, or butt base and moulding against the pedestal. In constructing open stairway where the stairs have a landing and the lower part of the stairway is open and the upper part is closed from the landing to the floor, how would you connect the stair string

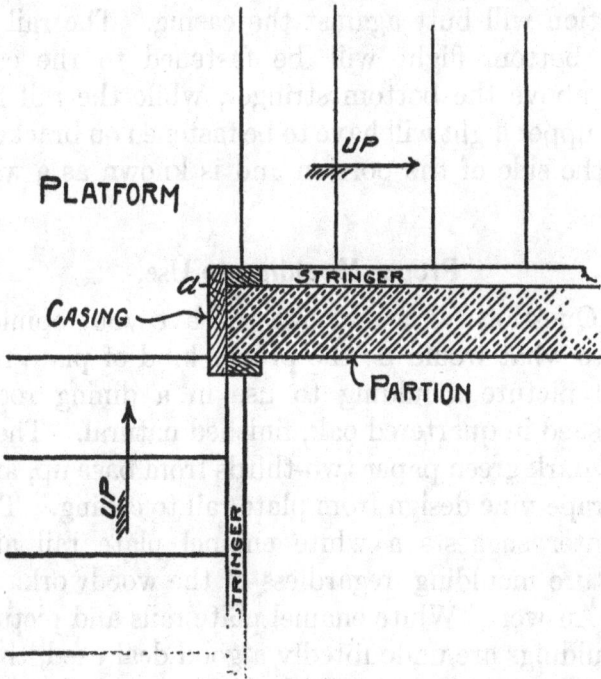

PLATFORM

UP

CASING

a

STRINGER

PARTION

STRINGER

UP

with the angle newel, as in this case the string would come on the inside of the angle newel.

Answer: We propose the solution exemplified in the accompanying sketch of the inquiry submitted in which it is shown that we omit the angle newel in the intersection of the two flights at the platform.   In place of which we continue the portion enclosing the upper flight far enough into the platform as shown at A to receive the stringer of the bottom flight. At the angle the portion is shown closed with stuff equal in thickness to the thickness of the stringers and the stringer of the closed upper flight which is on the inside of the portion will butt against the casing. The rail of the bottom flight will be fastened to the casing above the bottom stringer, while the rail for the upper flight will have to be fastened on brackets to the side of the portion and is known as a wall rail.

Proper Moulding to Use.

Question: I would like to have your opinion as to what would be the proper kind of plate rail and picture moulding to use in a dining room finished in quartered oak, finished natural. There is a dark green paper two-thirds from base up, and a grape-vine design from plate rail to ceiling. The painter suggests a white enamel plate rail and picture moulding, regardless of the woodwork.

Answer: White enamel plate rails and picture mouldings are undoubtedly a good deal used, they are, as a rule, suitable with dark paper only when all the woodwork of the room is finished in white, If the paper is light, then the plate rail may be white entirely, irrespective of the woodwork. In the present case, I should suggest using either an oak picture rail, or else to paint it in flat color (thinned with turpentine only) to match the green of the lower wall, or some shade of green in the upper-third. Another treatment would be to use flat black picture rail. The green, however, is most harmonious.

Effect of Frost on Paint.

Question: What effect has frost on paint? Will it freeze and come off or what causes paint in cold weather to blister?

Answer: There is no reason why painting done in cold weather should not be as durable as painting done at any other season of the year, providing the surface painted is dry before it is coated. But if paint is applied over a surface covered with frost, the dampness enclosed by the paint film, is sure to cause trouble and the paint will peel or blister. When painting is done in the winter time, the work should not be begun until all traces of frost have disappeared and the paint should have rather more driers than for summer painting and should be thoroughly brushed out with a "pound" or 6-inch brush. The flat wall brush frequently used for the purpose of saving labor, is not suitable for winter painting.

Roofing a Store Building.

Question: I have a flat roof to put on a double store building. Each store is twenty-one feet wide and sixty-four feet long with a center wall dividing them. I would like to have you advise as to which is the best way to put on a low pitched roof. Would you run all of the rafters to the center wall or to the outer walls?

Answer: The common way to roof a building of this kind is to slant the roof to the rear, giving a fall of about five-eighths of an inch to the foot, which in this case would be a fall of forty inches in the length of the building. Each joist is put on level, but set enough higher than the adjacent one

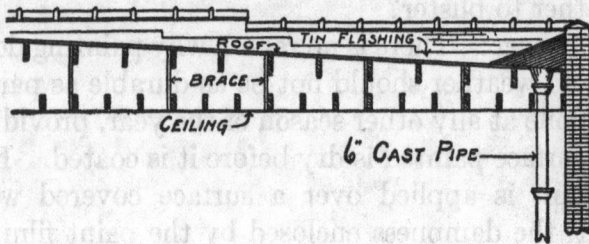

to give the required fall. The gutter is usually a hanging one, but may be objectionable in some sections of the country on account of ice formed from melting snow. In that case it is better to run the water to one place and empty into an internal down pipe as shown in the illustration.

Which is the Stronger?

Question: Which is the stronger for a barn, 2 by 6 studding set on 4-foot centers or two 2 by 6 spiked together on 8-inch centers?

Answer: Would prefer placing the studding on 4-foot centers. It is true the two studs spiked together would more than double the strength at that point, but in doing so the space between is robbed fo carrying strength for the hay load. The load, however, remains the same in either case, but where the carrying supports are on the lesser span the weight is more equalized.

Stair to Fit Circular Wall.

Question: I would like to know how to lay the string out so as to fit around a circular wall in the following plan of a stair?

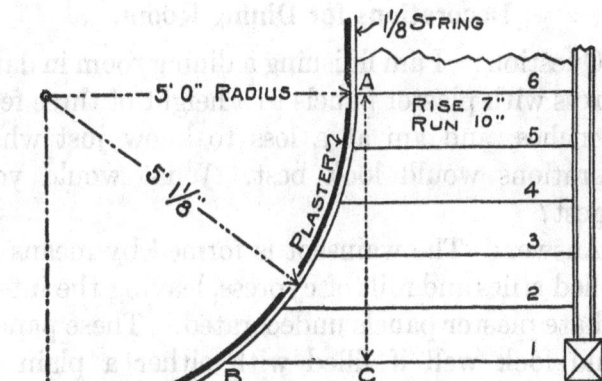

FIG. 1.

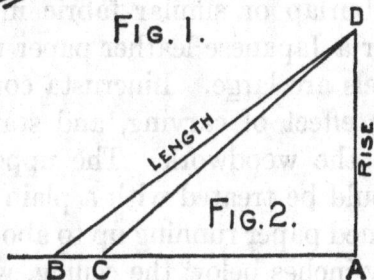

FIG. 2.

Answer: Measure the curve of the wall as from A to B taken at the floor line, as shown in Fig. 1, and this length will correspond with the natural run, as from A to C. To this set up the rise of the stairs, which in this case would be $6 \times 7 = 42$ inches,

as from A to D in Fig. 2, and D to B will be the
required length of the string. The back of the
string can then be kerfed same as for the ordinary
base, but the kerfs must be cut parallel with the
risers.

These diagrams should be laid off full size, from
which accurate measurement can be taken.

Decorations for Dining Room.

Question: I am finishing a dining room in dark
cypress with plaster panels to a height of three feet
six inches, and am at a loss to know just what
decorations would look best. What would you
suggest?

Answer: The wainscot is formed by means of
applied stiles and rails of cypress, leaving the intermediate plaster panels undecorated. These panels
would look well if filled with either a plain or
figured burlap or similar fabric in either red or
green; or a Japanese leather paper may be used if
the panels are large. Lincrusta could be used to
give the effect of carving, and stained the same
color as the woodwork. The upper part of the
wall should be treated with a plain fabric or with
a self-toned paper running up to about twenty-four
to thirty inches below the ceiling, where it should
be capped with a combined plate and picture rail,
or a wider shelf, on which steins, ornamental plates
or other bric-a-brac may be displayed. Above this
the frieze may be hung with a figured paper in
harmonious coloring or one of the pictorial friezes
could be used. These can be obtained in many

beautiful designs. Another treatment for this upper portion of the wall would be to use a plain ingrain paper of a lighter color than the side wall —for example, a light tan could be used with a red side wall—running this paper out some eighteen inches upon the ceiling, where it should be separated from the center panel of the ceiling by a narrow moulding. The ceiling could be tinted a deep ivory. Such a room would be refined in color effect, and would act as a pleasing background for any pictures or other decorations.

Cause of Plaster Cracking.

Question: Please tell me what is the cause of the cracking of the plastering along the lath in a house. The plastering has been on for two months and has been fine up to a short time ago. About three weeks ago the house was overheated.

Answer: There are several causes for plaster cracking. The real cause in some cases is hard to determine without a personal examination. We judge that in this case it is due to shrinking of the lath, as the plastering seems to crack along the lath lengthwise. If the house was overheated the ceiling would get the most of the heat, as the tendency of heat is upward, which would produce a greater shrinkage of the lath. The lath might have been put on wet or green, causing extra shrinkage when dried out with extreme heat. Often the lath are put on in a stretch without breaking joints, which would make the ceiling more susceptible to cracking. Sometimes the plaster cracks because of not getting the right proportions when mixing the mortar. So if there happened to be a combination of causes the cracks would be likely to be a permanent feature of the job.

Ventilating a Barn.

Question: Will you please give an illustration to ventilate a large barn without putting on the

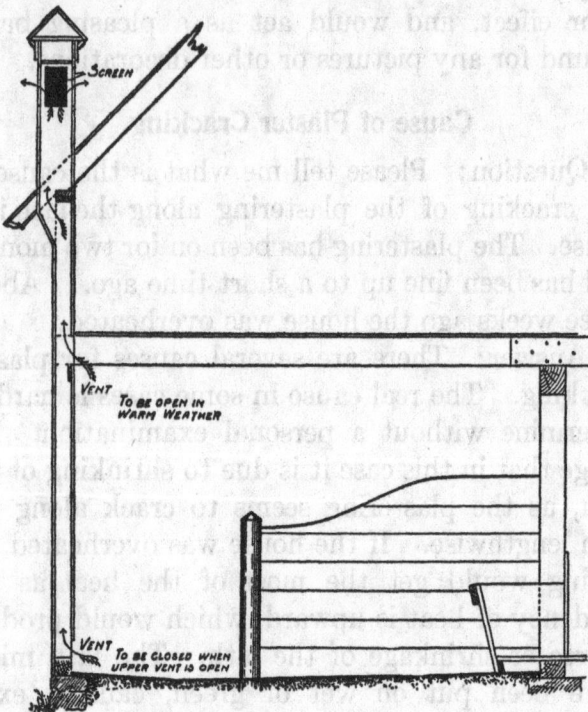

common roof ventilators, which are a nuisance in this part of the country on account of the sparrows and insects.

Answer: The nuisance caused by the sparrows and insects could be almost entirely avoided by screening the openings of the ventilator just the same as for the windows in a residence. As for ventilation from the stable part, this may be accomplished as shown in the illustration, which is simply done by boarding up the space between two studdings, boxing out at the cornice to clear the plate and finished with turret effect on the roof with screened openings on all sides. The interior openings should be as shown provided with slide shutters. Would have one of these vent shafts about every eight feet or opposite every other stall.

A Device for Holding Shingles.

We show herewith a drawing of the most useful article in a kit of tools. It is for shingling where the roof is sheathed tight, and especially at the

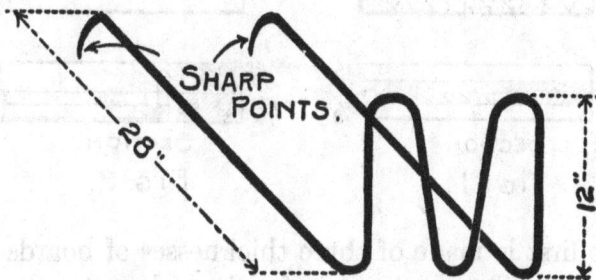

SHARP POINTS.

28"

12"

top in putting on the last rows of shingles. The object is to hold the loose shingles in convenient form for the workmen. The hooks are sharp and can be set anywhere on the roof or hooked over the comb. It is made out of ⅜-inch steel pump rod and any blacksmith can make them. A half dozen of these holders will be plenty for most anyone.

How to Make Barn Doors.

Question: I am building quite a large barn and am siding with drop siding. Will you give the best method of making the doors so as not to take up too much room in thickness?

Answer: We herewith produce two sectional drawings with elevations suitable for barn doors.

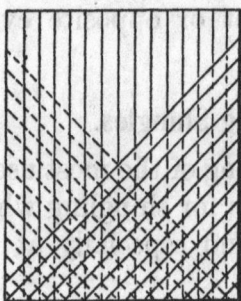
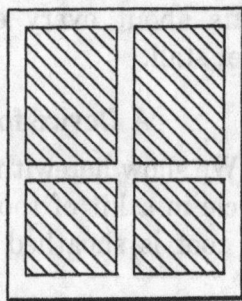

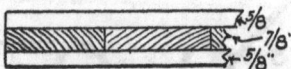
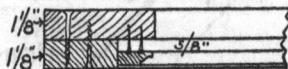

SECTION   SECTION.
FIG. 1.   FIG. 2.

The first is made of three thicknesses of boards as shown. The center is of ⅞-inch boards placed vertically and ⅝ ceiling placed diagonally on both sides, covering the whole space and well nailed. This will make a door 2⅜ inches thick.

The second is made of two 1⅛-inch pieces for

the frame work lapped and screwed together. The panel work is made of ⅝ ceiling cut in and nailed with a stop mould to cover the nail heads. Would paint all the laps and joints with white lead paint. This will make a door 2¼ inches thick.

How to Brace a Roof.

Question: How would you brace a flat roof building 28 feet wide? The joist will have to be spliced at the roof and also at ceiling, as it is not desired to use columns to support the joist.

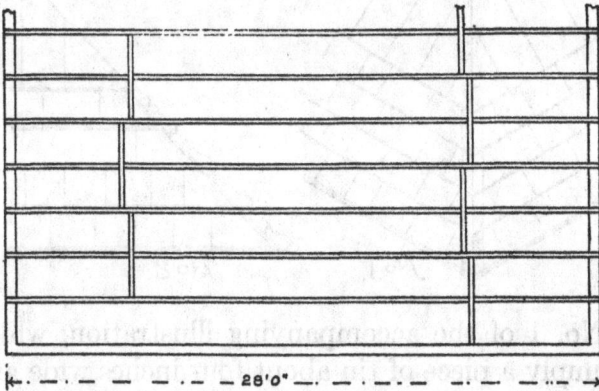

Answer: Judging from the above, there will be no other than its own weight to support, aside from snow that may collect on the roof. That being the case, would interlace the joist as shown, using joist at least 22 feet long. Placing the ceiling joist on 16-inch centers and the roof joist on 24-inch centers and spaced so that every other roof joist will be directly over a ceiling joist and brace with fencing alternately with a vertical and diagonal brace toward the center. Would use 2 by 10-inch for the roof joist and 2 by 6-inch for the ceiling.

How to Shingle Hips and Valleys.

Question: I would be pleased to have an illustration of the best way to shingle a hip or valley.

Answer: There are a number of ways of doing this work, but we think the best way is as shown

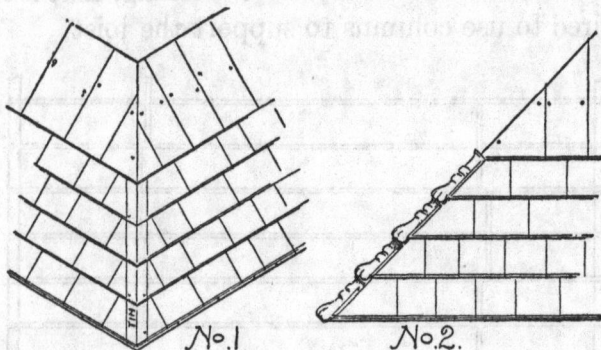

No. 1   No. 2.

in No. 1 of the accompanying illustration, which is simply a piece of tin about four inches wide and long enough to reach up about an inch and a half under the course of shingles above. The tin should be bent through the middle to fit over the hip and the corners trimmed to fit the angle of the butts of the shingles. These should be put in as each course of shingles is laid. As for the angle to cut the shingles to fit the hip, that is best obtained by taking a bunch of shingles, say five or six, and tack them together with a slender nail and then apply the square with the same figures that gives the side cut of the jack rafter, because the shingle laying in the same position, the angle must necessarily be the same. This also applies to the corresponding valley. The tin for the valley should be not less than fourteen inches wide and of best quality, because it will be cheapest in the long run. It should be well soldered and given a coat of oxide of iron paint on both sides several days in advance of using, so as to give it a chance to dry. All flashing tins should be treated in this way. In No. 2 we show an ornamental hip shingle put on as described above. It is made of tin or galvanized iron and pressed as indicated. This makes a very ornamental hip ridge. These shingles are kept in stock at most up-to-date hardware stores.

Roof Brackets for Shingling.

We print the following sketch of a bracket that has been used for the past twenty years, and always found to be safe and easily placed or re-

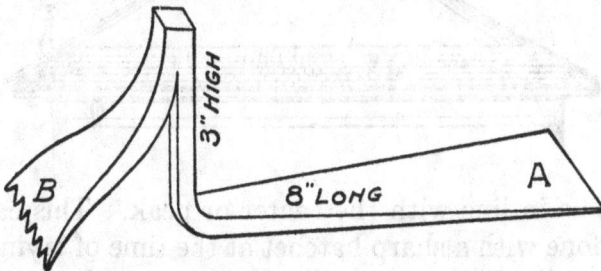

moved. They have been used on roofs of various pitches, and they have never yet failed to stay where they were put, even on the steepest roofs.

They can be made by a local blacksmith, the cost being about one dollar and a half per dozen. They are made from half-inch bar steel, shaped as per sketch, and drawn out thin at each end, the upper end, A, widened and thinned so as to slip under a course of shingles, while the lower end, or foot, B, is also drawn out and teeth filed so as to prevent slipping when in place. In these days of close competition, the best appliances are none too good, no matter what the work may be.

Shingling a Circular Roof.

Question: Please give best method of shingling a circle roof and how to run the chalk line. Also is paint considered better than shingle stain and how should it be applied?

Answer: Trim the shingles so that their edges

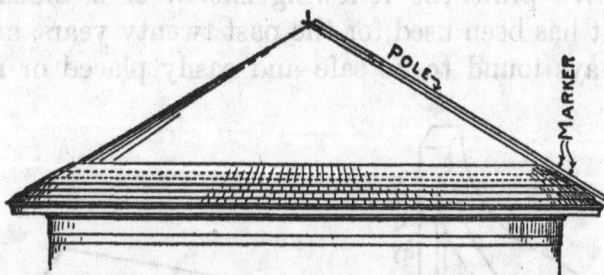

will be in line with the center or peak. This can be done with a sharp hatchet at the time of laying. As to the chalk or gauge line, this can be had with a pole pivoted at the peak as shown in the illustration, and with a marker set at the desired pacing, each circle can be easily described as needed. Or after the first course is laid, the spacing may be had with a gauged hatchet. Opinions differ widely as to whether paint or stain is the better shingle preservative. In either case the butts of the shingles should be dipped to a depth beyond where the second lap will be. Most dipping as a rule is simply a farce. The object seems to be not how much, but how little covering. We have seen dippers take a handful of shingles and dip them five or six inches deep as though they were practicing a sleight of hand performance, not giving the stain time to properly penetrate the wood. After they were laid bright wood at the joints was exposed. Thus the place which most needed protection had none at all. As a protection to the wood we prefer dipping in boiling linseed oil, dipping not more than three shingles at a time with each separated between the fingers, giving two or three seconds' time to each dip.

Finishing Bar Tops.

Question: How can I finish up bar tops both hard and soft wood, so they will have a finish that will not stain?

Answer: If the bar top is made of hard wood, it should first be filled with a good paste filler that should be stained to match the wood. If the natural color of the wood is not desired, it should be stained before it is filled. After the filler has become partially set, it should be wiped across the grain with burlaps to rub the filler into the pores of the wood and to remove any superfluous filler;

and after it has become dry it should be sand-papered lightly with the grain, and should then receive at least three coats of grain alcohol shellac. Each coat should be sandpapered lightly or rubbed with curled hair before applying the next succeeding coat. The final coat should be rubbed with pumice and oil until a perfectly smooth and level surface is obtained, and should then be polished with rotten stone and sweet oil. The treatment for a bar top made of soft, close grained wood is exactly similar to the above, except that the paste filler is omitted, the shellac being applied as soon as the stain has dried. Of course, if a water stain is used, the grain of the wood will be raised, re-quiring sandpapering after the stain is dry. To keep bar tops in good condition mix one part (by measure) of strong vinegar with two parts of boiled linseed oil, and after cleansing with luke-warm water, apply this mixture with a woolen cloth, well saturated, and rub briskly over all parts of the top until polished. Another method of finishing a bar top, whether of hard or soft wood, is to fill the wood with a paste filler, after first staining it if desired. And after the filler has been well rubbed into the grain of the wood and allowed to dry, the bar top is polished with any of the rubbing and polishing oils made for polishing furniture, or the mixture of vinegar and boiled oil may be used. It is well to apply this with a woolen cloth and then to polish by means of a piece of rubbing felt stretched over a wooden block that can be held comfortably in the hand.

Painting a Shingle Roof.

Question: I have just completed a shingle roofed house. These shingles were painted about two months ago with ready mixed red roofing paint two coats, the second being put on about three days later than the first. The water caught from this roof tastes so badly that it is unfit for drinking purposes. Please let me know what is the best way to go about remedying it.

Answer: The ready mixed red roofing paint referred to is in all probability a paint made by mixing a mineral red or metallic paint, red oxide of iron, with linseed oil (probably more or less adulterated) with rosin oil or mineral oil, since this is the usual composition of such paints. While such a paint would undoubtedly give a disagreeable taste to the water for some time and would discolor it to a certain extent, there is nothing poisonous about it, such as there would be in any white lead paint that could be used. As a rule, it is better to avoid painting a shingle roof if the water from it is to be used for drinking purposes. Any additional coats of paint applied to this roof would add to the difficulty, and moreover the paint would find its way into the crevices between the shingles, causing little dams, which would hold back the water and rot the shingles. The present condition will probably disappear in the course of a month or two at the most. When the paint becomes powdered on the surface, it may be given a coat of hot linseed oil, but other than that we should not advise any treatment. Care should be

taken in heating the oil to avoid fire. The best
way is to put the can containing it into a large
kettle of water, which is brought to a gentle boil
over a slow fire. Cold raw linseed oil will
answer the purpose, but will not penetrate the
wood as well as hot oil. It might be well to add
here that while dipping the shingles in creosote
stain is found to preserve them, painted shingles
do not last any longer than unpainted. Creosote,
however, will give a very disagreeable taste to
water taken from a roof where such stains are
used.

Filing a Saw

During an experience of many years at the
carpenter trade I have, very naturally, made a few
observations, and chief among them is the one
pertaining to filing a saw; and I have long ago
arrived at the conclusion that if there is any one
thing that the great majority of carpenters are
deficient in, it is the art of saw-filing.

It has always seemed to me that one of the first
things a carpenter should learn how to do is to
file his saw. I know that I had to file mine, although I nearly used up the first one that I tried.

Now, in regard to filing: When I see a man
using his file as though it were a "hack-saw," I set
him down as one who will never be a success at the
business. A file is made so as to cut but one way;
when you draw it back and forth you are not only
needlessly wearing out your file, but you are spoil-
ing the cutting edge of the teeth of your saw.

I have seen and tried several so-called "filing machines," but have never yet come across one that could compare favorably with hand filing by an expert. And why should not all carpenters, having good eyesight, be experts at filing? What would you think of a barber who could not sharpen his razor? You would not feel safe in trusting him to work for you, yet you will employ a man who calls himself a "carpenter," and allow him to haggle away at your lumber with a saw inadequate for the work for which it was intended, and wasting the time for which you are paying.

If the following instructions are noted, and carefully followed, I think that they may be of some benefit to perhaps a few of the readers of this book.

See to it that your saw has just enough set for the kind of work you are doing; just enough for the clearance, so that the saw will not bind, as too much set is nearly as bad as not enough. Next it should be jointed; by this I don't mean just running over the teeth with a flat file, but jointed with a "jointer." This tool may be purchased in most places at small cost, or can easily be made by fastening a flat mill-file in a hardwood piece, slotted to "straddle" the blade, thus insuring an evenness of the teeth. In jointing, all of the teeth should be touched, and if the saw is hollowing, it ought to be jointed till it is straight, even if some new teeth would have to be made. It would be far better if the saw was somewhat rounding, nevertheless, if straight, it would "pass inspection."

In filing a cut-off saw, much depends on the kind of wood to be cut; hardwood requires the teeth to have less "rake" and less bevel than soft woods. For use in soft woods I give the teeth quite a considerable rake, or pitch, and file at an angle of about thirty degrees, the handle of the file slightly depressed, and file toward the handle of the saw. There seems to be a difference of opinion among many as to which is the better way to file, some saying that they can make a saw cut better by filing to the point, I have tried both ways, and have found that I can do better work by filing toward the heel. But a good deal depends on the way a person commences; if he has become accustomed to file in any one direction he had better endeavor to become proficient in the manner that he accomplishes the best results. Care must be taken to hold the file at the same angle, and the same depression, throughout. Each tooth should be just brought to a point and not filed more; some of the teeth may require filing more than the one adjoining, and often the file will need to be pressed harder against that side of the tooth than the other. It requires "eternal vigilance" on the part of the filer to become an expert. Practice alone will teach the beginner how much he should file on either side. When he comes to file the other side of his saw he may have discovered that he filed too much on the first side; in such case the only proper thing to do is to file a like amount this time, even though it does cut some teeth more than just to a point, so that they will all have the same depth.

It may now be necessary to joint it again, perhaps to re-set it, and go over it carefully again.

After the saw has been filed there is a "wire edge" on the sides. The happy-go-lucky carpenter takes a slip, or an oil stone, and grinds off the sides of the teeth. The "practical" man takes his jackknife, or some such tool, and runs it along very lightly on the sides, trimming off that wire edge.

In filing a hand rip saw, I file square across, giving the teeth the proper rake, holding the file level. I file from both sides, after having set and jointed it as before stated. Having brought all the teeth to an even point, I then file the back of each opposite tooth at a slight bevel on the back, but being careful not to touch the cutting edge of the rear tooth. I hold the handle of the file slightly depressed during this operation. I have found that a rip saw filed in this way will cut well in nearly all kinds of wood.

In regard to files, I prefer a slim taper file, not too large, just a little larger than the depth of the tooth as the size of the file does not obstruct the points of the teeth so much. I have heard some men say that a large taper file makes a wider angle between the teeth; this is not so, as the angles of all taper files are the same, regardless of their size. The section of each is an equilateral triangle, and therefore the angles must be equal to each other.

There is another form of filing called "flem-tooth." To file those saws it requires a different operation, and a different vise, from ordinary filing. This method of filing is mostly used on fine saws, such as back saws, and requires but little set, and are used chiefly for sawing shoulders of tenons on fine work. In order to file a back-saw flem-tooth, I lay it flat on the work-bench, with a thin strip of wood under the blade, and secure in place by driving some small nails in the bench. I then file each side of the lower tooth, on both sides, to a "needle point," holding the file square across the blade, and depressed so that it will only just clear the back of the saw; after having filed one side, turn the saw over and repeat the operation. With a little practice one can file very nicely in this way, and have a saw that will cut keen when needed. No saws are better adapted to fine dovetail work than are these, and with care will last a long time without again filing.

Constructing an Ordinary Stair

Question: Will you tell me how to construct an ordinary stair, especially how to miter treads and risers?

Answer: In an open stair, and especially one in which the treads project over the face of the string, it is desirable to have the work rather well finished in order to present an attractive appearance, one that will harmonize with its surroundings. In the modern dwellings of to-day the front hall or the stair hall is made larger than is necessary to accommodate merely the stair. The reception hall and stair hall are combined and appropriately so, but it is necessary then to finish the room more elaborately than if it were used as a stair hall only.

The object in the stair we are taking up is to avoid having the end of any piece of wood show. In order to accomplish this in the riser, the rise in the string is mitered and the end of the riser is cut on the same miter.

In Fig. 1 the different ways of mitering are shown. At (a) is a miter of forty-five degrees cut on both the string and the riser. This is the simplest method and the one more often used because of the saving in time. In (b) the riser has a

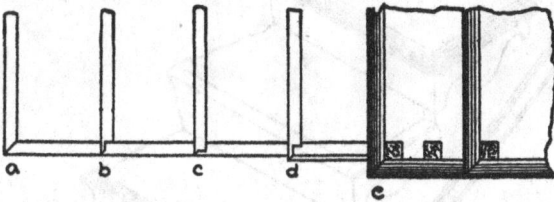

FIG. 1.

shoulder to fit against the string and only the outside is mitered. This makes a more rigid joint. In (c) the miter is cut at the front as in (b) and the string is cut out to receive the remainder of the riser. Here the riser gets a stronger bearing upon the string, while in (b) only the front of the riser gets a bearing.

Where it is desired to make the face of the string more ornamental, a thin bracket is placed against the string, as shown in Fig. 3 at (g). When this is done, the riser must be longer than the thickness of the bracket where no bracket is used. This is necessary because the bracket is mitered to the riser. The cove under the nosing is placed upon the bracket just as it is returned upon the face of the string in the case where the bracket is not used. The lower front part of the bracket rests upon the returned nosing of the tread. In the best grade of work the brackets are glued upon the string, but ordinarily they are nailed on

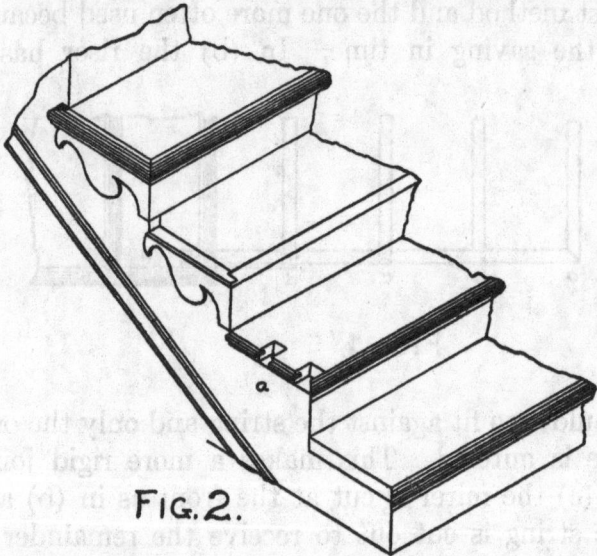

FIG. 2.

with brads, which are then set and the holes filled with putty.

The return nosing is mitered at the front of the tread to fit the nosing over the riser. At the back of the tread, a return is cut as seen at (m) in Fig. 3, which is a plan of (h) in Fig. 3.

After the balusters are in place the return nosing is nailed to its proper place and nail-holes filled

with putty or a groove may be cut into it, as shown
in Fig. 4.   On the end of the tread, as seen in Fig.
2 at (a), a similar groove is cut and a thin piece of
wood or tongue glued into the groove in the end of
the tread.   This tongue should properly have
the grain of the wood run in the same direction as
the grain in the tread.   The return nosing is then

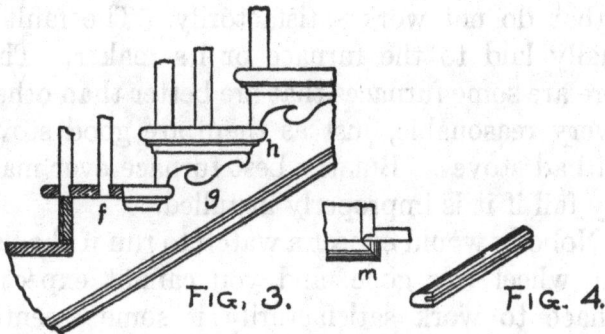

FIG. 3.   FIG. 4.

fastened into place by gluing the tongue and the
groove and driving the nosing to a tight fit.

Another way to fasten the nosing is to cut
notches on the underside of the tread and putting
wood screws through into the return nosing.

A glance at Fig. 1 and 2 will show how the balusters are dove-tailed into the tread.   The outside
of the baluster should be flush with the face of the
string and where a bracket is used, this must be
considered the face of the string.

Installing Furnace Pipes

We notice in the ordinary residence that the
contractor or foreman makes no provision to take

care of furnace pipes until the furnace man comes
to put in the pipes, and then perhaps there is twice
as much cutting to do as there would have been
had the matter been give a little attention when
it was most needed. The piping of a house for a
furnace is a matter that deserves more care and
consideration than it usually gets. It is a well-
known fact that there are more or less furnaces put
in that do not work satisfactorily. The fault is
usually laid to the furnace or its maker. That
there are some furnaces that are better than others
is very reasonable, just as there are good stoves
and bad stoves. But the best furnace ever made
may fail if it is improperly installed.

Nobody would expect a watch to run if the balance wheel was gone, and you cannot expect a
furnace to work satisfactorily if some essential
point has been left out in its installation. By
installation, we do not mean simply setting the
furnace, but we mean the entire plant and system
of pipes. A furnace may not be quite as delicate
as a watch, but there are essential things about
the working of a furnace that must not be overlooked if satisfactory results are to be expected.
The free and easy circulation of air is undoubtedly
the balance wheel that makes the furnace work
satisfactorily.

Before going into further details regarding the
furnace, we want to offer a few suggestions to carpenters and contractors in regard to setting studding and joists to accommodate the pipes. First,
get all the registers located and marked on the

plan. The architect should mark the location of the registers on the plan and the location of the furnace and the registers should be duly considered. They should be placed so as to keep out of the way of beds and not come too close to doors, and yet be where they can be reached by the most direct line possible from the furnace, and also avoid cutting the timbers of the house to any great extent. This makes quite a bit to look after, and enough to need some care in the location of pipes and registers.

No furnace pipe should be run up in a two by four partition; nothing less than two by six should be used, and if a large pipe, use two by eight studding where the pipe is put in. Many contractors use partitions with two by two where the furnace pipes are put in, but we do not believe this is good judgment—might as well get the studding wide enough in one piece and save the extra work of furring, and also the extra nails that the furring requires. Where there are partitions running parallel with the floor joists, it is common for the architect to specify double joists for such places. This, of course, is all right in its way, but if there are furnace pipes to go up in the partition they cannot be put in without cutting the joists half in two and perhaps more. A better way is to know just exactly the space taken up by the partition and set a joist up close to each side of the partition, so the furnace pipe can go between the joists and up through the partition. Pieces can be cut in so a plate can be put on top of the joists to set the studding on, and the plate would only have to be cut where the pipes come through, and this would give full strength to the joists, and if looked after at the right time could be put in quicker than any carpenter could cut out the holes for the pipes in the old haphazard way.

A very serious blunder in putting in a furnace is the lack of providing for the escape of the cold air from the rooms. Cold air is drawn into the furnace through the cold air duct, heated by the furnace and sent through the pipes to the different rooms in the house wherever it is desired. Stand over a register when the heat is turned on and you can feel a strong current of warm air coming into the room. It is the universal opinion and conclusion that the warm air coming into the room forces the cold air out, and by this process the room is heated. Now is it not reasonable to suppose that there should be a suitable place provided for the cold air in the room to escape? You cannot continually force air into a room unless there is some place for it to escape. Without this, when the room is full of air, the air in the room will hold back the warm air from the furnace, and the room will not heat satisfactorily and possibly not at all. We have seen houses with rooms that would not heat even when the pipes to all other parts of the house were shut off. Such a condition cannot be anything else but faulty construction.

If you want to heat a house nicely, provide a cold air register of almost the same capacity as the warm air register for each room, and locate them in the best possible positions. Do not be satisfied with just an opening into a partition with a plate over it, for it is not right. See that each cold air register is of proper size and that it has a pipe of proper size leading either to a vent flue in a chimney or to a main cold air return pipe leading to the furnace well. It is a good thing to run a large cold air return pipe to the furnace well and connect the cold air return pipes from each cold air register with this. We mean this pipe to be entirely separate from the cold air duct to the furnace.

The main trouble is, people want things too cheap, and in order to cut down expenses they leave out all the flues and pipes that they can, and thus hundreds of furnaces are condemned every day, and all through the false idea of economy.

The furnace is all right; it's the way they are allowed to be put in that leads up to the trouble.

Forming an Octagon

Enclosed find drawing and description of how to form an octagon from a square timber, which is

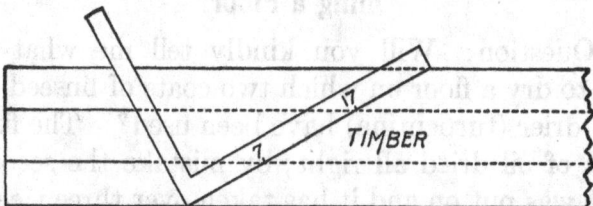

obtained by laying the square as shown and follow back seven inches from either end, and from that point to the corner of the timber set your gauge and size to that, and you will have an octagon timber.

If it be a tapering column, get the size at each end and apply the square by the same rule and take your points at each end and strike a line instead of gauging.

Remember it does not make any difference as to size of stick up to twenty-four inches square.

Filing Saws

I take a cut-off saw, say No. 6 or 7, file it square across with the file level, then I start on the back corner of the tooth, say one-third to one-half way down, and file it to a point on the set side, which gives it a bevel point, but leaves the face square. I don't think it can be beat for the cuts I have mentioned. I file all my rip saws the same way, as it only takes a trifle longer, and if the timber is curly or knotty it cuts just as easy as if it was all straight grain, and it cuts just as fast and much smoother.

Oiling a Floor

Question: Will you kindly tell me what to use to dry a floor on which two coats of linseed oil and drier (turpentine) have been used? The first coat of oil dried all right; by mistake the second coat was put on and it has taken over three weeks to dry; in fact, it is sticky and tacky yet. Can you advise me how to make a good finish?

Answer: It is practically impossible to get two coats of linseed oil to dry hard on a floor, in any reasonable time, if it will ever dry entirely. The only really satisfactory thing to do with the floor in question would be to remove the oil with ammonia or soda, afterward thoroughly washing the floor to remove the alkali, and neutralizing any traces that remain with vinegar. Then the floor should be sandpapered and the work begun anew. For an oil finished floor the best method to pursue is first to fill it with a good paste filler, which is allowed to set and then rubbed with burlap across the grain of the wood, to make it smooth. The best way is to take a strip of, say, five or six boards wide, and apply the filler; then go back and rub well with the burlap. After the filler has hardened for at least twenty-four hours, the floor should be oiled with crude oil or one of the specially prepared floor oils or rubbing oils, applying it with a rag and allowing it to remain for at least thirty minutes and then rubbing it well with a dry cloth. Such a floor can be maintained in good condition by an occasional oiling, in the same manner, which can be done by anyone and is no more trouble than wiping up the floor with water. If it is considered undesirable to remove the linseed oil with alkali, or some of the paint removers that are on the market, a coat of grain alcohol shellac might be tried, but this is at best only an expensive experiment.

Proper Floor Draining

Floor drains, when used in cellar or basement, should be connected to leader side of a rain leader trap wherever it is possible. Some sanitary engineers go so far as to say that floor drains should never be used, their objection to them being that the floor is not washed often enough to furnish sufficient water to maintain a water seal at all times against sewer gas ingress, and their argument is well taken; but floor drains in a basement are very convenient, and are part of a well-installed sanitary sewer system.

In case of a seepage of water through the foundation walls, during a rainy period, it is well to be provided with some means to carry the water away

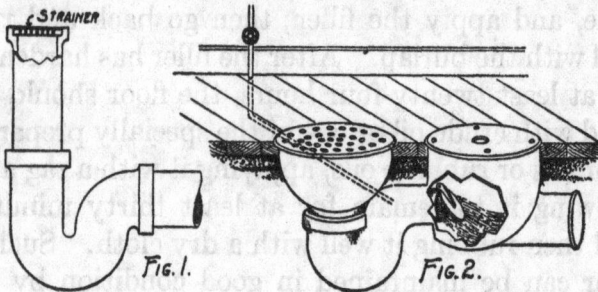

quickly, without having to resort to the laborious (and oftentimes expensive) practice of pumping.

The evils of a floor drain are not so much due to their inefficiency as they are to the care taken of them. The cemented floor basement of the modern home to-day is just as important to be kept clean and sweet as the bathroom, and the thorough housekeeper takes just as much pride in it, and realizes the necessity for having it so from a sanitary standpoint at least.

The old method of installing a floor drain or

floor outlet which consisted of placing a running
trap in the line of drain pipe to the catch-basin,
and running a piece of pipe to the floor level and
simply closing the opening with a bar strainer
grate, as shown in Fig. 1, is wrong. The grate,
even when cemented into the hub end of the pipe,
will in time become loosened, and dirt, sticks and
other rubbish will soon clog up the trap and render
it useless.

As we said before, the one great objection to a
floor drain in the ordinary house, is that there is

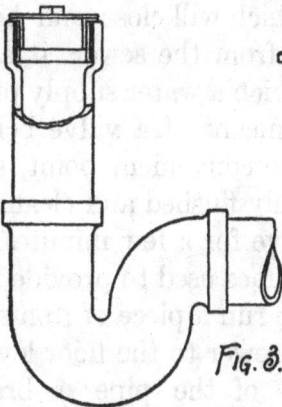
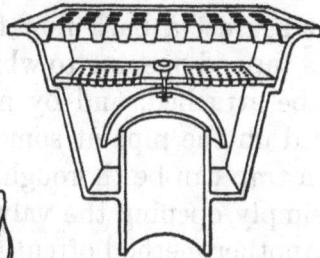

Fig. 3.   Fig. 4.

seldom sufficient water used on the basement floor
to maintain a perfect water seal in the trap. To
neglect to see that the floor drain trap is not always
filled with water and to argue against its installation on that point only is wrong. Neglect renders
many valuable comforts of life valueless.

Floor drains should never be used without a
back-water or tide valve, which will prevent sewer

water from backing up into the basement. We show a number of different styles of floor drains in this article, which are built on the proper lines. The one shown in Fig. 2 is a combination floor drain and back-water gate valve. This accessible cleanout cellar drain flushing cesspool and back-water gate trap valve combination has much to be commended. It has a hinged strainer, through which seeping and floor waste water finds a direct outlet and sewer. The trap has a deep water seal, which is always desirable, and is always provided with a brass back-water gate valve or flap valve which will not rust and which will close and hold tight against a back flow from the sewer; it also has a tapped opening to which a water supply pipe can be attached, and by means of a valve being placed on the pipe at some convenient point, the drain trap can be thoroughly flushed and cleansed by simply opening the valve for a few minutes.

Another method oftentimes used to provide for a floor outlet to sewer is to run a piece of iron soil pipe from the trap on the sewer to the floor level, and to calk into the hub of the pipe a brass ferrule or thimble with a brass screwed cover, which is screwed down tight against a rubber gasket, as shown in Fig. 3. An outlet of this character is only opened when occasion demands, by unscrewing and removing the cover until its need is past.

In Fig. 4 we show an extra heavy cesspool suitable for barns, carriage rooms and places of like nature. The top is sixteen inches square, the body ten inches deep and has a four-inch outlet, suitable for calking into the hub of a four-inch iron sewer pipe; the top cover or grating is heavy enough to permit of horses, wagons and carriages passing over it. The second grating or strainer is of finer mesh, which catches any obstacles which might clog up the sewer; it can be lifted out by

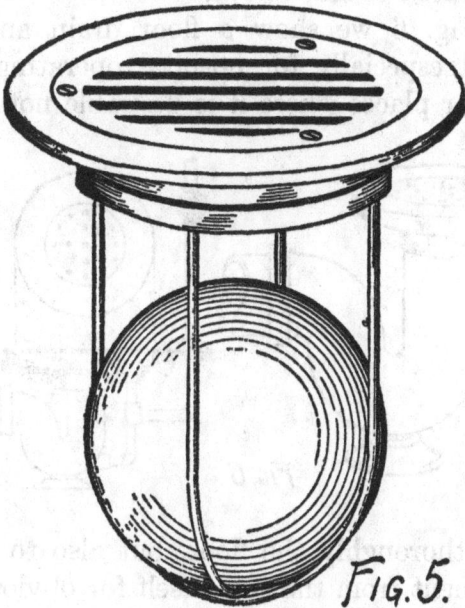

FIG. 5.

the knob and easily cleaned at any time. The deep water seal in this trap is one of its good features, the bell or hood not only serves to maintain a water seal, but where used in stables is a shield over the outlet to prevent oats or grain of any description which might fall through the second strainer from getting into the sewer.

Care should be taken to prevent the bottom of the cesspool from filling up with fine strainings.

Fig. 5 is a combination floor strainer and back-water seal and is used in the hub of a sewer pipe which extends down to the trap placed in the sewer run. The rubber ball prevents the flooding of the basement from backing up of water, by being floated to seat above.

In Fig. 6 we show a floor drain and trap, designed especially for hospital operating rooms and other places where it is desirable not only to

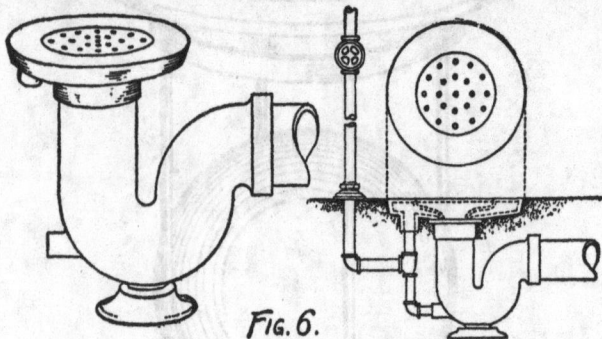

Fig. 6.

cleanse thoroughly the floor, but also to remove all sediment from the trap itself for obvious sanitary reasons. The trap is of cast iron, and is enameled inside. This gives it an impervious and smooth surface and prevents the trap from becoming coated and slimy. This trap is provided with heavy brass cast flushing rim and has a brass removable strainer.

In the sectional cut, we show the method by which the water supply is connected to both the rim and trap, by means of which not only every portion of the body may be cleansed, but also all sediment removed from the jet inlet at the bottom.

The trap is built especially to maintain a deep seal and is three inches in diameter.

Suggestions for Modern Decoration

Modern decorators are not content to follow the old and so-called historical styles, but have broadened out and are no longer bound by conventionalities. If they believe a certain combination of lines or colors has intrinsic merit in itself, it need not have the sanction of Greek or Renaissance taste, but it must stand or fall upon its own merits. It is true that in those of our public buildings in which the architect has followed some historical more or less closely, the decorator usually aims to keep his work in the same period, but when he comes to the homes of the people, it is no longer necessary for him to restrict himself to so limited a field, but he can search wherever he will for beautiful forms that may be adapted to decorative use.

The modern English wall paper designers are particularly happy in this respect, and they adapt the most commonplace things to decorative purposes. For example, one of the recent pictorial friezes produced by a leading English wall paper manufacturer represents a view of the River Thames above London, with one of the typical English river steamboats as a prominent part of the picture. Who but an Englishman would ever have thought of considering a steamboat as a fit subject for decorative design?

In modern house decoration the decorator avails himself very freely of mouldings, applying them over the plastered wall simply for ornament and employing them to separate the various

divisions of his decorative scheme. Not only are these mouldings used for the side wall, but they are applied to the ceiling as well, and one manufacturer of decorative room mouldings has offered to the trade the necessary mouldings to produce the effect of a deeply beamed ceiling, or a heavy paneled wooden ceiling, the whole thing being merely tacked to the plaster by means of thin wire nails. Such moulding as picture mouldings, plate rails and the like, which at one time were made a part of the carpenter's specifications and were put up by him with no regard whatever to what was to follow in the way of decoration, are now almost always left to the decorator and the carpenter has nothing whatever to do with them.

In this connection we offer a couple of suggestions for decorations that may prove useful either carried out as they are, or may be used as suggestions for other treatments.

The first is well adapted for a hall treatment and is intended to be carried out in fabrics in combination with mouldings and stenciling. For about three-fourths the height of the room the walls are to be paneled, after a rich, warm, brown burlap has first been hung upon them. Wide, flat mouldings are used to form the stiling of the of the panels. Then the stencil ornament shown in the illustration, or some other suitable design, is stenciled upon the burlaps in a deep red. The frieze is carried out in natural burlaps, the stenciling being done in a light yellow brown. The ceiling angle is broken by a heavy wooden cornice. The mouldings may either be finished in fumed oak, or in the dark, almost black, Flemish oak, or they may be stained a forest green or sealing wax red, with good effect. One curious feature of the stenciling is that the pattern is carried right through as though it were continued back of the stiling. A similar idea would be quite effective in the case of stenciled ornaments on doors, a treatment which deserves more recognition than it usually gets at the hands of decorators, who leave the doors as great blank spaces of monochrome in the midst of a highly decorated wall.

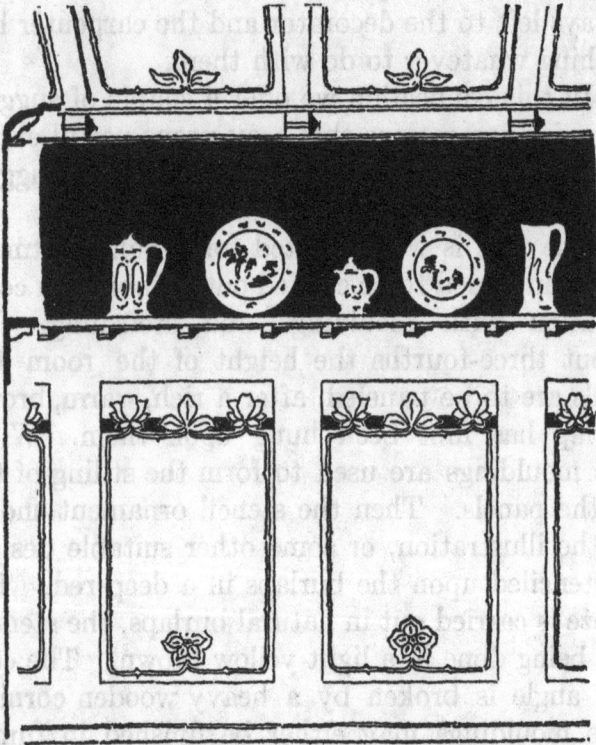

Another effective method of carrying out a hall decoration of this character would be to use a figured burlap or a lincrusta or some similar material that is made with ornament in low relief, for the panel fillings. In that case it is not neces-

sary to stencil any ornaments. The upper part
of the wall could then be hung with a paper or a
burlap having a small set figure powdered upon
the background.

The second suggestion is for a dining room
decoration. Unfortunately the necessity for ren-
dering this sketch in black and white, in order that
it might be engraved, takes away from its effective-
ness. In carrying out this design, the lower portion of the wall is paneled with wide, flat oak
boards, using a rich bright red for the panels.
The border design is stenciled on in shades of
green, with the flowers in a dull yellow tone, taking
care that it does not clash with the red background.
The oak woodwork is stained a forest green, or
may be finished in black Flemish oak if desired.
A shelf rail, supported by brackets, serves as a
resting place for odd bits of pottery, which stand
out against a plain background of dark green burlaps. The ceiling is also paneled, a bracketed
cornice breaking the angle. The ceiling panels
are filled with bright red buckram, lighter in tone
than the dado panels, and the border is stenciled
in green bronze. Other color treatments will suggest themselves according to the amount and
character of the lighting which the room receives
and the colors desired for the draperies and furniture coverings. Another color scheme, for example, would be to stain the oak mouldings a rich
red, making the dado panels a light green and the
upper wall a warm brown. A very effective treatment, although somewhat expensive, would be to

gild the oak on the unfilled wood with leaf gold, using the brush to force the gold well down into the grain of the wood. This gilded surface is then given a glaze coat of asphaltum to soften down its brightness. The dado panels and the ceiling are strong red and the upper wall is a dull blue.

Suggestions of this kind are intended primarily to show the possibilities that are open to the modern decorator, who finds himself absolutely unhampered by precedent in the choice of decorative forms, colors and materials with which his ideas are to be carried out. Recent years have brought so many beautiful decorative fabrics, such as burlaps, grass cloths, finely woven matting, buckram, imitation leathers and similar materials, that it is no wonder that the wide-awake decorators who are alive to the possibilities of all these new things should have struck out for themselves along new lines and should be producing decorative effects that are not only new and novel, but have every element of good taste as well.

Cellar Drainer or Water Elevator

Very often it is necessary—and it is always a good practice—to drain the soil around the outside of a foundation wall. Sometimes it is thought impossible to do this, owing to the fact that the sewer is higher than the level of the drain.

To drain cellars, wheel pits, furnace, cesspools, etc., removing waste water from kitchens below the level of the sewer, drippings from iceboxes and for any purpose where it is necessary to remove water economically from one level to a higher one, the automatic cellar drainer (of which there are several makes on the market) seems to answer the purpose on a small scale very satisfactorily.

Taking the minimum pressure outside of Chicago, i. e., 40 pounds to the square inch, a small

Fig. 1.

cellar drainer has a capacity of four hundred gallons per hour, and is made in larger sizes up to a capacity of fifteen hundred gallons per hour. With additional pressure, it will throw, proportionately, a greater amount of water.

A cellar drainer of this type will elevate seepage water one foot for every five pounds of city water pressure.

In Fig. 1 we show the arrangement of valve and unions, and it should always be installed in this way, so that the drainer can be disconnected and taken out of the pit any time to be repaired and cleansed. A swinging check valve should be placed at the discharge pipe, should there be any possibility of back water getting into the same while in the cut we show a brick pit which is substantial and lasting. A very good strainer pit is made of an old barrel, with a few holes bored in the side and bottom of same. In operation, a drainer of this type performs its functions by passing water or steam under pressure, through the drainer point or jet, thus creating a suction which draws the water from the barrel or pit in which the drainer is placed into the discharge pipe, and both the jet water and the cellar water are discharged together. As long as the city water or steam passes through the drainer point or jet, this suction and discharge is continuous.

The accumulation of water in the barrel or pit raises the float ball gradually, and when the water has accumulated about eight inches in depth, the float ball opens the supply valve. When the water has been removed from the barrel or pit, the valve is closed by the ball dropping back and closing the valve, and the drainer becomes inactive until the water again accumulates, when the valve is opened and the water is discharged as before.

This drainer can also be purchased without the automatic attachment and is used where there is always somebody to attend to same, and is not to be recommended for cellar or any place where water accumulates unexpectedly or irregularly.

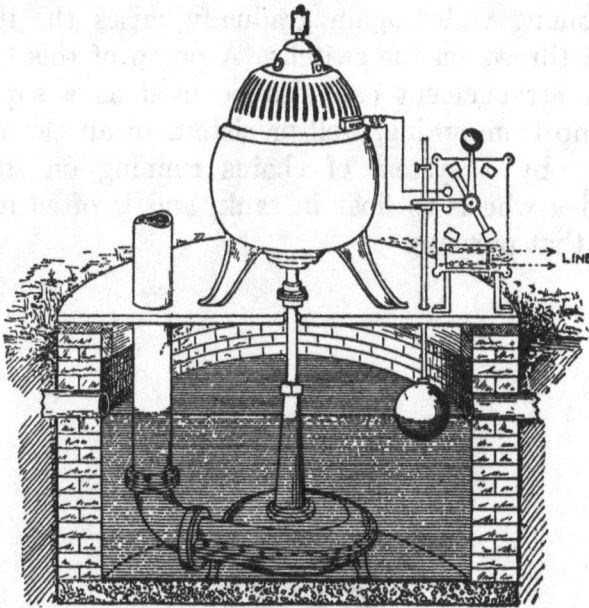

FIG. 2.

This type of drainer or elevator is especially adapted for light work.

In Fig. 2 we show an electric ejector or Bilge pump, which can be had to meet any conditions or capacity necessary, while the first cost, including cost of installation, is greater. The expense of running it is very slight. The operation of this device is that the water rising in the pit raises the copper float, which in turn throws the switch on, starting the motor and pump immediately, the ball receding as the water is pumped out, gradually opens the switch, and when the water is pumped to the level at which the ball is set to stop the pump, the switch is wide open and remains so until the incoming water again gradually raises the float and throws on the switch. A pump of this type and arrangement can also be used as a supply pump from spring, well or cistern to an elevated tank by a system of chains running on small pulley wheels to float in tank, and is often used for that purpose.

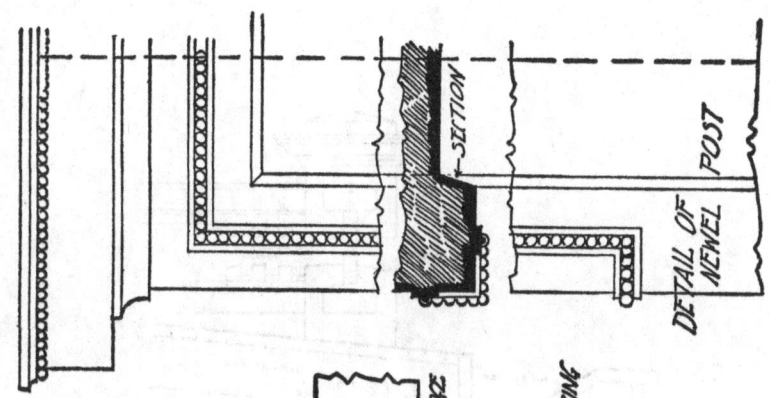

DETAIL OF NEWEL POST

SECTION

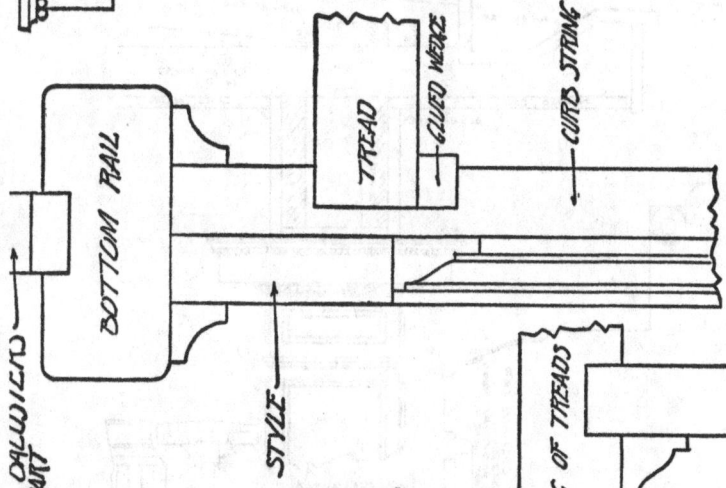

BOTTOM RAIL

TREAD

CURB WEDGE

CURB STRING

STYLE

1" X 1" BALUSTERS 7" APART

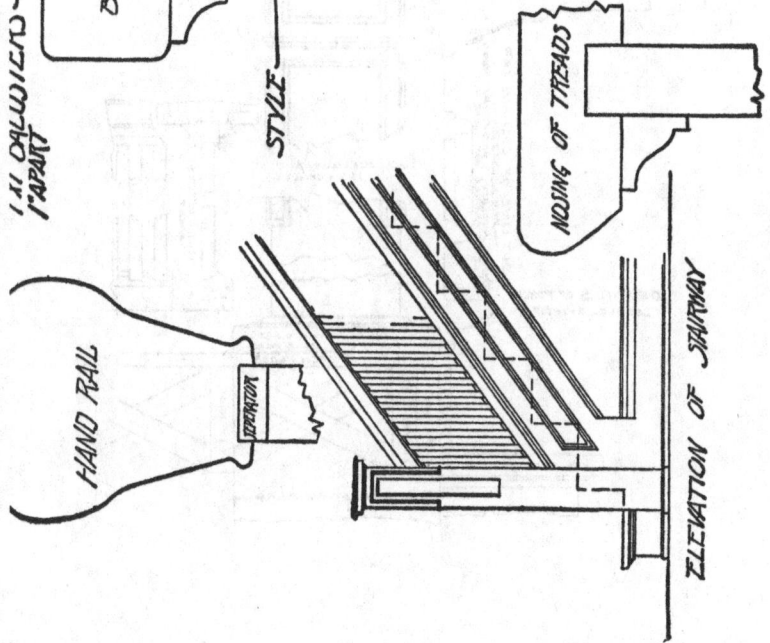

HAND RAIL

BALUSTER

NOSING OF TREADS

ELEVATION OF STAIRWAY

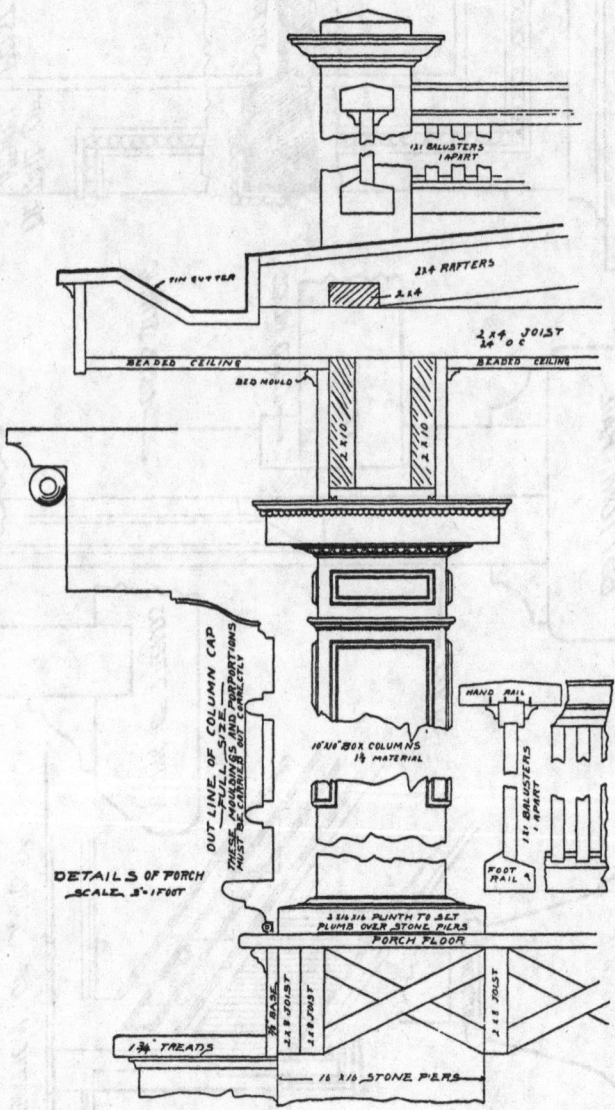

DETAILS OF PORCH
SCALE 3" = 1 FOOT

131 BALUSTERS 1 APART

TIN GUTTER

2X4 RAFTERS

2X4

2X4 JOIST
16" O C

BEADED CEILING   BEADED CEILING

BED MOULD

2X10   1X10

16"X16" BOX COLUMNS
1½ MATERIAL

OUT LINE OF COLUMN CAP
FULL SIZE
THESE MOULDINGS AND PROPORTIONS
MUST BE CARRIED OUT CORRECTLY

HAND RAIL

131 BALUSTERS
1 APART

FOOT
RAIL

3 2X2 X16 PLINTH TO SET
PLUMB OVER STONE PIERS
PORCH FLOOR

1¾ TREADS

¾ BASE
1 X 8 JOIST
1 X 8 JOIST

2 X 8 JOIST

16 X 16 STONE PIERS

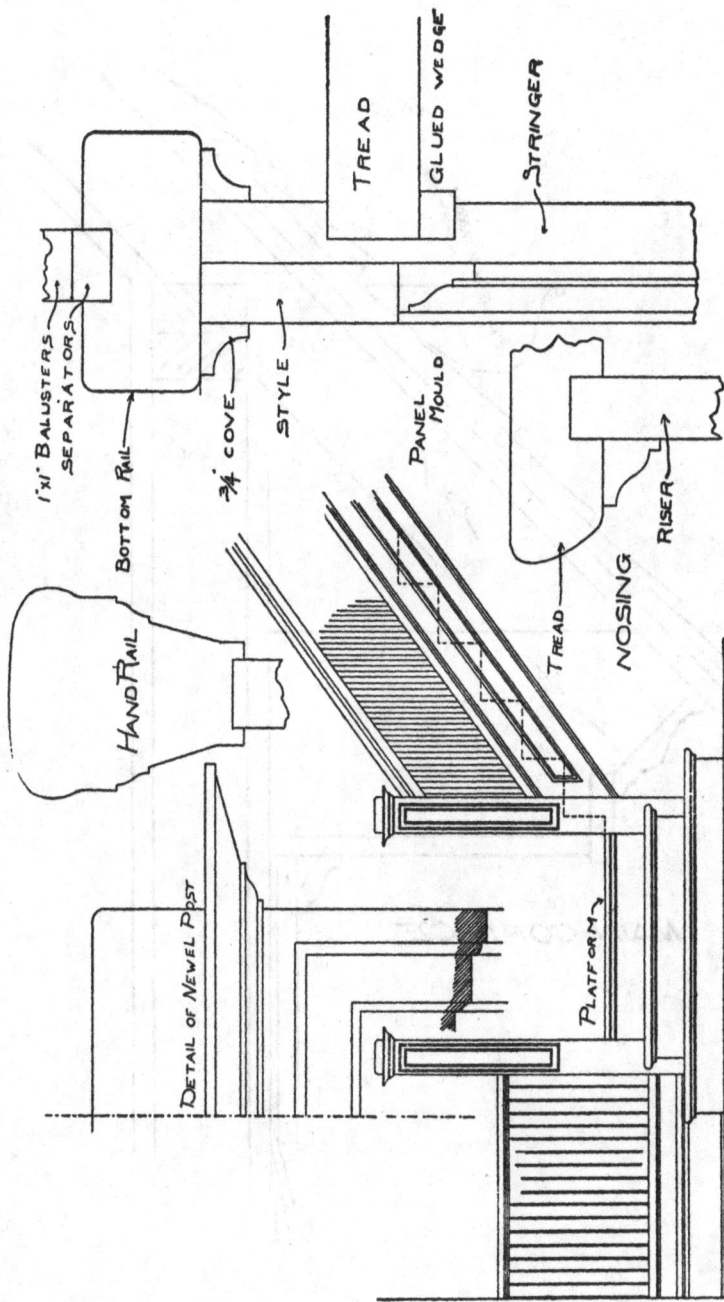

TREAD

GLUED WEDGE

STRINGER

1"x1" BALUSTERS
SEPARATORS

BOTTOM RAIL

¾" COVE

STYLE

PANEL MOULD

TREAD

NOSING

RISER

HAND RAIL

DETAIL OF NEWEL POST

PLATFORM

DETAIL OF STAIRS
Scale ½ = 1 Foot

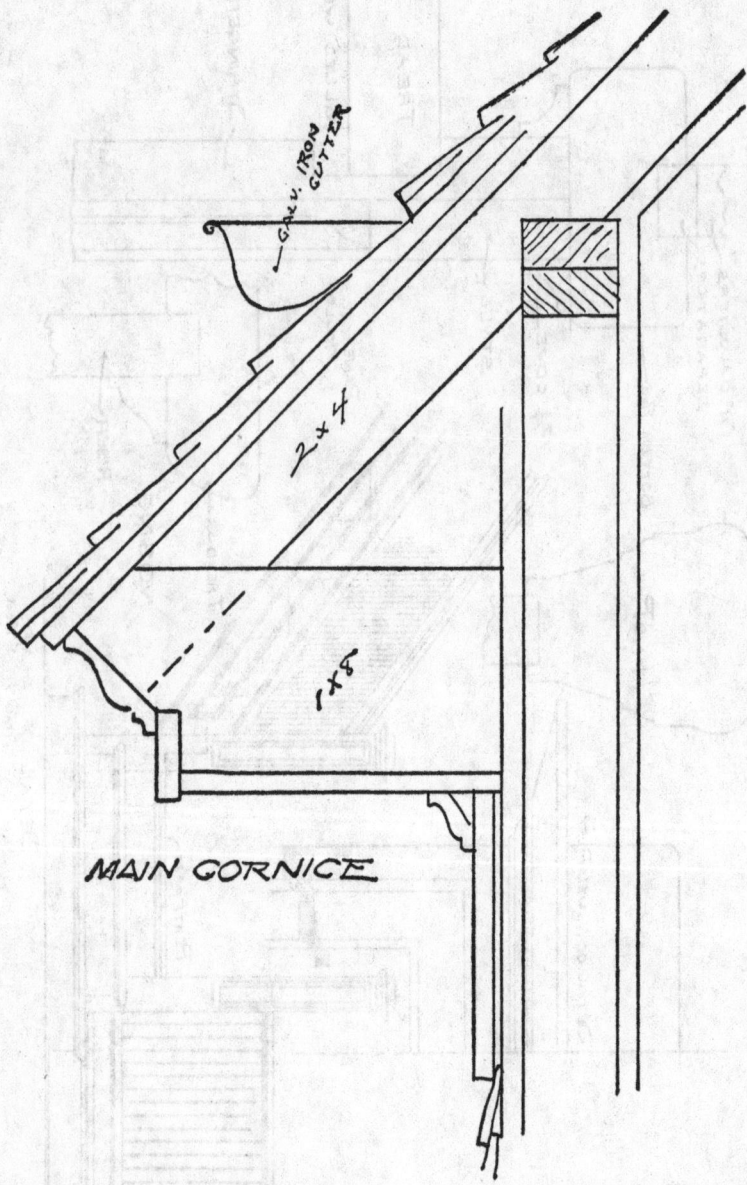

GALV. IRON GUTTER

2 x 4

1 x 8

MAIN CORNICE

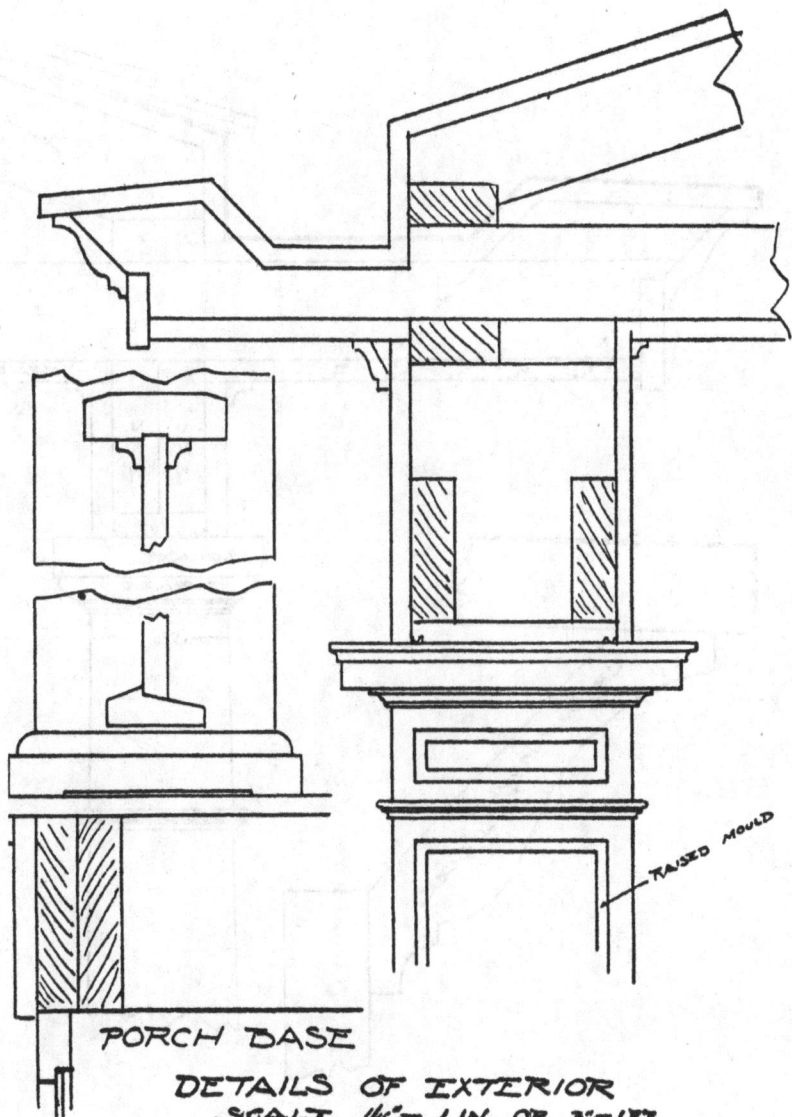

RAISED MOULD

PORCH BASE

DETAILS OF EXTERIOR
SCALE 1/4"= 1 IN OR 3"=1FT.

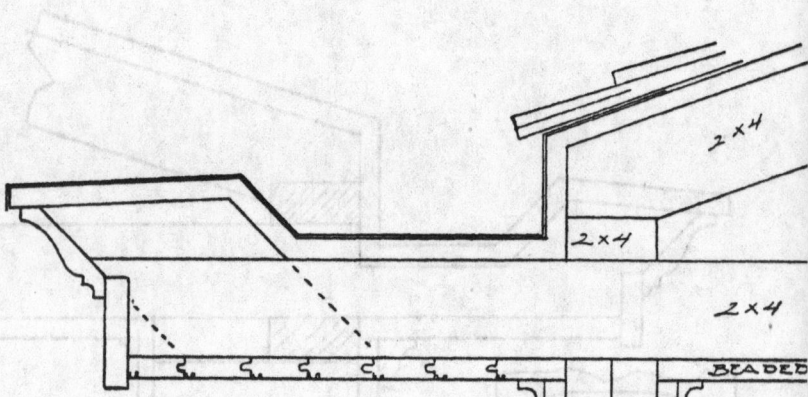

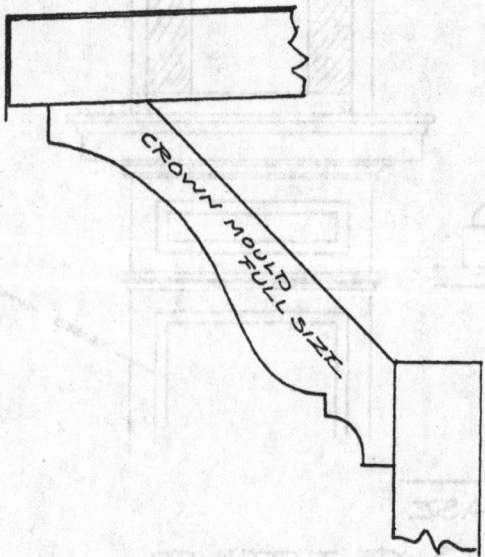

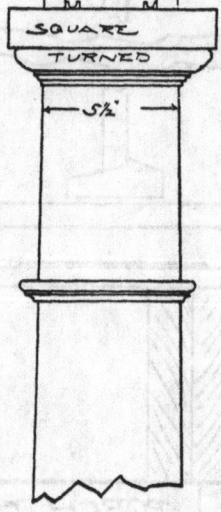

2 × 4

2 × 4

2 × 4

BEADED

2 × 6   2 × 6

SQUARE

TURNED

5½"

CROWN MOULD FULL SIZE

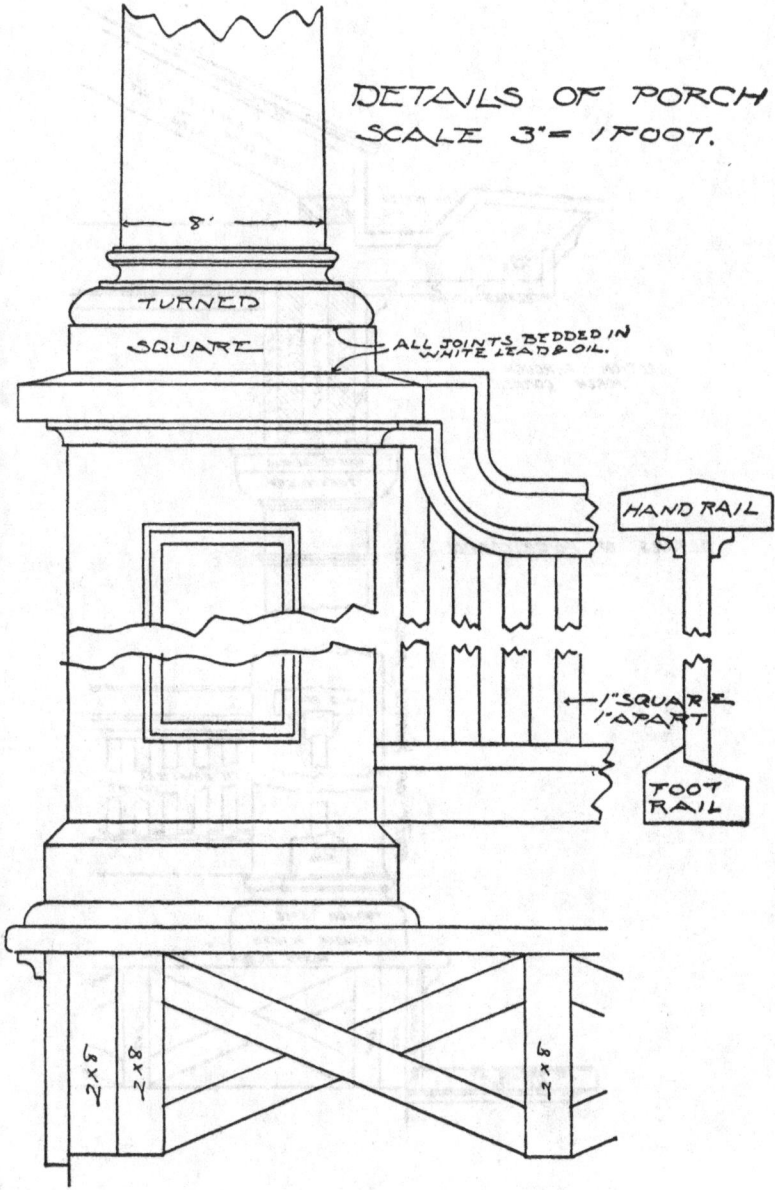

DETAILS OF PORCH
SCALE 3" = 1 FOOT.

8'

TURNED

SQUARE

ALL JOINTS BEDDED IN
WHITE LEAD & OIL.

HAND RAIL

1" SQUARE
1" APART

FOOT
RAIL

2 X 8

2 X 8

2 X 8

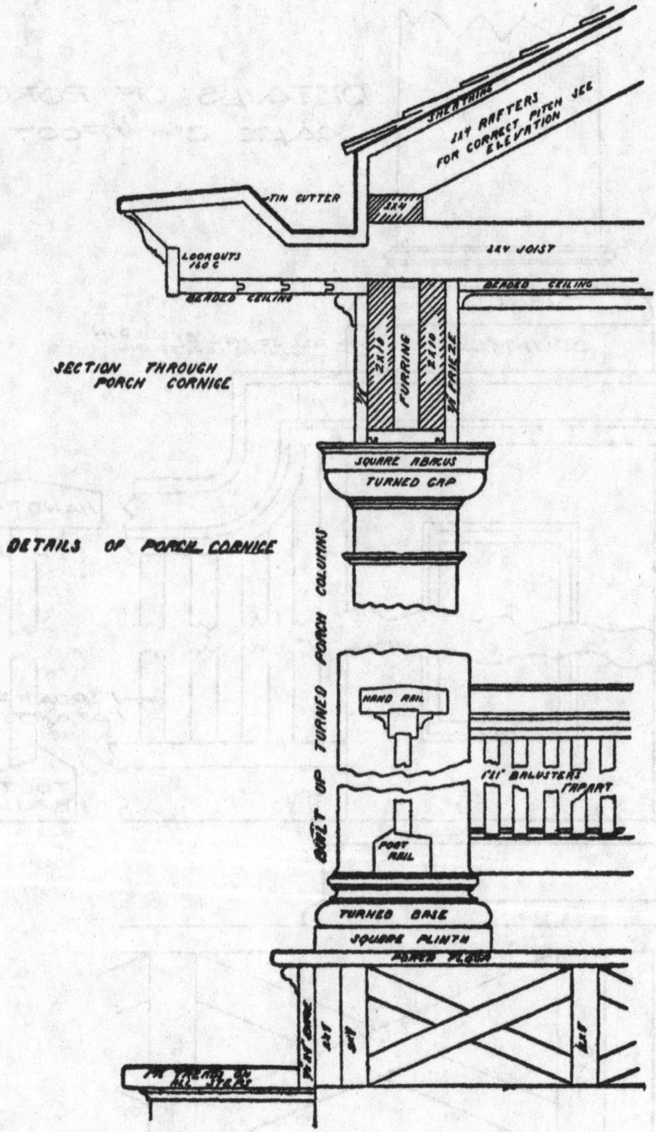

SECTION THROUGH
PORCH CORNICE

DETAILS OF PORCH CORNICE

SHEATHING

2x4 RAFTERS
FOR CORRECT PITCH SEE
ELEVATION

TIN GUTTER

LOOKOUTS
16.0 C

BEADED CEILING

2x4

2x4 JOIST

BEADED CEILING

2x10
FURRING
2x10

FRIEZE

SQUARE ABACUS

TURNED CAP

BUILT UP TURNED PORCH COLUMN

HAND RAIL

1¾" BALUSTERS
7" APART

POST RAIL

TURNED BASE

SQUARE PLINTH

PORCH FLOOR

PORCH CEILING

2x8

2x8

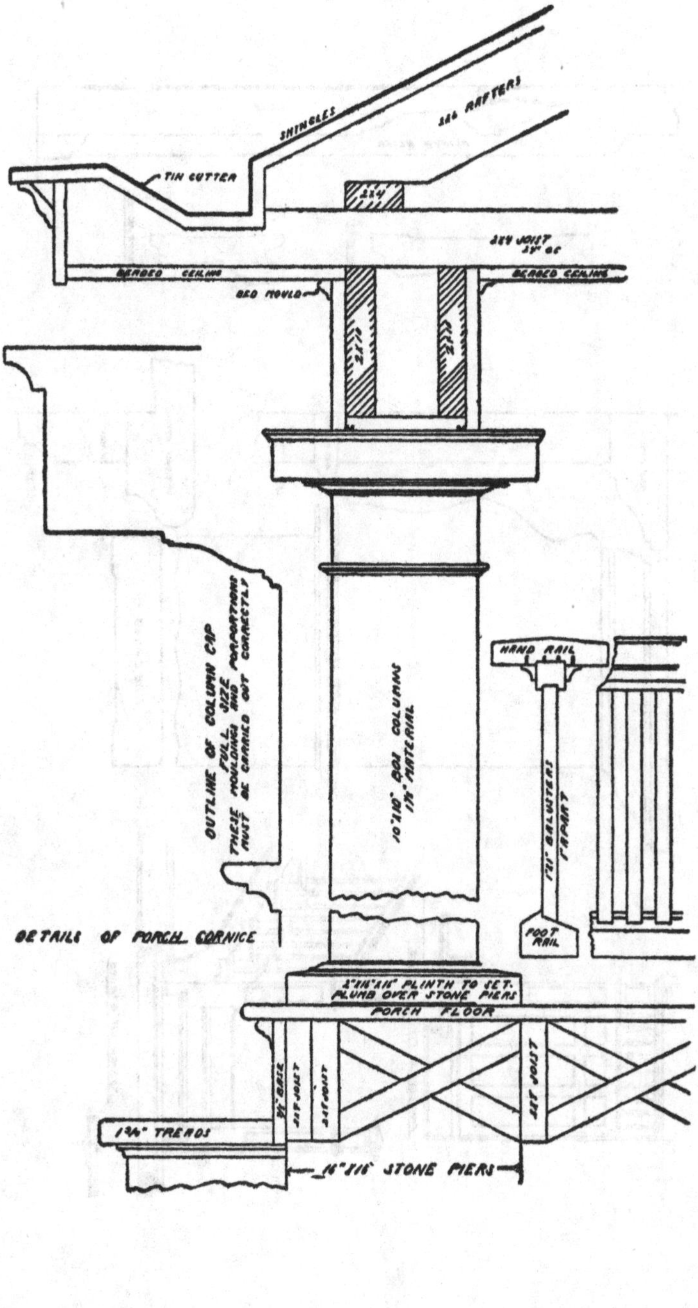

SHINGLES

2x6 RAFTERS

TIN GUTTER

2x4

2x4 JOIST 24" O.C.

BEADED CEILING

BED MOULD

BEADED CEILING

2x10

OUTLINE OF COLUMN CAP
FULL SIZE
THESE MOULDINGS AND PROPORTIONS
MUST BE CARRIED OUT CORRECTLY

HAND RAIL

1¾" BALUSTERS 4" APART

10"x10" BOX COLUMN 1 IN. MATERIAL

FOOT RAIL

DETAILS OF PORCH CORNICE

2"x18"x18" PLINTH TO SET PLUMB OVER STONE PIERS

PORCH FLOOR

2x8 JOIST

2x8 JOIST

2x8 SILL

2x8 JOIST

1¾" TREADS

16"x16" STONE PIERS

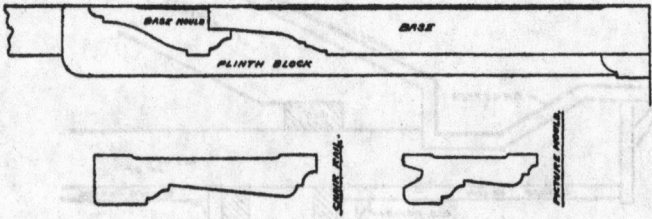

BASE MOULD · BASE

PLINTH BLOCK

STOOL MOULD · PICTURE MOULD

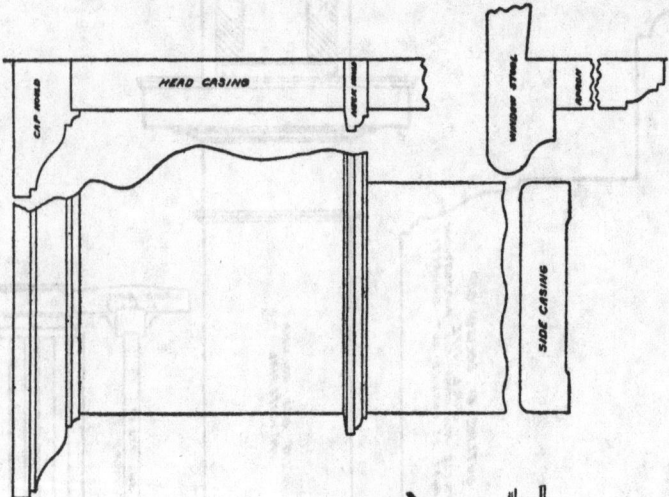

CAP MOULD

HEAD CASING

CHAIR MOULD

WINDOW STOOL

APRON

SIDE CASING

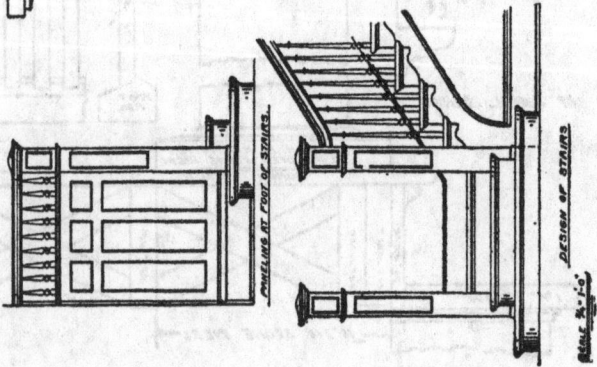

PANELING AT FOOT OF STAIRS

DESIGN OF STAIRS

SCALE ¾″ = 1′-0″

DETAIL OF CHINA CLOSET

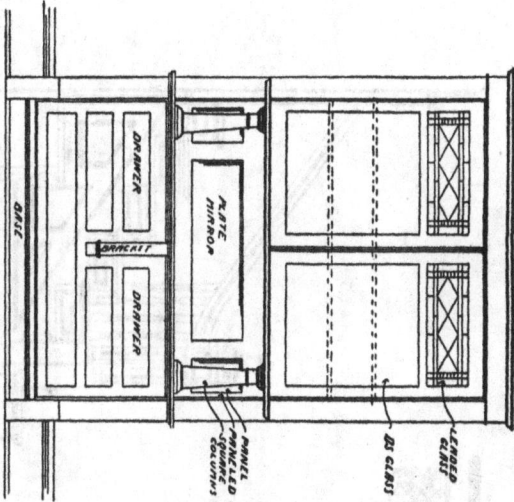

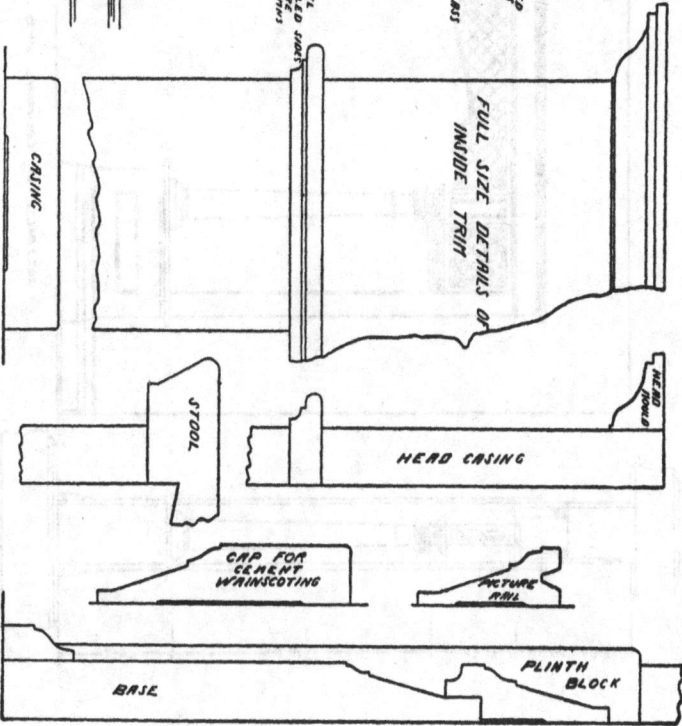

BASE

DRAWER

BRACKET

DRAWER

PLATE MIRROR

PANEL

PANELLED SQUARE COLUMNS

DS GLASS

LEADED GLASS

FULL SIZE DETAILS OF INSIDE TRIM

CASING

STOOL

HEAD CASING

HEAD MOULD

CAP FOR CEMENT WAINSCOTING

PICTURE RAIL

BASE

PLINTH BLOCK

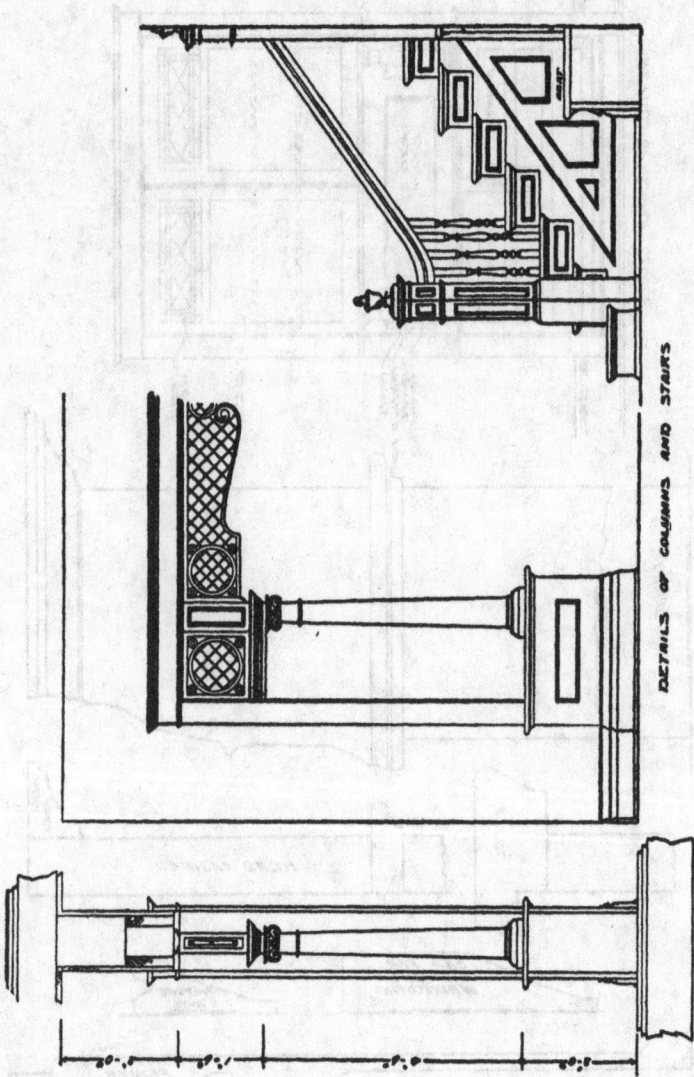

DETAILS OF COLUMNS AND STAIRS

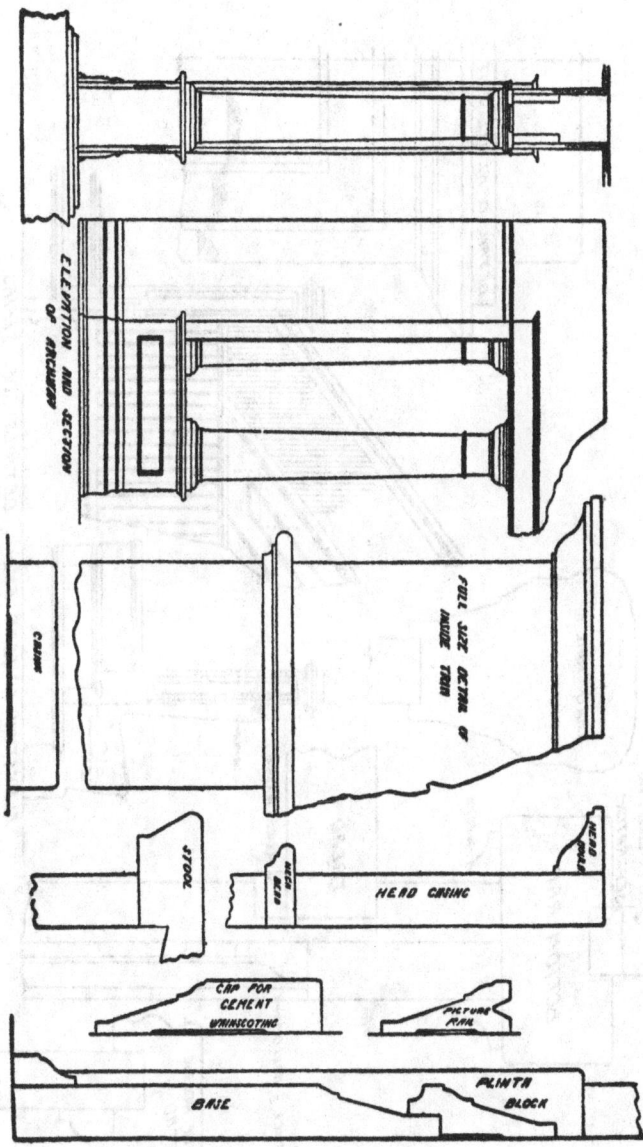

ELEVATION AND SECTION OF ARCHWAY

FULL SIZE DETAIL OF INSIDE TRIM

CASING

TOOLS

HEAD

HEAD CASING

HEAD CASING

CAP FOR CEMENT WAINSCOTING

PICTURE RAIL

BASE

PLINTH BLOCK

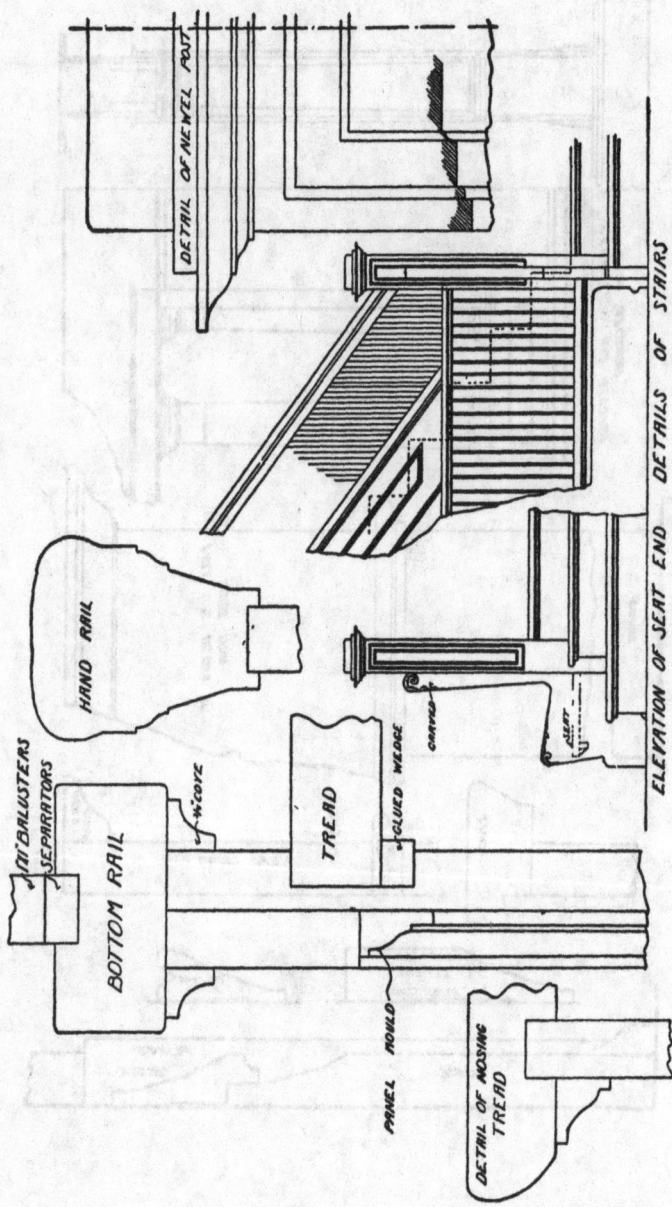

DETAIL OF NEWEL POST

ELEVATION OF SEAT END   DETAILS OF STAIRS

HAND RAIL

BALUSTERS
SEPARATORS

BOTTOM RAIL

TREAD

COVE

GLUED WEDGE

CHAMF

SEAT

PANEL MOULD

DETAIL OF NOSING TREAD

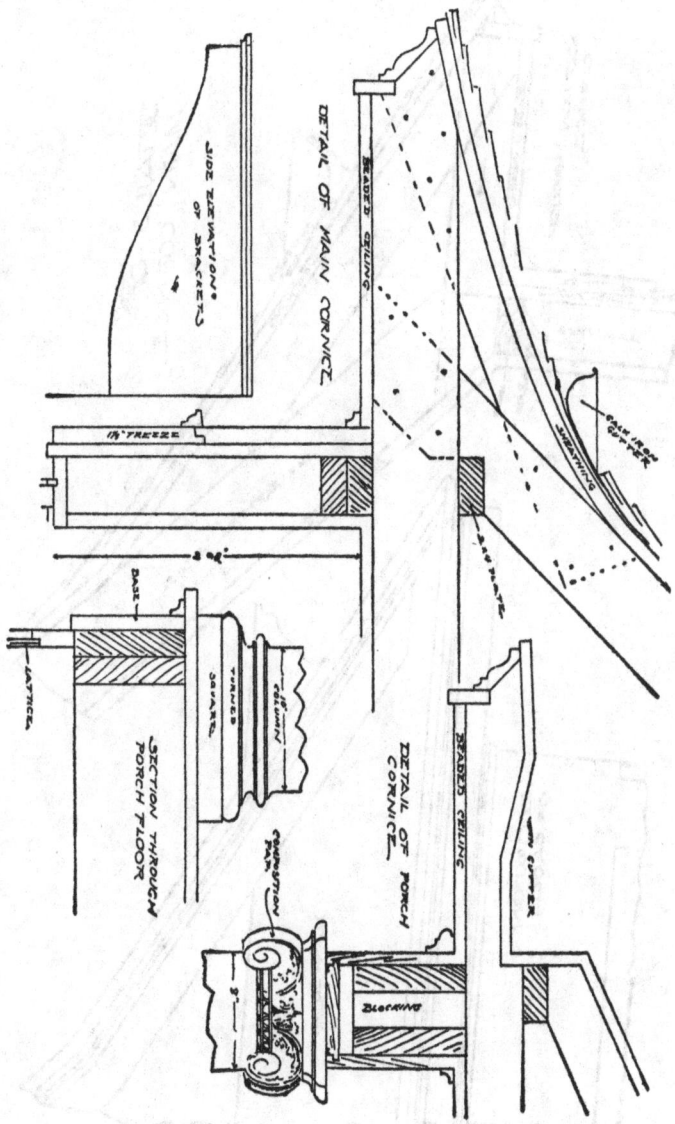

SIDE ELEVATION
OF BRACKETS

DETAIL OF MAIN CORNICE.

BEADED CEILING

GALV. IRON GUTTER

SHEATHING

1 1/2" FREEZE

RAFTER

BASE

4 0"

LATTICE

36" COLUMN

SECTION THROUGH
PORCH FLOOR

TURNED
SPIRAL

DETAIL OF PORCH
CORNICE.

BEADED CEILING

TIN GUTTER

COMPOSITION
CAP.

BLOCKING

8"

DETAILS OF
MAIN CORNICE.

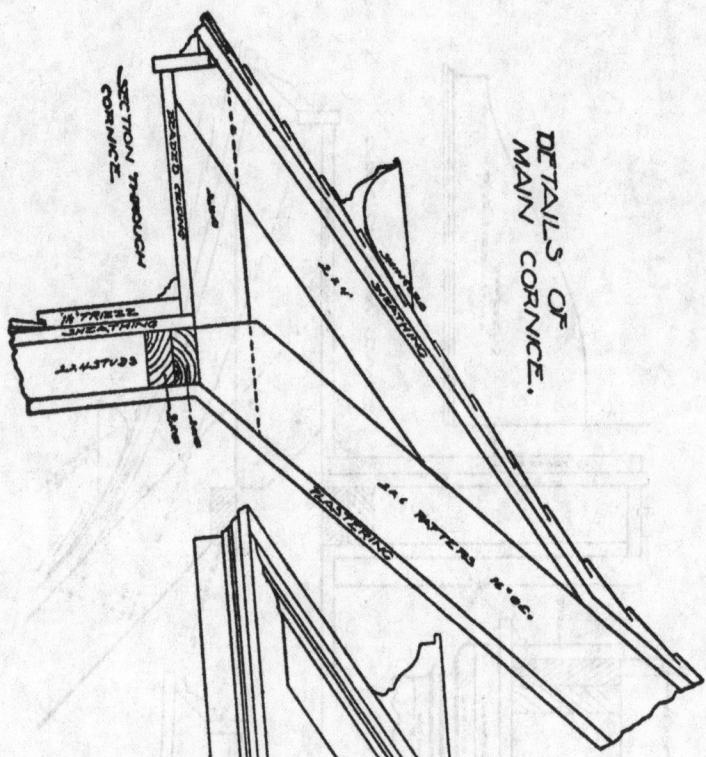

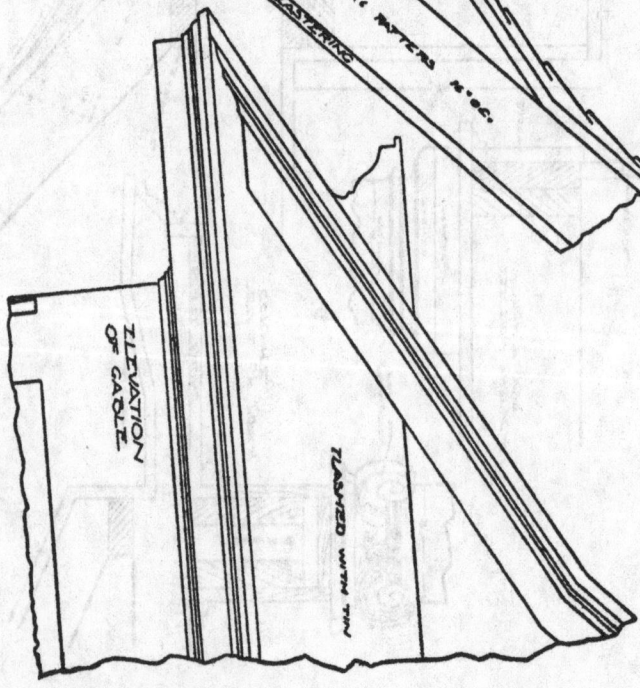

SECTION THROUGH
CORNICE.

BEADED CEILING

1/2 FRIEZE
SHEATHING

2x4 STUDS

SHINGLES
SHEATHING

2x4

2x4 RAFTERS 16" OC

RAFTERS

ELEVATION
OF GABLE

FLASHED WITH TIN

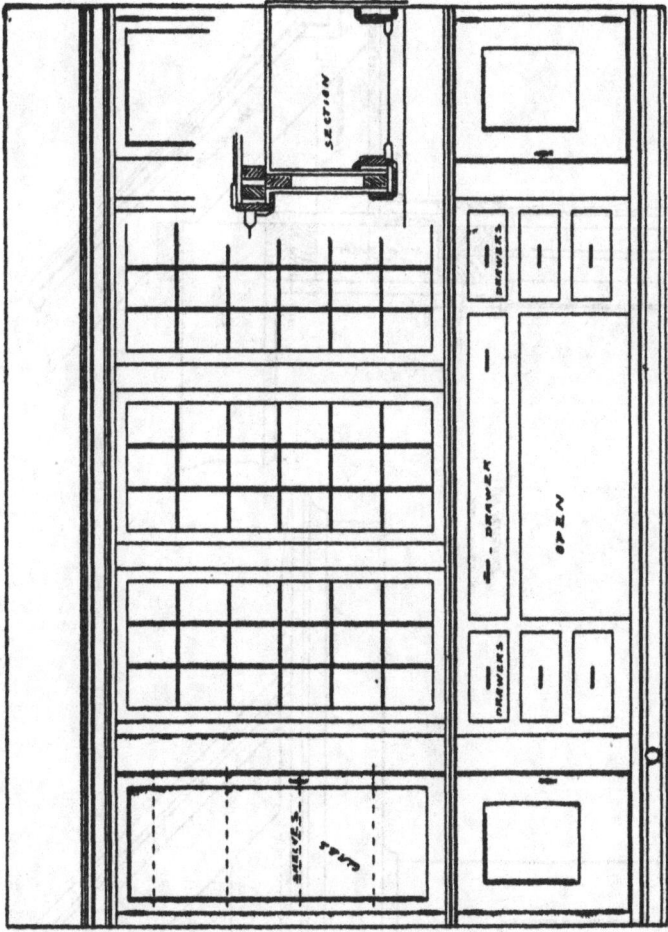

ELEVATION OF DRAWER CASE

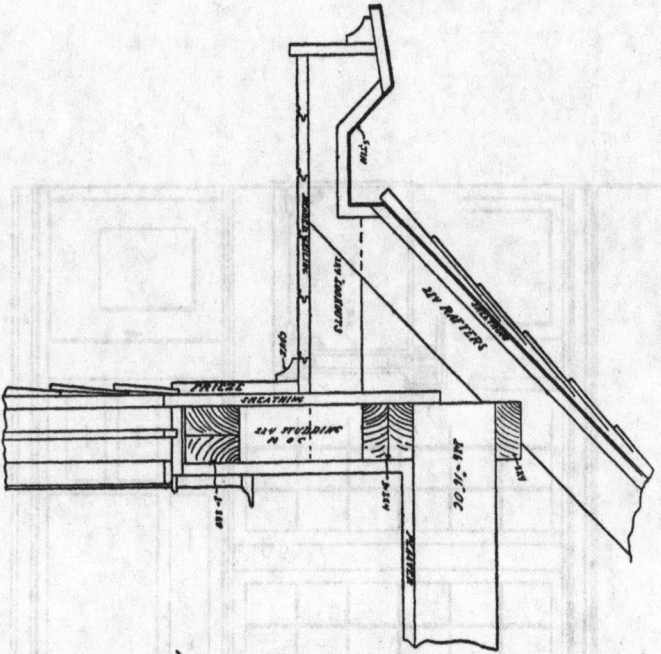

DETAILS OF MAIN CORNICE

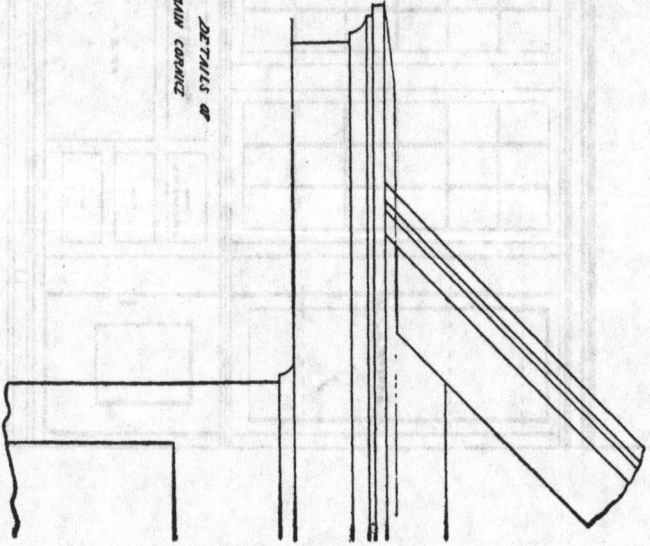

2x4 RAFTERS

SHEATHING

2x4 STUDDING 16" O.C.

FRIEZE

SHEATHING

SIDING

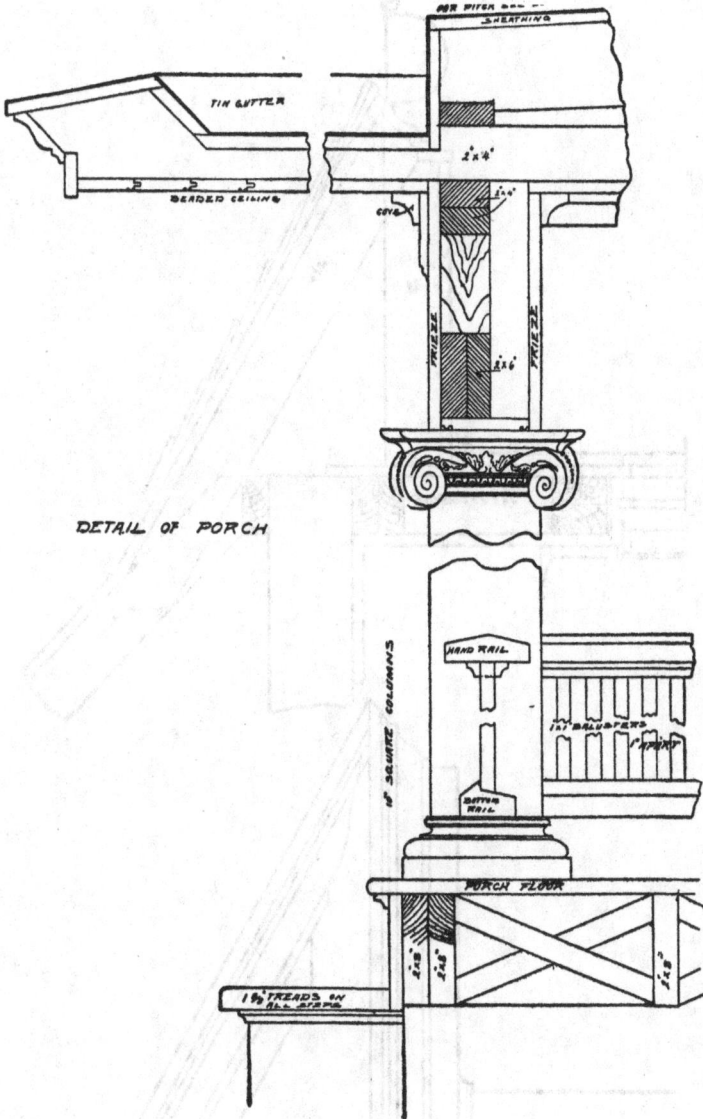

DETAIL OF PORCH

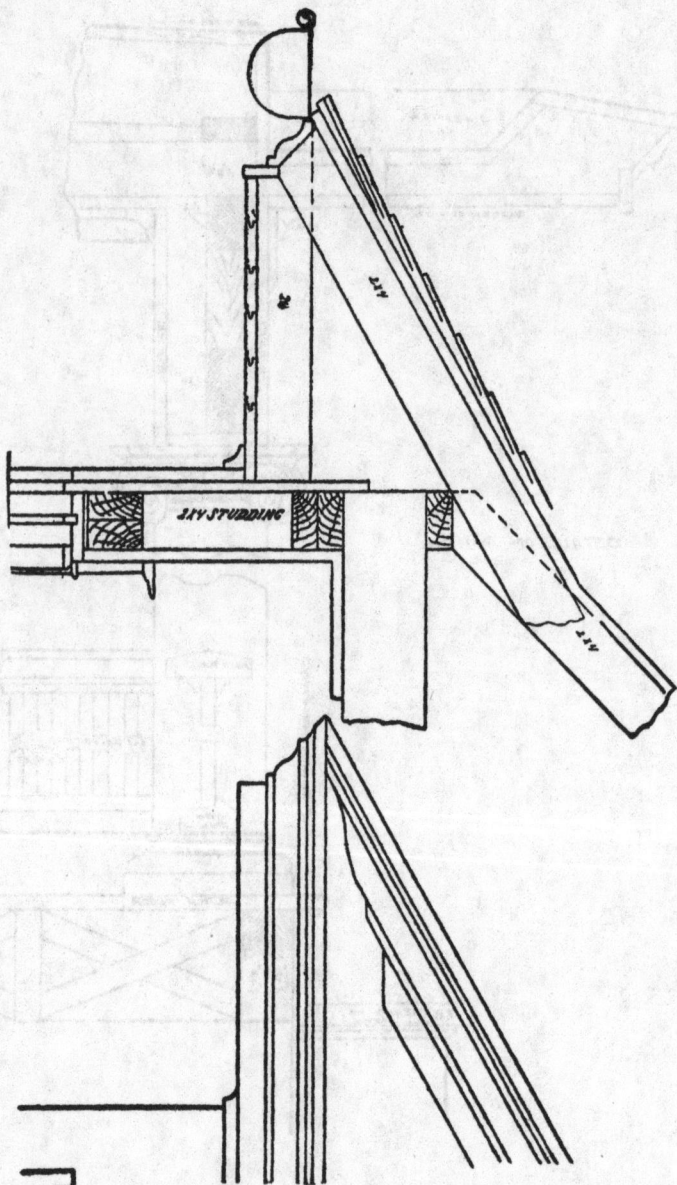

JIB STUDDING

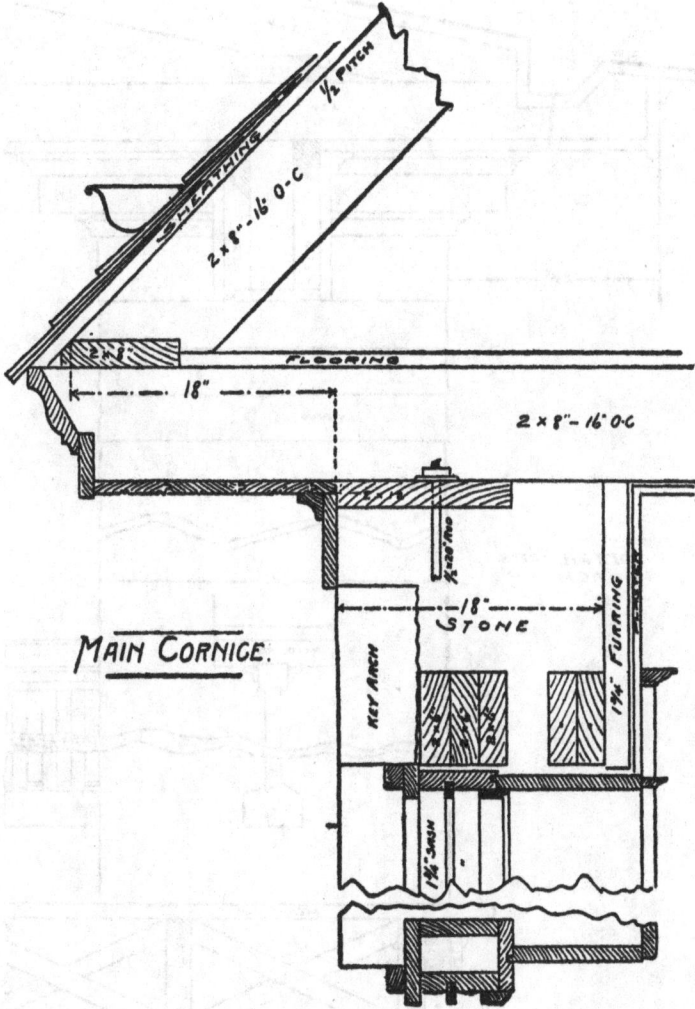

MAIN CORNICE.

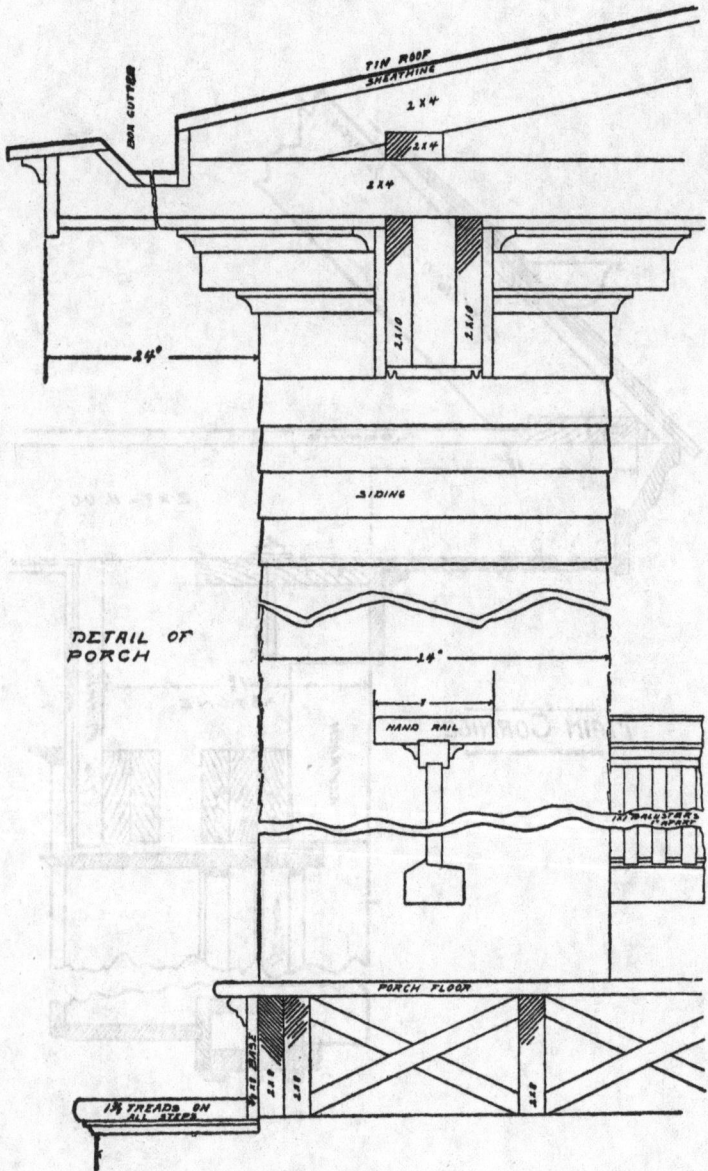

TIN ROOF

SHEATHING

1 X 4

2 X 4

BOX GUTTER

2 X 4

24°

1 X 10

1 X 10

SIDING

DETAIL OF
PORCH

14°

HAND RAIL

1 X 1 BALUSTERS
4" APART

PORCH FLOOR

1¼ TREADS ON
ALL STEPS

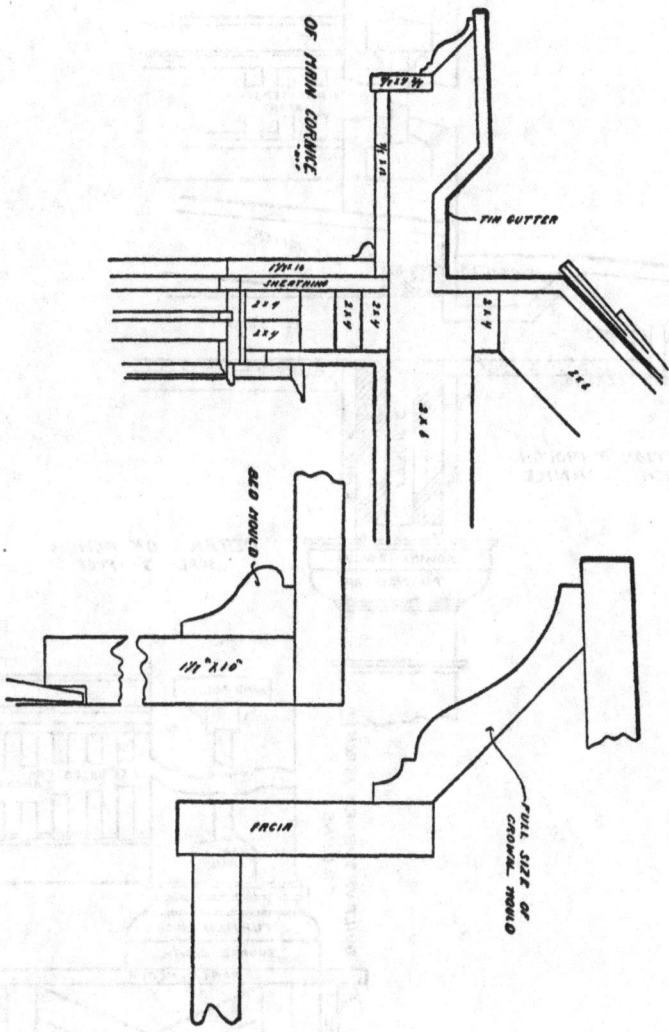

DETAIL OF MAIN CORNICE

4"x3/4"

3/4"x10

TIN GUTTER

1/4"x10 SHEATHING

2"x4"

4"x4"

3"x4" 3"x4"

4"x4"

3"x8"

3"x3"

BED MOULD

1 1/2"x10"

FACIA

FULL SIZE OF CROWN MOULD

1x1 BALUSTERS
1 APART

2 X 4 RAFTERS

TIN GUTTER

2 X 4 JOIST

LOOK OUTS
16" O C

BEADED CEILING

BEADED CEILING

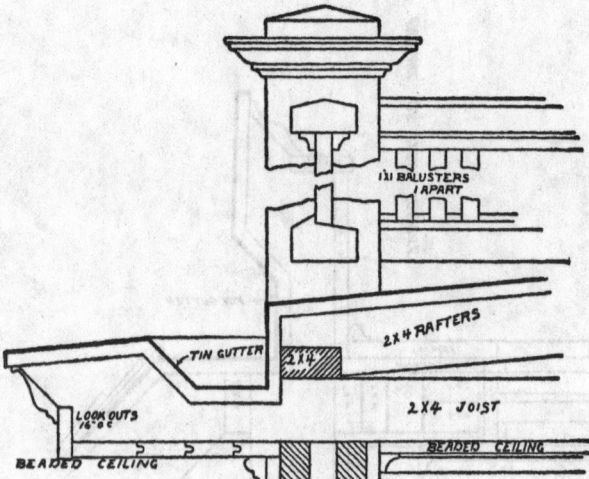

SECTION THROUGH
PORCH CORNICE

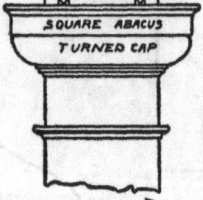

2x10   FURRING   2x10

7/8   7/8 FRIEZE

DETAILS OF PORCH
SCALE 3"=1 FOOT

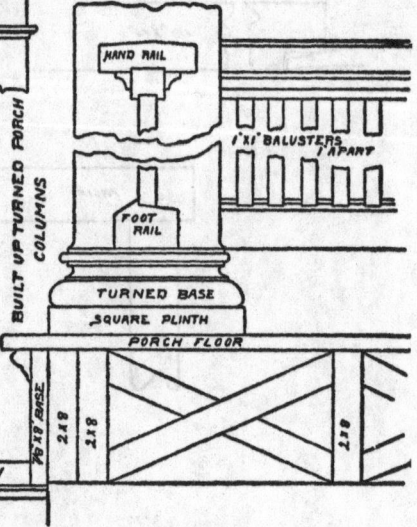

SQUARE ABACUS
TURNED CAP

HAND RAIL

1"x1" BALUSTERS
1 APART

BUILT UP TURNED PORCH
COLUMNS

FOOT
RAIL

TURNED BASE

SQUARE PLINTH

PORCH FLOOR

7/8 x1" BASE

2x8

2x8

2x8

1 3/8 TREADS ON
ALL STEPS

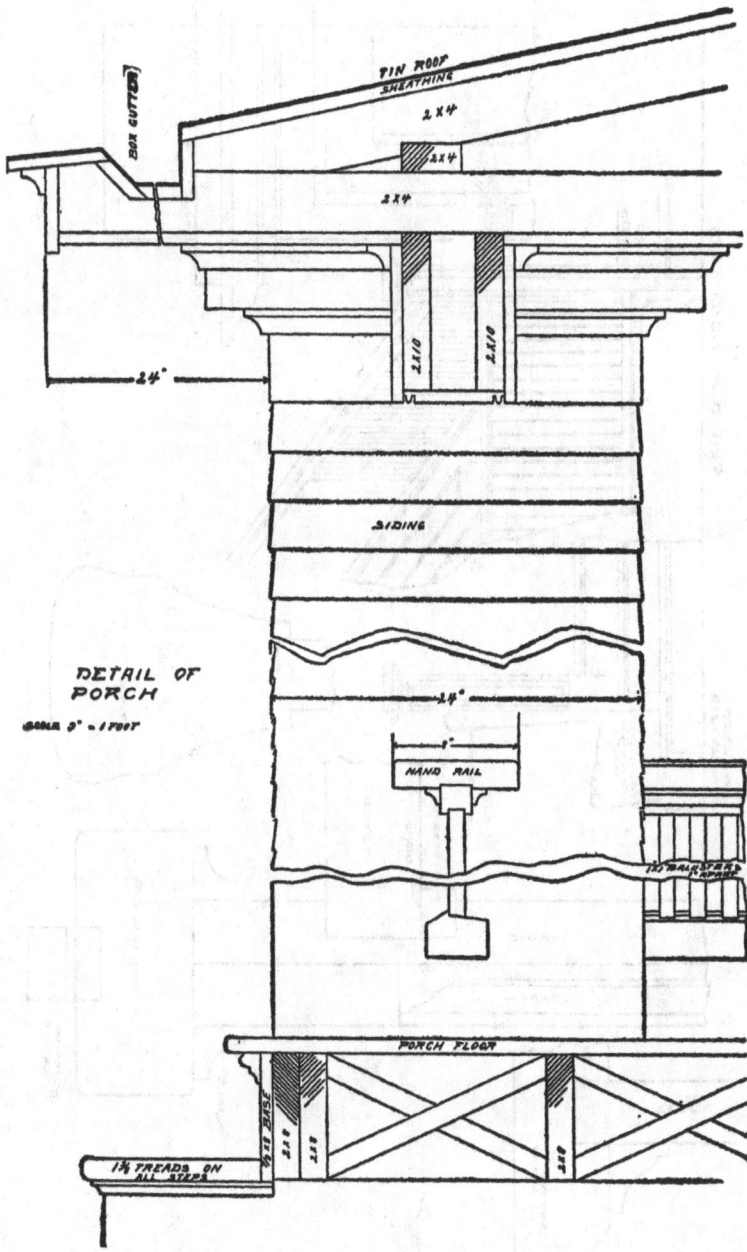

DETAIL OF
PORCH
SCALE 3" = 1 FOOT

TIN ROOF
SHEATHING
2 X 4
2 X 4
2 X 4

BOX GUTTER

24"

2 X 10
2 X 10

SIDING

24"

5"

HAND RAIL

1 X 1 BALUSTERS
3" APART

PORCH FLOOR

4 X 4 BASE

4 X 4

4 X 4

1 1/2 TREADS ON
ALL STEPS

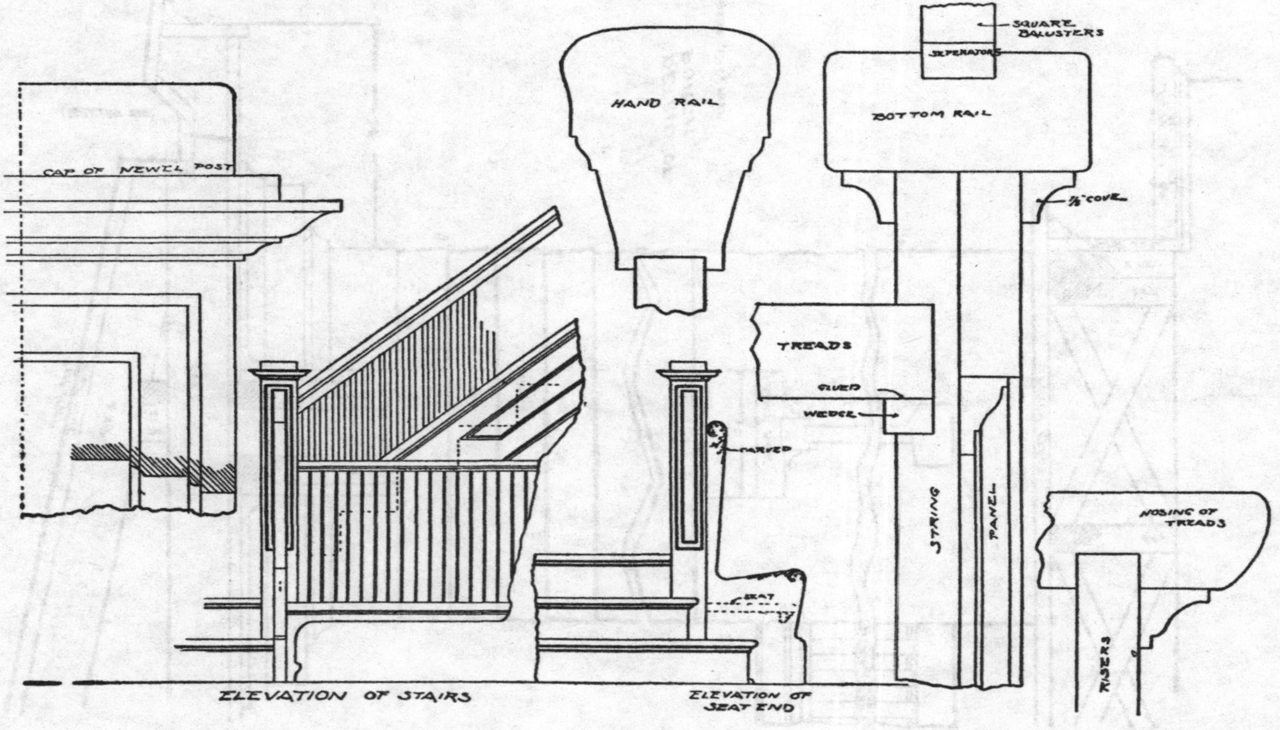
CAP OF NEWEL POST
HAND RAIL
SQUARE BALUSTERS
SEPERATORS
BOTTOM RAIL
⅞" COVE
TREADS
GLUED
WEDGE
CARVED
SEAT
STRING
PANEL
NOSING OF TREADS
RISERS
ELEVATION OF STAIRS
ELEVATION OF SEAT END

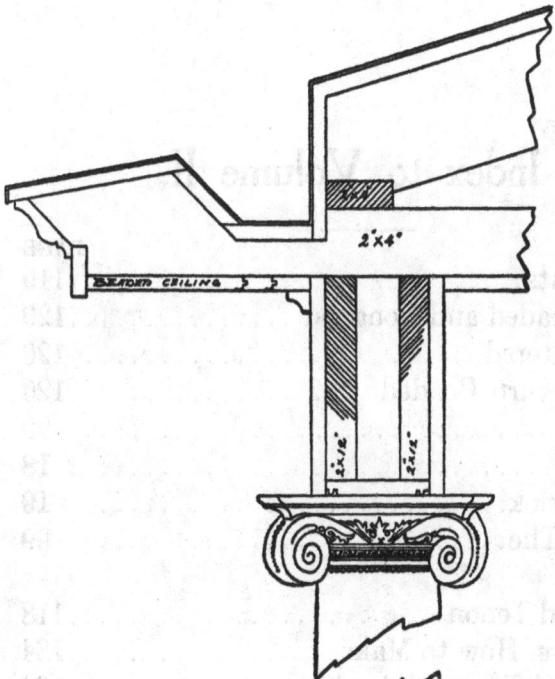

2'x4'

BEADED CEILING

2'x4' 2'x4'

DETAILS OF PORCH
SCALE ¼" = 1 FOOT

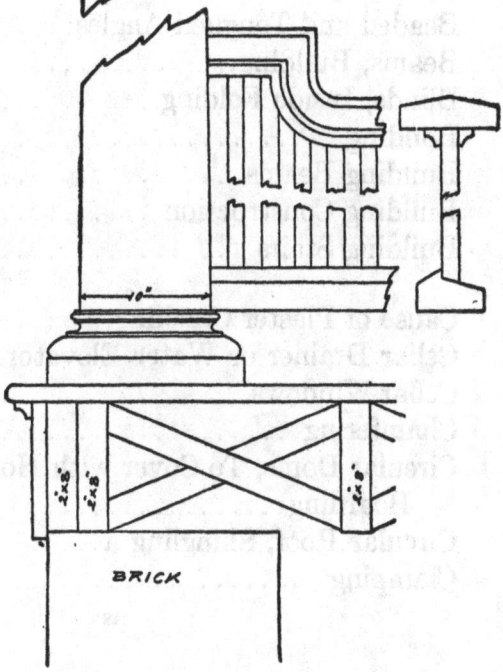

1'0"

BRICK

2'x8' 2'x8' 2'x8'

Index to Volume II

Practical House Plans

WE ILLUSTRATE IN THIS BOOK the perspective view and floor plans of 50 low and medium-priced houses. In the preparation of this work great care has been exercised in the selection of original, practical and attractive house designs, such as seventy-five to ninety per cent of the people to-day wish to build. In drawing these plans special effort has been made to provide for the MOST ECONOMICAL CONSTRUCTION, thereby giving the home builder and contractor the benefit of the saving of many dollars; for in no case have we put any useless expense upon the building simply to carry out some pet idea. Every plan illustrated will show, by the complete working plans and specifications, that we give you designs that will work out to the best advantage and will give you the most for your money; besides every bit of space has been utilized to the best advantage.

$50.00 PLANS FOR ONLY $5.00 This department has for its foundation the best equipped architectural establishment ever maintained for the purpose of furnishing the public with complete working plans and specifications at the remarkably low price of only $5.00 per set. Every plan we illustrate has been designed by a licensed architect, who stands at the head of his profession in this particular class of work and has made a specialty of low and medium-priced houses. The price usually charged for this work is from $50.00 to $75.00.

WHAT WE GIVE YOU The first question you will ask is, "What do we get in these complete working plans and specifications? Of what do they

257

consist? Are they the cheap printed plans on tissue paper without details or specifications?" We do not blame you for wishing to know what you will get for your money.

BLUE PRINTED WORKING PLANS
The plans we send out are the regular blue printed plans, drawn one-quarter inch scale to the foot, showing all the elevations, floor plans and necessary interior details. All of our plans are printed by electricity on an electric circular blue-printing machine, and we use the very best grade of electric blue-printing paper; every line and figure showing perfect and distinct.

FOUNDATION AND CELLAR PLANS
This sheet shows the shape and size of all walls, piers, footings, posts, etc., and of what materials they are constructed; shows the location of all windows, doors, chimneys, ash-pits, partitions, and the like. The different wall sections are given, showing their construction and measurements from all the different points.

FLOOR PLANS
These plans show the shape and size of all rooms, halls and closets; the location and size of all doors and windows; the position of all plumbing fixtures, gas lights, registers, pantry work, etc., and all the measurements that are necessary are given.

ELEVATIONS
A front, right, left and rear elevation are furnished with all the plans. These drawings are complete and accurate in every respect. They show the shape, size and location of all doors, windows, porches, cornices, towers, bays, and the like; in fact, give you an exact scale picture of the house as it should be at completion. Full wall sections are given showing the construction from foundation to roof, the height of stories between the joists, height of plates, pitch of roof, etc.

ROOF PLAN This plan is furnished where the roof construction is at all intricate. It shows the location of all hips, valleys, ridges, decks, etc. All the above drawings are made to scale one-quarter inch to the foot.

DETAILS All necessary details of the interior work, such as door and window casings and trim, base, stools, picture moulding, doors, newel posts, balusters, rails, etc., accompany each set of plans. Part is shown in full size, while some of the larger work, such as stair construction, is drawn to a scale of one and one-half inch to the foot. These blue prints are substantially and artistically bound in cloth and heavy water-proof paper, making a handsome and durable covering and protection for the plans.

SPECIFICATIONS The specifications are typewritten on Lakeside Bond Linen paper and are bound in the same artistic manner as the plans, the same cloth and water-proof paper being used. They consist of from about sixteen to twenty pages of closely typewritten matter, giving full instructions for carrying out the work. All directions necessary are given in the clearest and most explicit manner, so that there can be no possibility of a misunderstanding.

BASIS OF CONTRACT The working plans and specifications we furnish can be made the basis of contract between the home builder and the contractor. This will prevent mistakes, which cost money, and they will prevent disputes which are unforeseen and never settled satisfactorily to both parties. When no plans are used the contractor is often obliged to do some work he did not figure on, and the home builder often does not get as much for his money as he expected, simply because there was no basis on which to work and upon which to base the contract.

NO MISUNDERSTANDING CAN ARISE when a set of our
plans and specifica-
tions is before the contractor and the home builder, showing the interior and exterior construction of the house as
agreed upon in the contract.   Many advantages may be
claimed for the complete plans and specifications.   They
are time savers and, therefore, money savers.   Workmen
will not have to wait for instructions when a set of plans is
left on the job.   They will prevent mistakes in cutting lumber, in placing door and window frames, and in many other
places where the contractor is not on the work and the men
have received only partial or indefinite instructions.   They
also give instructions for the working of all material to the
best advantage.

FREE PLANS FOR FIRE INSURANCE ADJUSTMENT You
take
every precaution to have your house covered by insurance;
but do you make any provision for the adjustment of the
loss, should you have a fire?   There is not one man in ten
thousand who will provide for this embarrassing situation.
You can call to mind instances in your own locality where
settlements have been delayed because the insurance companies wanted some proof which could not be furnished.
They demand proof of loss before paying insurance money,
and they are entitled to it.   We have provided for this and
have inaugurated the following plan, which cannot but
meet with favor by whoever builds a house from our plans.

IMMEDIATELY UPON RECEIPT OF INFORMATION from
you
that your house has been destroyed by fire, either totally
or partially, we will forward you, free of cost, a duplicate
set of plans and specifications, and in addition we will furnish an affidavit giving the number of the design and the
date when furnished, to be used for the adjustment of the
insurance.

WITHOUT ONE CENT OF COST TO YOU and without
one particle
of trouble. We keep a record of the number of the house
design and the date it was furnished, so that, in time of
loss, all it will be necessary for you to do is to drop us a line
and we will furnish the only reliable method of getting a
speedy and satisfactory adjustment. This may be the means
of saving you hundreds of dollars, besides much time and
worry.

OUR LIBERAL PRICES Many have marveled at our
ability to furnish such excellent
and complete working plans and specifications at such low
prices. We do not wonder at this, because we charge but
$5.00 for a more complete set of working plans and specifi-
cations than you would receive if ordered in the ordinary
manner, and when drawn especially for you, at a cost of
from fifty to seventy-five dollars. On account of our large
business and unusual equipment, and owing to the fact
that WE DIVIDE THE COST of these plans among so many,
it is possible for us to sell them at these low prices. The
margin of profit is very close, but it enables us to sell
thousands of sets of plans, which save many times their
cost to both the owner and the contractor in erecting even
the smallest dwelling.

OUR GUARANTEE Perhaps there are many who feel that
they are running some risk in ordering
plans at a distance. We wish to assure our customers that
there is no risk whatever. If, upon receipt of these plans,
you do not find them exactly as represented, if you do not
find them complete and accurate in every respect, if you
do not find them as well prepared as those furnished by any
architect in the country, or any that you have ever seen,
we will refund your money upon the return of the plans
from you in perfect condition. All of our plans are prepared
by architects standing at the head of their profession, and

the standard of their work is the very highest   We could
not afford to make this guarantee if we were not positive
that we were furnishing the best plans put out in this
country, even though our price is not more than one-seventh
to one-tenth of the price usually charged.

BILL OF MATERIAL   We do not furnish a bill of material.
We state this here particularly, as
some people have an idea that a bill of material should accompany each set of plans and specifications. In the first place,
our plans are gotten up in a very comprehensive manner,
so that any carpenter can easily take off the bill of materia'
without any difficulty. We realize that there are hardly two
sections of the country where exactly the same kinds of
materials are used, and, moreover, a bill which we might
furnish would not be applicable in all sections of the
country. We furnish plans and specifications for houses
which are built as far north as the Hudson Bay and as far
south as the Gulf of Mexico. They are built upon the
Atlantic and Pacific Coasts, and you can also find them in
Australia and South Africa. Each country and section of
a country has its peculiarities as to sizes and qualities;
therefore, it would be useless for us to make a list that
would not be universal. Our houses, when completed, may
look the same whether they are built in Canada or Florida,
but the same materials will not be used, for the reason that
the customs of the people and the climatic conditions will
dictate the kind and amount of materials to be used in
their construction.

ESTIMATED COST   It is impossible for anyone to estimate
the cost of a building and have the
figures hold good in all sections of the country. We do not
claim to be able to do it. The estimated cost of the houses
we illustrate is based on the most favorable conditions in
all respects and includes everything but the plumbing and

heating. We are not familiar with your local conditions, and, should we claim to know the exact cost of a building in your locality, a child would know that our statement was false. We leave this matter in the hands of the reliable contractors, for they, and they alone, know your local conditions.

WE WISH TO BE FRANK WITH YOU and therefore make no statement that we cannot substantiate in every respect. If a plan in this book pleases you; if the arrangement of the rooms is satisfactory, and if the exterior is pleasing and attractive, then we make this claim—that it can be built as cheaply as if any other architect designed it, and we believe cheaper.

WE HAVE STUDIED ECONOMY in construction, and our knowledge of all the material that goes into a house qualifies us to give you the best for your money. We give you a plan that pleases you, one that is attractive, and one where every foot of space is utilized at the least possible cost. Can any architect do more, even at seven to ten times the price we charge you for plans?

REVERSING PLANS We receive many requests from our patrons for plans exactly according to the designs illustrated, with the one exception of having them reversed or placed in the opposite direction. It is impossible for us to make this change and draw new plans, except at a cost of about eight times our regular price. We see no reason why our regular plans will not answer your purpose. Your carpenter can face the house exactly as you wish it, and the plans will work out as well facing in one direction as in another. We can, however, if you wish, and so instruct us, make you a reversed blue print and turnish it at our regular price; but in that case all the figures and letters will be reversed and, therefore, liable to cause as much confusion as if your carpenter reversed the plan himself while constructing the house.

WE WOULD ADVISE however, in all cases where the plan is to be reversed, and there is the least doubt about the contractor not being able to work from the plans as we have them, that two sets of blue prints be purchased, one regular and the other reversed, and in such cases we will furnish two sets of blue prints and one set of specifications for only fifty per cent added to the regular cost, making the $5.00 plan cost only $7.50.

Design No. 1083

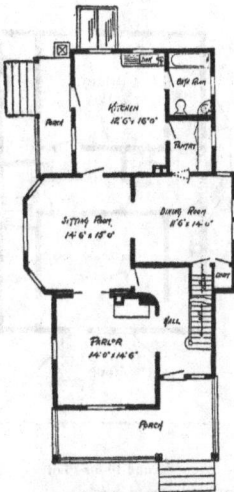

First Floor Plan

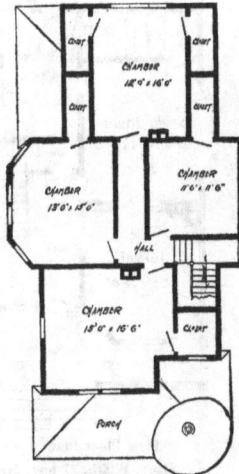

Second Floor Plan

Size: Width, 30 feet; length, 48 feet, exclusive of porch

Blue prints consist of cellar and foundation plan; first and second floor plans; front, rear, two side elevations; wall sections and all necessary interior details. Specifications consist of about fifteen pages of typewritten matter.

Full and complete working plans and specifications of this house will be furnished for $5.00. Cost of this house is from about $2,750.00 to about $3,000.00, according to the locality in which it is built.

Design No. 1163

PRICE
of Blue
Prints, to-
gether with
a complete
set of type-
written
specifica-
tions is

ONLY

$5.00

We mail
Plans and
Specifica-
tions the
same day
order is re-
ceived.

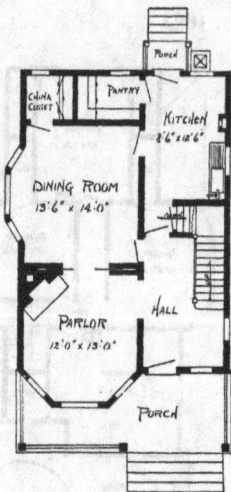

DINING ROOM
13'6" x 14'0"

KITCHEN
8'6" x 14'6"

CHINA CLOSET

PANTRY

PARLOR
12'0" x 13'0"

HALL

PORCH

PORCH

CLOSET

CHAMBER
11'0" x 13'0"

CLOSET

BATH ROOM
6'0" x 8'0"

HALL

CLOSET

CHAMBER
11'6" x 13'0"

CLOSET

ROOF

First Floor Plan   **Second Floor Plan**

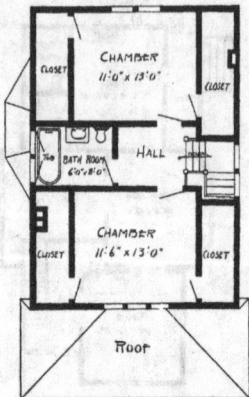

Size: Width, 22 feet; length, 34 feet, exclusive of porches

Blue prints consist of cellar and foundation plan; first and second floor plans;
front, rear, two side elevations; wall sections and all necessary interior details.
Specifications consist of about twenty
pages of typewritten matter.

Full and complete working plans and
specifications of this house will be furnished for $5.00. Cost of this house is
from about $1,050.00 to about $1,200.00,
according to the locality in which it is
built.

Design No. 1508

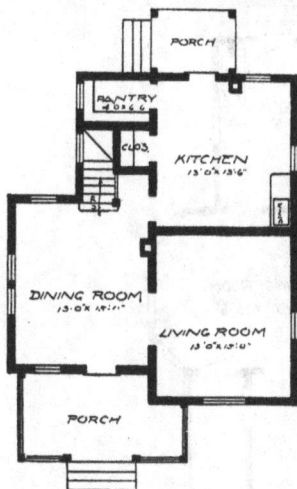

First Floor Plan

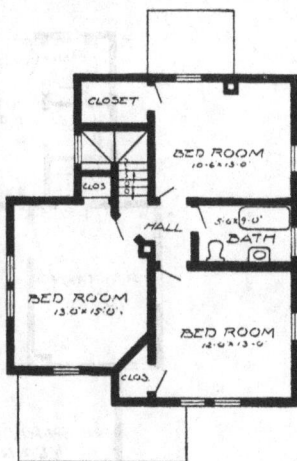

Second Floor Plan

PRICE

of Blue Prints, together with a complete set of typewritten specifications is

ONLY

$5.00

We mail Plans and Specifications the same day order is received.

Size: Width, 27 feet 6 inches; length, 30 feet, exclusive of porches

Blue prints consist of cellar and foundation plan; roof plan; first and second floor plans; front, rear, two side elevations, wall sections and all necessary interior details. Specifications consist of about twenty pages of typewritten matter.

Full and complete working plans and specifications of this house will be furnished for $5.00. Cost of this house is from about $1,150.00 to about $1,300.00, according to the locality in which it is built.

Design No. 1032

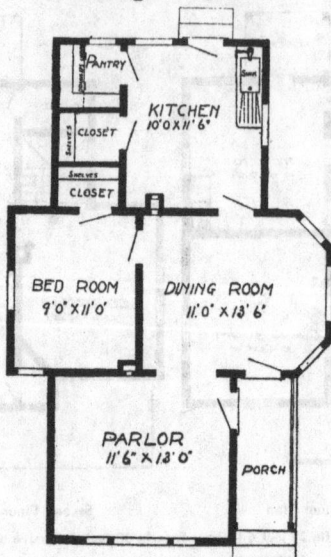

Floor Plan

Size: Width, 24 feet; length, 36 feet

Blue prints consist of foundation plan; floor plan; front, rear, two side elevations; wall sections and all necessary interior details. Specifications consist of about twelve pages of typewritten matter.

Full and complete working plans and specifications of this house will be furnished for $5.00. Cost of this house is from about $1,050.00 to about $1,200.00, according to the locality in which it is built.

Design No. 1157

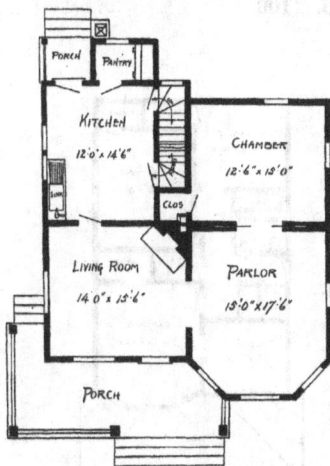

First Floor Plan

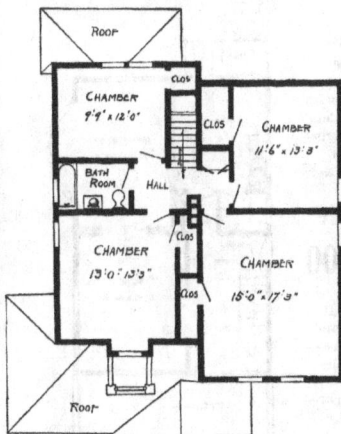

Second Floor Plan

Size: Width, 32 feet; length 34 feet 6 inches, exclusive of porch

Blue prints consist of cellar and foundation plan; first and second floor plans; front, rear, two side elevations; wall sections and all necessary interior details. Specifications consist of about twenty pages of typewritten matter.

Full and complete working plans and specifications of this house will be furnished for $5.00. Cost of this house is from about $1,750.00 to about $1,900.00, according to the locality in which it is built.

Design No. 1100

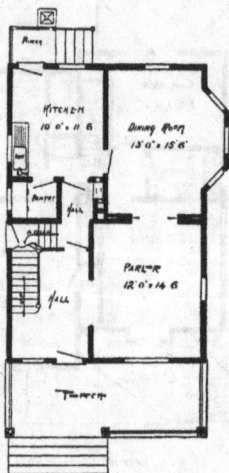

First Floor Plan   Second Floor Plan

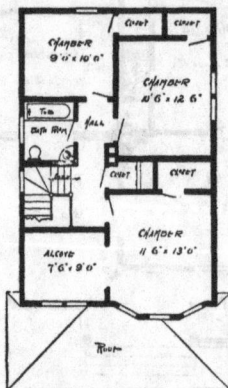

Size: Width, 24 feet; length, 32 feet, exclusive of porches

Blue prints consist of cellar and foundation plan; first and second floor plans, front, rear, two side elevations; wall sections and all necessary interior details. Specifications consist of about twenty pages of typewritten matter.

Full and complete working plans and specifications of this house will be furnished for $5.00. Cost of this house is from about $2,100.00 to about $2,300.00, according to the locality in which it is built.

Design No. 1038

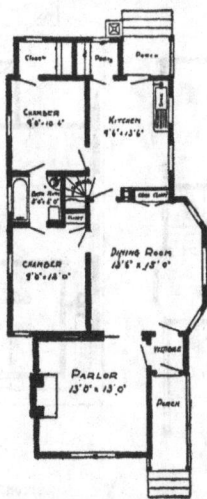

Floor Plan

Size: Width, 24 feet; length, 48 feet, exclusive of porches

Blue prints consist of cellar and foundation plan; floor plan; front, rear, two side elevations; wall sections and all necessary interior details. Specifications consist of about fifteen pages of typewritten matter.

Full and complete working plans and specifications of this house will be furnished for $5.00. Cost of this house is from about $1,650.00 to about $1,800.00, according to the locality in which it is built.

Design No. 1161

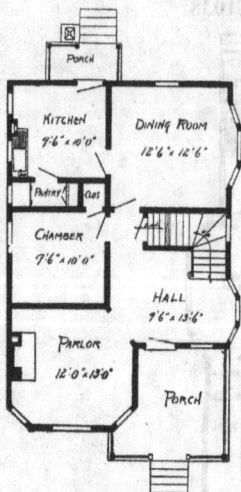

First Floor Plan

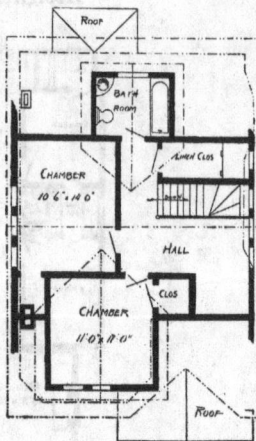

KITCHEN
7'6" x 10'0"

DINING ROOM
12'6" x 12'6"

PANTRY

CLOS

CHAMBER
7'6" x 10'0"

HALL
7'6" x 13'6"

PARLOR
12'0" x 15'0"

PORCH

PORCH

Second Floor Plan

ROOF

BATH ROOM

CHAMBER
10'6" x 14'0"

LINEN CLOS

HALL

CLOS

CHAMBER
11'0" x 11'0"

ROOF

First Floor Plan   **Second Floor Plan**

Size: Width, 25 feet; length, 36 feet, exclusive of porches

Blue prints consist of cellar and foundation plan; first and second floor plans; front, rear, two side elevations; wall sections and all necessary interior details. Specifications consist of about fifteen pages of typewritten matter.

Full and complete working plans and specifications of this house will be furnished for $5.00. Cost of this house is from about $1,150.00 to about $1,300.00, according to the locality in which it is built.

Design No. 1018

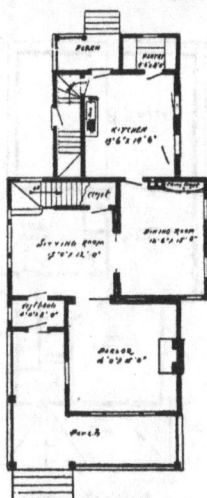

First Floor Plan

Second Floor Plan

Size: Width, 30 feet; length, 48 feet, exclusive of porch

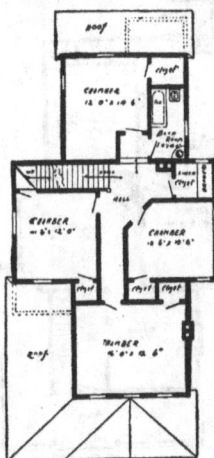

Blue prints consist of cellar and foundation plan; roof plan; first and second floor plans; front, rear, two side elevations; wall sections and all necessary interior details. Specifications consist of about twenty pages of typewritten matter.

Full and complete working plans and specifications of this house will be furnished for $5.00. Cost of this house is from about $2,250.00 to about $2,500.00, according to the locality in which it is built.

PRICE
of Blue Prints, together with a complete set of typewritten specifications is

ONLY

$5.00

We mail Plans and Specifications the same day order is received.

Design No. 1019

First Floor Plan   Second Floor Plan

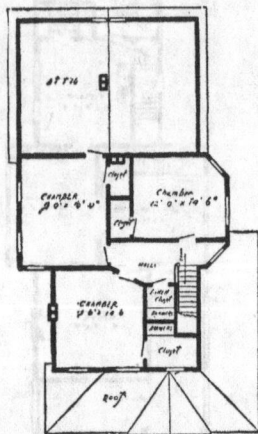

Size: Width, 32 feet; length, 52 feet, exclusive of porch

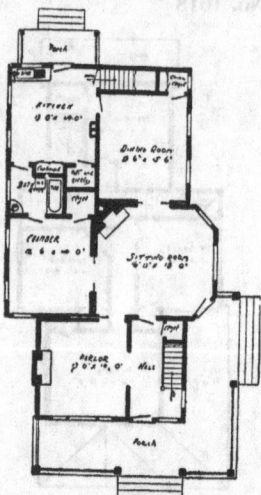

Blue prints consist of cellar and foundation plan; first and second floor plans; front, rear, two side elevations; wall sections and all necessary interior details. Specifications consist of about fifteen pages of typewritten matter.

Full and complete working plans and specifications of this house will be furnished for $5.00. Cost of this house is from about $2,750.00 to about $3,000.00, according to the locality in which it is built.

Design No. 1064

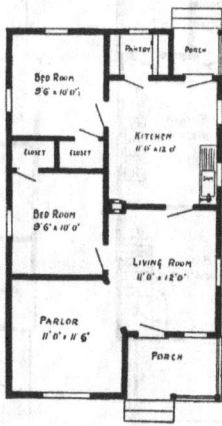

Floor Plan

BED ROOM
9'6" x 10'0"

CLOSET CLOSET

BED ROOM
9'6" x 10'0"

PARLOR
11'0" x 11'6"

PANTRY PORCH

KITCHEN
11'0" x 12'0"

LIVING ROOM
11'0" x 12'0"

PORCH

PRICE

of Blue Prints, together with a complete set of typewritten specifications is

ONLY

$5.00

We mail Plans and Specifications the same day order is received.

Size: Width, 22 feet; length, 36 feet

Blue prints consist of foundation plan; floor plan; front, rear, two side elevations; wall sections and all necessary interior details. Specifications consist of about fifteen pages of typewritten matter.

Full and complete working plans and specifications of this house will be furnished for $5.00. Cost of this house is from about $950.00 to about $1,100.00, according to the locality in which it is built.

Design No. 1013

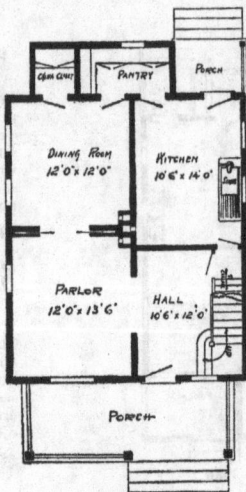

CEM CLOST — PANTRY — PORCH

DINING ROOM
12'0" x 12'0"

KITCHEN
10'6" x 14'0"

PARLOR
12'0" x 13'6"

HALL
10'6" x 12'0"

PORCH

First Floor Plan

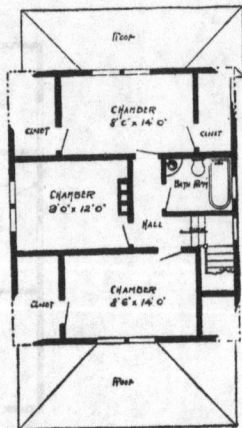

ROOF

CHAMBER
8'6" x 14'0"

CLOST — CLOST

CHAMBER
8'0" x 12'0"

BATH RM

HALL

CLOST

CHAMBER
8'6" x 14'0"

ROOF

Second Floor Plan

Size: Width, 24 feet; length, 32 feet 6 inches

Blue prints consist of cellar and foundation plan; first and second floor plans;
front, rear, two side elevations; wall sections and all necessary interior details.
Specifications consist of about twenty
pages of typewritten matter.

Full and complete working plans and
specifications of this house will be furnished for $5.00. Cost of this house is
from about $1,750.00 to about $2,000.00,
according to the locality in which it is
built.

Design No. 1149

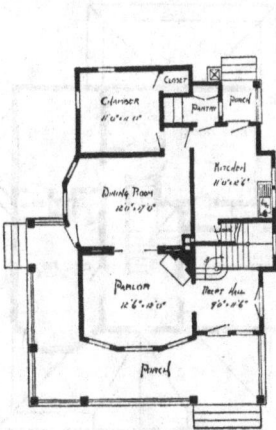

First Floor Plan

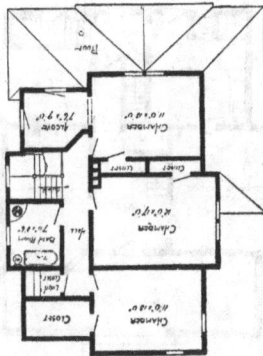

Second Floor Plan

Size: Width, 29 feet 6 inches; length, 38 feet, exclusive of porch

Blue prints consist of cellar and foundation plan; first and second floor plans; front, rear, two side elevations; wall sections and all necessary interior details. Specifications consist of about fifteen pages of typewritten matter.

Full and complete working plans and specifications of this house will be furnished for $5.00. Cost of this house is from about $2,500.00 to about $2,750.00, according to the locality in which it is built.

Design No. 1520

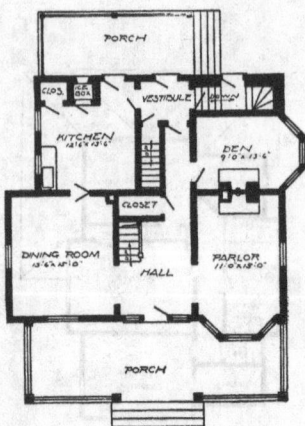

First Floor Plan

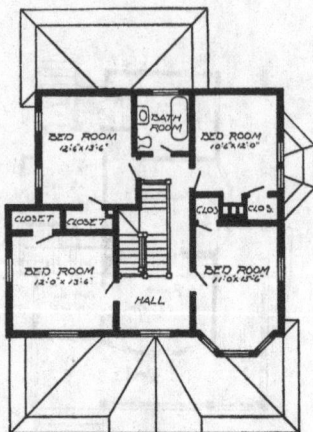

Second Floor Plan

Size: Width, 36 feet; length, 30 feet, exclusive of porches

Blue prints consist of cellar and foundation plan; roof plan; first and second floor plans, front, rear, two side elevations; wall sections and all necessary interior details. Specifications consist of about twenty pages of typewritten matter.

Full and complete working plans and specifications of this house will be furnished for $5.00. Cost of this house is from about $2,500.00 to about $2,750.00, according to the locality in which it is built.

Design No. 1096

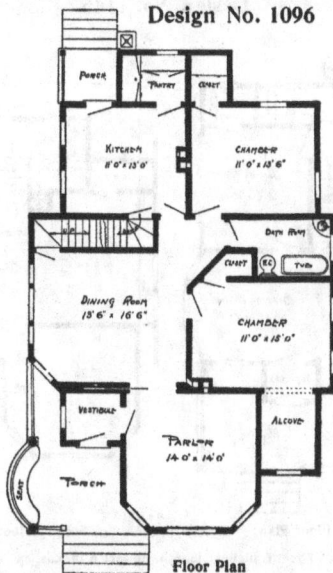

Floor Plan

Size: Width, 33 feet; length, 50 feet

Blue prints consist of foundation plan; floor plan; front, rear, two side elevations; wall sections and all necessary interior details. Specifications consist of about fifteen pages of typewritten matter.

Full and complete working plans and specifications of this house will be furnished for $5.00. Cost of this house is from about $1,850.00 to about $2,000.00, according to the locality in which it is built.

Design No. 1146

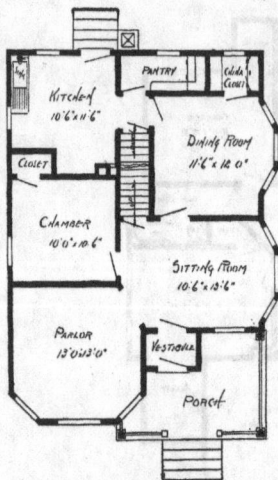

First Floor Plan

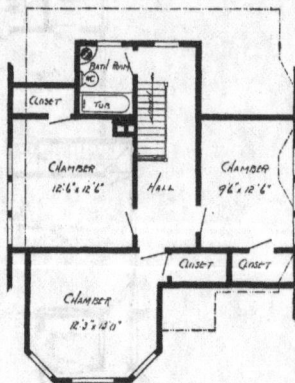

Second Floor Plan

Size: Width, 27 feet 6 inches; length, 36 feet 6 inches, exclusive of porch

Blue prints consist of cellar and foundation plan; roof plan; first and second
floor plans; front, rear, two side elevations,
wall sections and all necessary interior
details. Specifications consist of about
twenty pages of typewritten matter.

Full and complete working plans and
specifications of this house will be furnished for $5.00. Cost of this house is
from about $1,150.00 to about $1,300.00,
according to the locality in which it is
built.

Design No. 1154

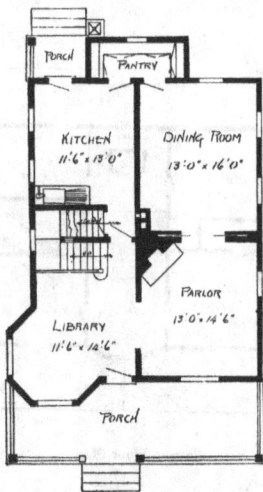

First Floor Plan

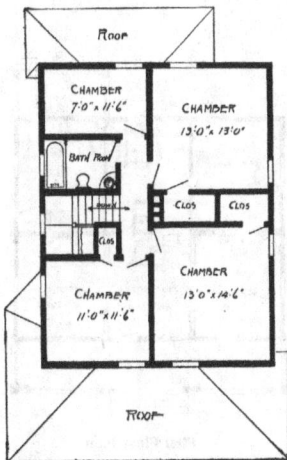

Second Floor Plan

Size: Width, 28 feet 6 inches; length, 37 feet, exclusive of porch

Blue prints consist of cellar and foundation plan; first and second floor plans; front, rear, two side elevations; wall sections and all necessary interior details. Specifications consist of about twenty pages of typewritten matter.

Full and complete working plans and specifications of this house will be furnished for $5.00. Cost of this house is from about $1,950.00 to about $2,100.00, according to the locality in which it is built.

Design No. 1148

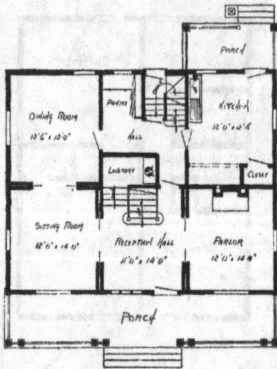

First Floor Plan

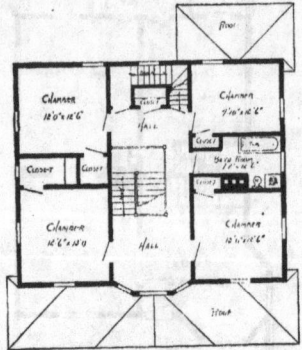

Second Floor Plan

Size: Width, 38 feet; length, 31 feet, exclusive of porches

Blue prints consist of cellar and foundation plan; first and second floor plans; front, rear, two side elevations; wall sections and all necessary interior details. Specifications consist of about twenty pages of typewritten matter.

Full and complete working plans and specifications of this house will be furnished for $5.00. Cost of this house is from about $2,500.00 to about $2,750.00, according to the locality in which it is built.

Design No. 1165

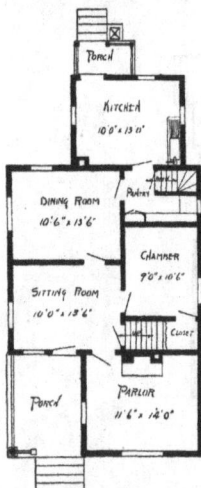

First Floor Plan

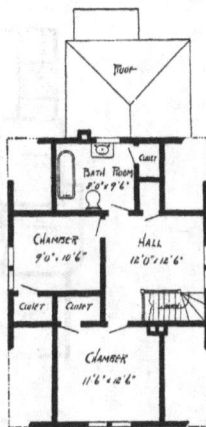

- Porch
- KITCHEN 10'0" x 13'0"
- Pantry
- DINING ROOM 10'6" x 13'6"
- CHAMBER 9'0" x 10'6"
- SITTING ROOM 10'0" x 13'6"
- Closet
- Porch
- PARLOR 11'6" x 13'0"

Second Floor Plan

- Roof
- BATH ROOM 8'0" x 9'6"
- Closet
- CHAMBER 9'0" x 10'6"
- HALL 12'0" x 13'6"
- Closet
- Closet
- CHAMBER 11'6" x 13'6"

Size: Width, 24 feet; length, 45 feet 6 inches, exclusive of porch

Blue prints consist of cellar and foundation plan; first and second floor plans; front, rear, two side elevations; wall sections and all necessary interior details. Specifications consist of about fifteen pages of typewritten matter.

Full and complete working plans and specifications of this house will be furnished for $5.00. Cost of this house is from about $1,050.00 to about $1,200.00, according to the locality in which it is built.

Design No. 1001

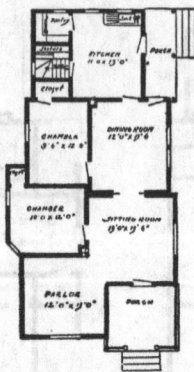

Floor Plan

Size: Width, 27 feet; length, 50 feet, exclusive of porch

Blue prints consist of cellar and foundation plan; first and second floor plans; front, rear, two side elevations, wall sections and all necessary interior details. Specifications consist of about twenty pages of typewritten matter.

Full and complete working plans and specifications of this house will be furnished for $5.00 Cost of this house is from about $1,550.00 to about $1,700.00, according to the locality in which it is built.

Design No. 1040

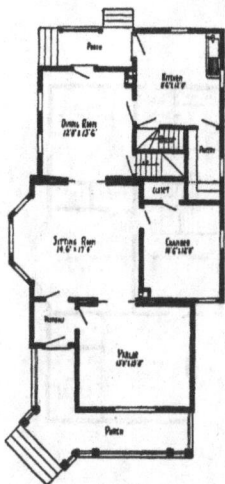

First Floor Plan   Second Floor Plan

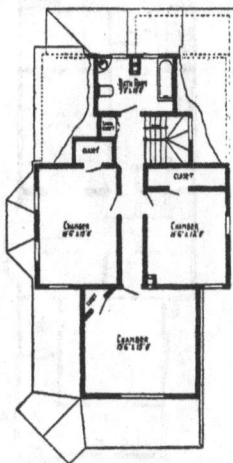

Size: Width, 29 feet; length, 49 feet, exclusive of porches

Blue prints consist of cellar and foundation plan; first and second floor plans; front, rear, two side elevations; wall sections and all necessary interior details. Specifications consist of about twenty pages of typewritten matter.

Full and complete working plans and specifications of this house will be furnished for $5.00. Cost of this house is from about $2,250.00 to about $2,500.00, according to the locality in which it is built.

Design No. 1011

First Floor Plan

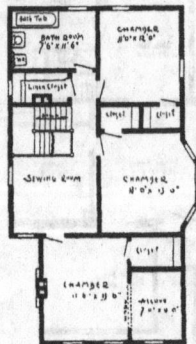

Second Floor Plan

PRICE

of Blue
Prints, to-
gether with
a complete
set of type-
written
specifica-
tions is

ONLY

$5.00

We mail
Plans and
Specifica-
tions the
same day
order is re-
ceived.

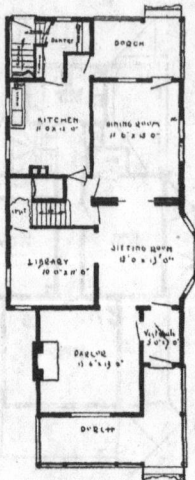

Size: Width, 26 feet; length, 52 feet, exclusive of porch

Blue prints consist of cellar and foundation plan; first and second floor plans; front, rear, two side elevations; wall sections and all necessary interior details. Specifications consist of about twenty pages of typewritten matter.

Full and complete working plans and specifications of this house will be furnished for $5.00. Cost of this house is from about $2,750.00 to about $3,000.00, according to the locality in which it is built.

Design No. 1024

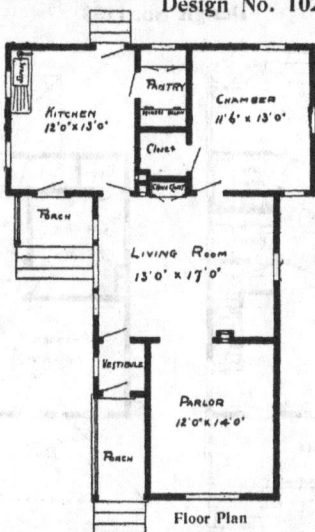

Floor Plan

Size: Width, 30 feet; length, 42 feet

Blue prints consist of foundation plan; or plan; front, rear, two side vations; wall sections and all neces- ry interior details. Specifications nsist of about twelve pages of type- itten matter.

Full and complete working plans and specifications of this house will be furnished for $5.00. Cost of this house is from about $1,350.00 to about $1,500.00, according to the locality in which it is built.

Design No. 1153

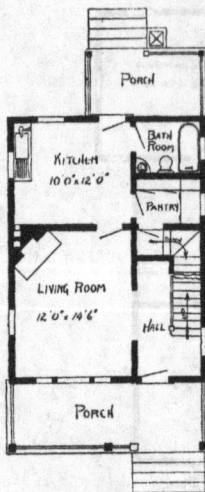

First Floor Plan

Porch

KITCHEN 10'0" x 12'0"

BATH ROOM

PANTRY

LIVING ROOM 12'0" x 14'6"

HALL

PORCH

Second Floor Plan

Roof

CLOS   CLOS

CHAMBER 9'0" x 10'0"

CHAMBER 9'6" x 10'0"

CHAMBER 11'0" x 14'0"

CLOS

Roof

Size: Width, 20 feet; length, 26 feet, exclusive of porches

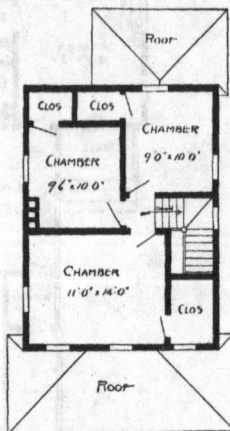

Blue prints consist of cellar and foundation plan; first and second floor plans; front, rear, two side elevations; wall sections and all necessary interior details. Specifications consist of about twenty pages of typewritten matter.

Full and complete working plans and specifications of this house will be furnished for $5.00. Cost of this house is from about $1,150.00 to about $1,300.00, according to the locality in which it is built.

Design No. 1147

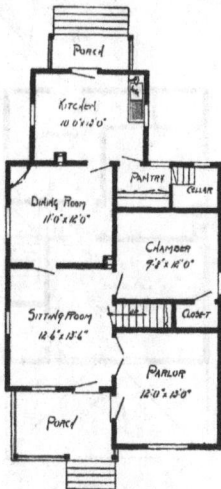

First Floor Plan

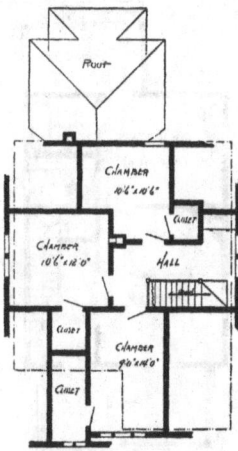

Second Floor Plan

Size: Width, 26 feet; length, 43 feet, exclusive of porches

PRICE of Blue Prints, together with a complete set of type-written specifications is

ONLY

$5.00

We mail Plans and Specifications the same day order is received.

Blue prints consist of cellar and foundation plan; first and second floor plans; front, rear, two side elevations; wall sections and all necessary interior details. Specifications consist of about fifteen pages of typewritten matter.

Full and complete working plans and specifications of this house will be furnished for $5.00. Cost of this house is from about $1,150.00 to about $1,300.00, according to the locality in which it is built.

Design No. 1151

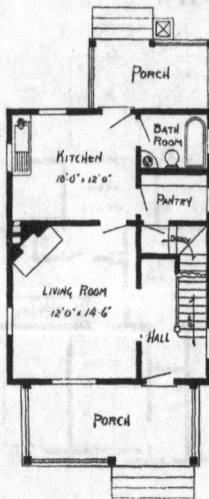

Porch

Kitchen
10'0" x 12'0"

Bath Room

Pantry

Living Room
12'0" x 14'6"

Hall

Porch

Roof

Closet   **Closet**

Chamber
9'0" x 10'0"

Chamber
9'6" x 10'0"

Chamber
11'0" x 12'0"

Closet

Roof

First Floor Plan   **Second Floor Plan**

Size: Width, 20 feet; length, 26 feet, exclusive of porches

Blue prints consist of cellar and foundation plan; first and second floor plans; front, rear, two side elevations; wall sections and all necessary interior details. Specifications consist of about twenty pages of typewritten matter.

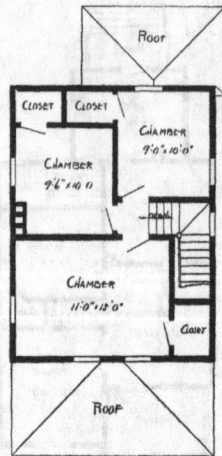

Full and complete working plans and specifications of this house will be furnished for $5.00. Cost of this house is from about $1,150.00 to about $1,300.00, according to the locality in which it is built.

Design No. 1509

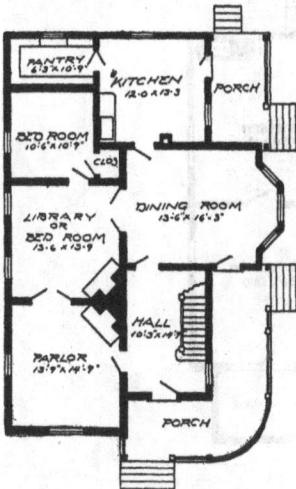

First Floor Plan labels:
PANTRY 6'3"x10'9" · KITCHEN 12'0"x13'3" · PORCH · BED ROOM 10'6"x10'9" · CLOS · DINING ROOM 13'6"x16'3" · LIBRARY OR BED ROOM 13'6"x13'9" · HALL 10'5"x14'9" · PARLOR 13'9"x14'9" · PORCH

Second Floor Plan labels:
BATH ROOM 8'x12' · BED ROOM 13'0"x13'0" · BED ROOM 11'6"x13'0" · CLOSET CLOSET · HALL · CLOSET · BED ROOM 15'0"x15'0" · CLOS

First Floor Plan   Second Floor Plan

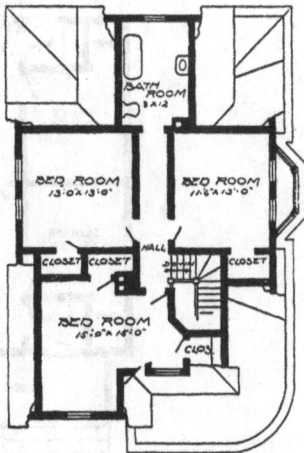

Size: Width, 30 feet; length, 45 feet, exclusive of porches

Blue prints consist of cellar and foundation plan; roof plan; first and second floor plans; front, rear, two side elevations; wall sections and all necessary interior details. Specifications consist of about twenty pages of typewritten matter.

Full and complete working plans and specifications of this house will be furnished for $5.00. Cost of this house is from about $1,400.00 to about $1,600.00, according to the locality in which it is built.

Design No. 1046

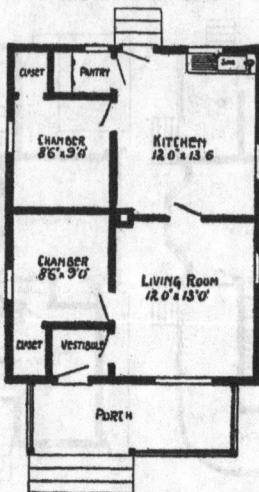

CLOSET | PANTRY

CHAMBER
8'6" x 9'0"

KITCHEN
12'0" x 13'6"

CHAMBER
8'6" x 9'0"

LIVING ROOM
12'0" x 13'0"

CLOSET | VESTIBULE

PORCH

Floor Plan

Size: Width, 22 feet; length, 28 feet, exclusive of porch

Blue prints consist of foundation plan; floor plan; front, rear, two side elevations; wall sections and all necessary interior deatils. Specifications consist of about fifteen pages of typewritten matter.

Full and complete working plans and specifications of this house will be furnished for $5.00. Cost of this house is from about $900.00 to about $1,000.00, according to the locality in which it is built.

Design No. 1021

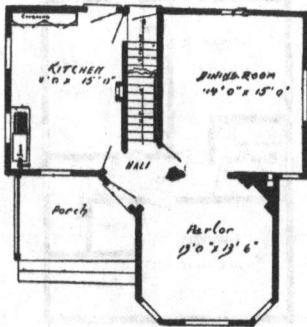

First Floor Plan

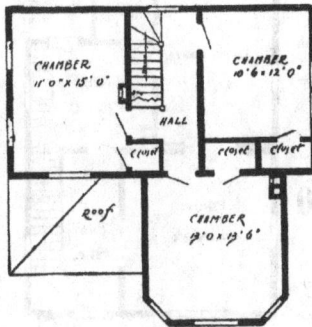

Second Floor Plan

Size: Width, 30 feet; length, 30 feet

Blue prints consist of cellar and foundation plan; first and second floor plans; front, rear, two side elevations; wall sections and all necessary interior details. Specifications consist of about twenty pages of typewritten matter.

Full and complete working plans and specifications of this house will be furnished for $5.00. Cost of this house is from about $1,350.00 to about $1,500.00, according to the locality in which it is built.

Design No. 1158

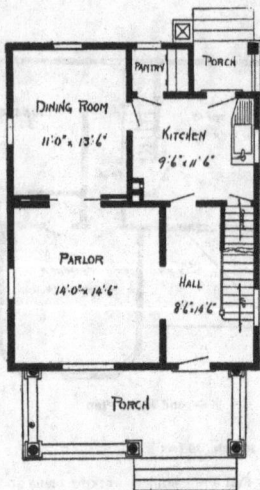

First Floor Plan

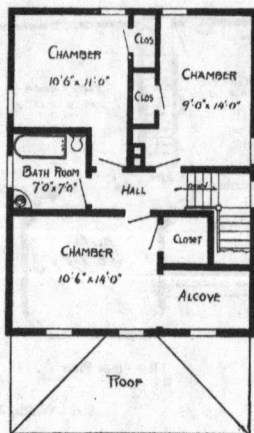

Second Floor Plan

Size: Width, 24 feet; length, 30 feet, exclusive of porch

Blue prints consist of cellar and foundation plan; first and second floor plans; front, rear, two side elevations; wall sections and all necessary interior details. Specifications consist of about fifteen pages of typewritten matter.

Full and complete working plans and specifications of this house will be furnished for $5.00. Cost of this house is from about $1,450.00 to about $1,600.00, according to the locality in which it is built.

Design No. 1121

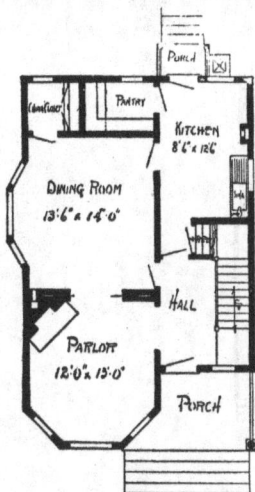

First Floor Plan

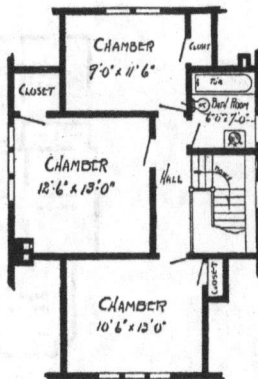

CLOSET | PANTRY | Porch | [sink]
KITCHEN
8'6" x 18'6"

DINING ROOM
13'6" x 15'0"

HALL

PARLOR
12'0" x 15'0"

PORCH

Second Floor Plan

CHAMBER
9'0" x 11'6"

CLOSET | CLOSET

CHAMBER
12'6" x 13'0"

Bath Room
5'0" x 7'6"

HALL

CHAMBER
10'6" x 13'0"

CLOSET

First Floor Plan   **Second Floor Plan**

Size: Width, 23 feet 6 inches; length, 34 feet

Blue prints consist of cellar and foundation plan; first and second floor plans; front, rear, two side elevations; wall sections and all necessary interior details. Specifications consist of about fifteen pages of typewritten matter.

Full and complete working plans and specifications of this house will be furnished for $5.00. Cost of this house is from about $1,150.00 to about $1,350.00, according to the locality in which it is built.

Design No. 1076

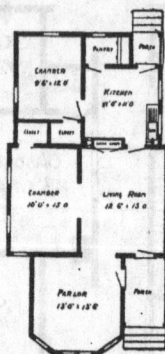

Floor Plan

Size: Width 24 feet; length, 46 feet

Blue prints consist of foundation plan; floor plan; front, rear, two side elevations; wall sections and all necessary interior details. Specifications consist of about fifteen pages of typewritten matter.

Full and complete working plans and specifications of this house will be furnished for $5.00. Cost of this house is from about $1,250.00 to about $1,400.00, according to the locality in which it is built.

Design No. 1008

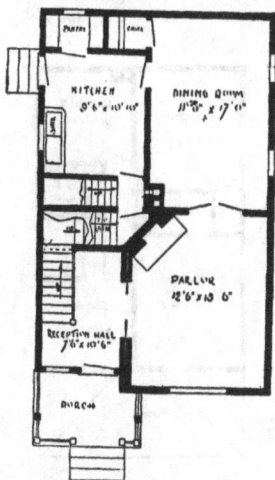

First Floor Plan

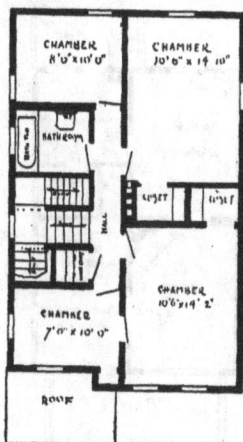

Second Floor Plan

Size: Width, 22 feet; length, 34 feet, exclusive of porch

Blue prints consist of cellar and foundation plan; first and second floor plans; front, rear, two side elevations; wall sections and all necessary interior details. Specifications consist of about twenty pages of typewritten matter.

Full and complete working plans and specifications of this house will be furnished for $5.00. Cost of this house is from about $1,850.00 to about $2,000.00, according to the locality in which it is built.

Design No. 1160

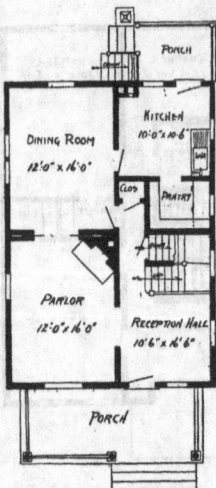

First Floor Plan

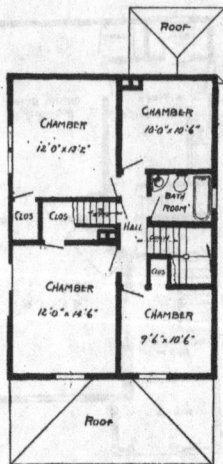

Second Floor Plan

Size: Width, 24 feet; length, 34 feet, exclusive of porches

Blue prints consist of cellar and foundation plan; first and second floor plans;
front, rear, two side elevations; wall
sections and all necessary interior details. Specifications consist of about
twenty pages of typewritten matter.

Full and complete working plans and
specifications of this house will be furnished for $5.00. Cost of this house is
from about $1,750.00 to about $1,900.00,
according to the locality in which it is
built.

Design No. 1002

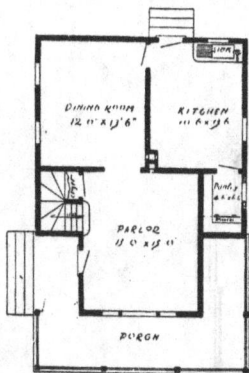

First Floor Plan

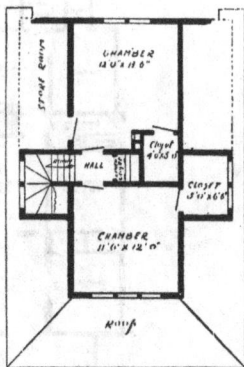

Second Floor Plan

Size: Width, 24 feet; length, 30 feet, exclusive of porches

Blue prints consist of cellar and foundation plan; first and second floor plans; front, rear, two side elevations; wall sections and all necessary interior details. Specifications consist of about 20 pages of typewritten matter.

Full and complete working plans and specifications of this house will be furnished for $5.00. Cost of this house is from about $1,250.00 to about $1,400.00, according to the locality in which it is built.

Design No. 1016

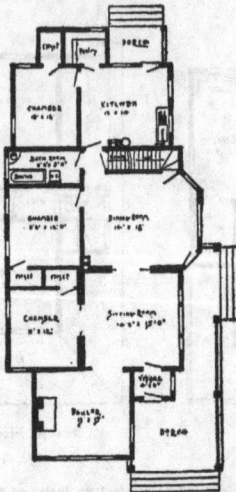

Floor Plan

Size: Width, 31 feet; length, 60 feet, exclusive of porches

Blue prints consist of cellar and foundation plan; floor plan; front, rear, two side elevations; wall sections, and all necessary interior details. Specifications consist of about twenty pages of typewritten matter.

Full and complete working plans and specifications of this house will be furnished for $5.00. Cost of this house is from about $1,850.00 to about $2,000.00, according to the locality in which it is built.

Design No. 1150

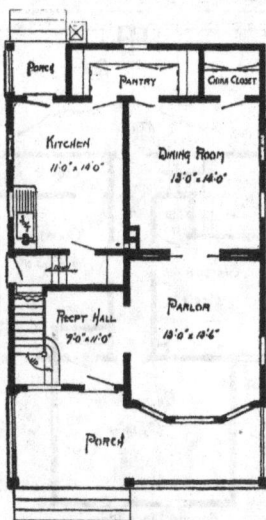

First Floor Plan

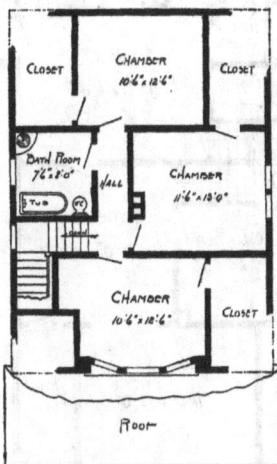

Second Floor Plan

Size: Width, 25 feet 6 inches; length, 36 feet, exclusive of porch

Blue prints consist of cellar and foundation plan; roof plan; first and second floor plans; front, rear, two side elevations; wall sections and all necessary interior details. Specifications consist of about twenty pages of typewritten matter

Full and complete working plans and specifications of this house will be furnished for $5.00. Cost of this house is from about $1,500.00 to about $1,650.00, according to the locality in which it is built.

Design No. 1123

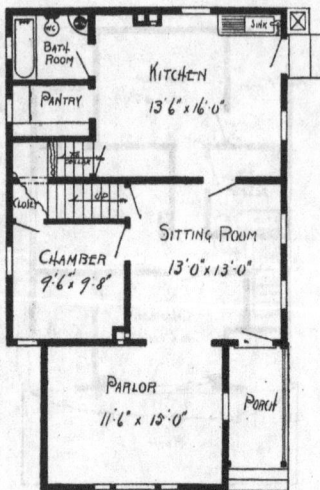

BATH ROOM

PANTRY

KITCHEN
13'6" x 16'0"

Closet

CHAMBER
9'6" x 9'8"

SITTING ROOM
13'0" x 13'0"

PARLOR
11'6" x 15'0"

PORCH

ROOF

ATTIC

Closet HALL

CHAMBER
9'6" x 9'8"

CHAMBER
9'6" x 13'0"

Closet

CHAMBER
11'6" x 15'0"

ROOF

First Floor Plan   **Second Floor Plan**

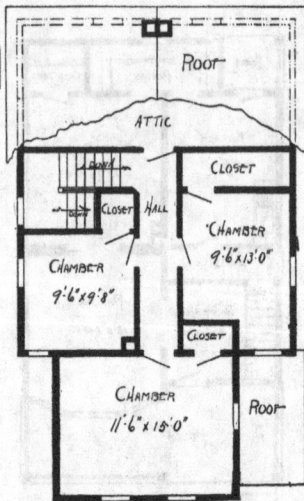

Size: Width, 24 feet; length, 39 feet 6 inches

Blue prints consist of cellar and foundation plan; first and second floor plans;
front, rear, two side elevations; wall sections and all necessary interior details.
Specifications consist of about twenty
pages of typewritten matter.

Full and complete working plans and
specifications of this house will be furnished for $5.00. Cost of this house is
from about $1,225.00 to about $1,450.00,
according to the locality in which it is
built.

Design No. 1156

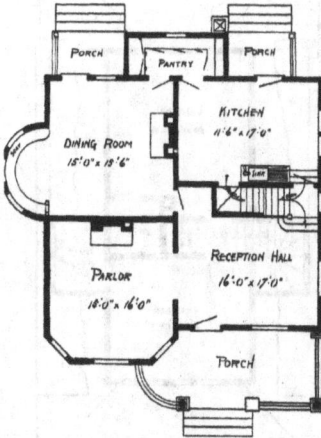

First Floor Plan

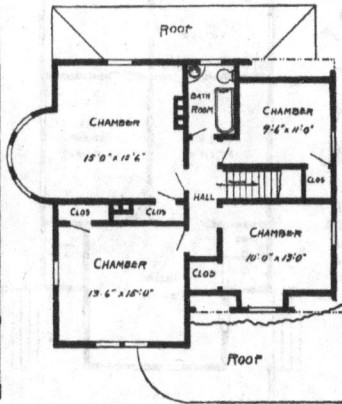

Second Floor Plan

PRICE of Blue Prints, together with a complete set of typewritten specifications is **ONLY**

$5.00

We mail Plans and Specifications the same day order is received.

Size: Width, 39 feet; length, 38 feet, exclusive of porches

Blue prints consist of cellar and foundation plan; first and second floor plans; front, rear, two side elevations; wall sections and all necessary interior details. Specifications consist of about twenty pages of typewritten matter.

Full and complete working plans and specifications of this house will be furnished for $5.00. Cost of this house is from about $3,250.00 to about $3,500.00, according to the locality in which it is built.

Design No. 1072

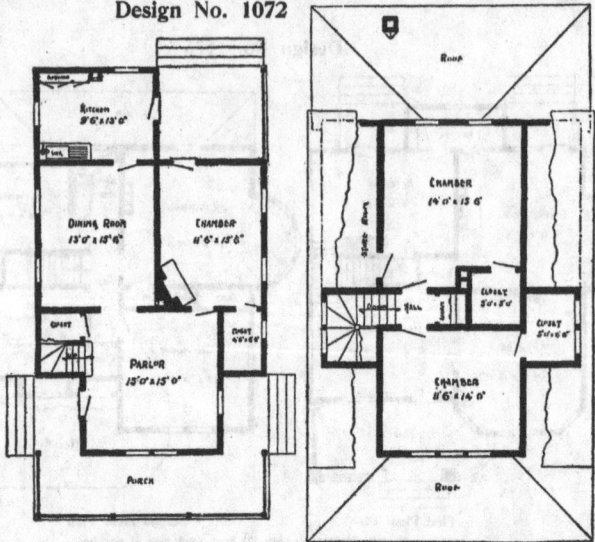

First Floor Plan

Kitchen
9'6" x 13'0"

Dining Room
13'0" x 13'6"

Chamber
11'6" x 13'6"

Closet

Bath 4'6" x 8'6"

Parlor
13'0" x 15'0"

Porch

Second Floor Plan

Roof

Chamber
14'0" x 13'6"

Closet 5'0" x 5'0"

Hall

Closet 5'0" x 6'6"

Chamber
11'6" x 14'0"

Roof

First Floor Plan   **Second Floor Plan**

Size: Width, 26 feet; length, 42 feet, exclusive of porch

Blue prints consist of foundation plan; first and second floor plans; front, rear, two side elevations; wall sections and all necessary interior details. Specifications consist of about twenty pages of typewritten matter.

Full and complete working plans and specifications of this house will be furnished for $5.00. Cost of this house is from about $1,250.00 to about $1,400.00, according to the locality in which it is built.

Design No. 1159

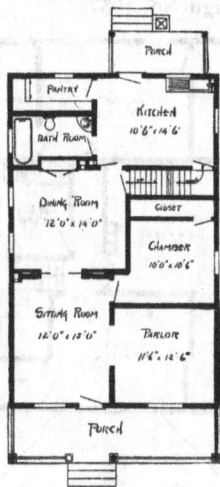

First Floor Plan

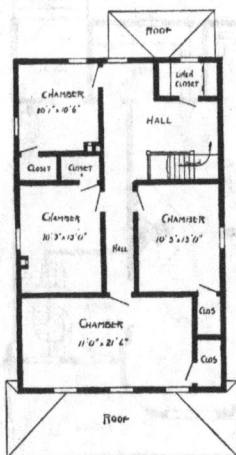

Second Floor Plan

PRICE of Blue Prints, together with a complete set of typewritten specifications is

ONLY

$5.00

We mail Plans and Specifications the same day order is received.

Size: Width, 26 feet; length, 40 feet, exclusive of porches

Blue prints consist of cellar and foundation plan; first and second floor plans; front, rear, two side elevations; wall sections and all necessary interior details. Specifications consist of about fifteen pages of typewritten matter.

Full and complete working plans and specifications of this house will be furnished for $5.00. Cost of this house is from about $2,150.00 to about $2,350.00, according to the locality in which it is built.

Design No. 1057

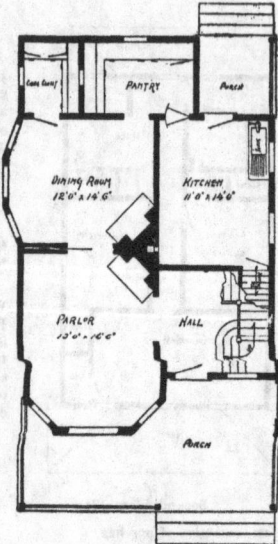

First Floor Plan

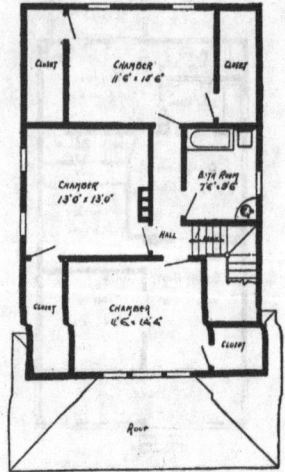

Second Floor Plan

Size: Width, 27 feet; length, 38 feet

Blue prints consist of cellar and foundation plan; first and second floor plans; front, rear, two side elevations; wall sections and all necessary interior details. Specifications consist of about twenty pages of typewritten matter.

Full and complete working plans and specifications of this house will be furnished for $5.00. Cost of this house is from about $1,850.00 to about $2,000.00, according to the locality in which it is built.

Design No. 1152

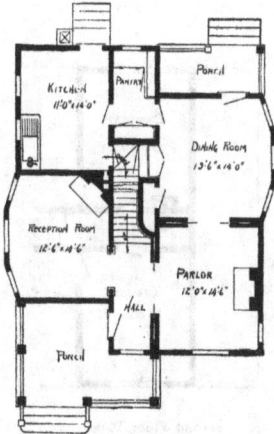

First Floor Plan

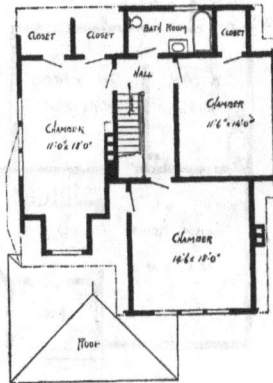

Second Floor Plan

Size: Width, 33 feet; length, 36 feet, exclusive of porch

Blue prints consist of cellar and foundation plan; first and second floor plans; front, rear, two side elevations; wall sections and all necessary interior details. Specifications consist of about fifteen pages of typewritten matter.

Full and complete working plans and specifications of this house will be furnished for $5.00. Cost of this house is from about $1,650.00 to about $1,850.00, according to the locality in which it is built.

Design No. 1155

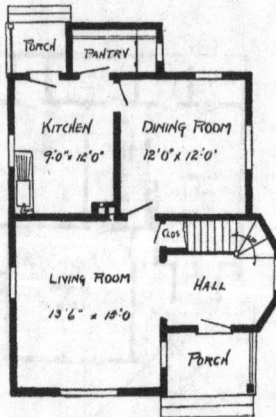

PORCH

PANTRY

KITCHEN
9'0" x 12'0"

DINING ROOM
12'0" x 12'0"

CLOS

LIVING ROOM
13'6" x 15'0

HALL

PORCH

CHAMBER
10'0" x 12'0"

CLOSET

HALL

CHAMBER
10'0" x 11'6"

CLOSET

First Floor Plan   **Second Floor Plan**

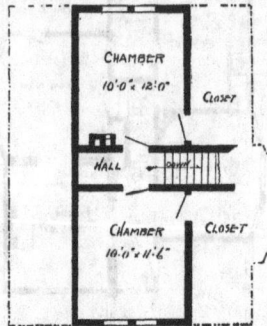

Size: Width, 25 feet; length, 33 feet, exclusive of porch

Blue prints consist of foundation plan; first and second floor plans; front, rear, two side elevations; wall sections and all necessary interior details. Specifications consist of about fifteen pages of typewritten matter.

Full and complete working plans and specifications of this house will be furnished for $5.00. Cost of this house is from about $950.00 to about $1,100.00, according to the locality in which it is built.

Design No. 1034

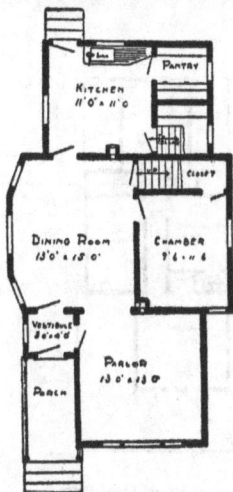

First Floor Plan

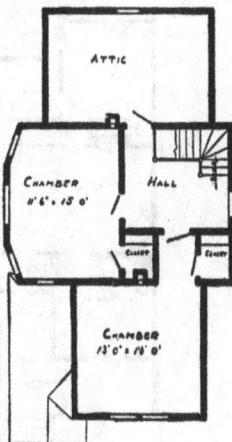

Second Floor Plan

PRICE

of Blue Prints, together with a complete set of typewritten specifications is

ONLY

$5.00

We mail Plans and Specifications the same day order is received.

Size: Width, 24 feet; length, 41 feet

Blue prints consist of cellar and foundation plan; first and second floor plans; front, rear, two side elevations; wall sections and all necessary interior details. Specifications consist of about twenty pages of typewritten matter.

Full and complete working plans and specifications of this house will be furnished for $5.00. Cost of this house is from about $1,750.00 to about $1,900.00, according to the locality in which it is built.

Design No. 1059

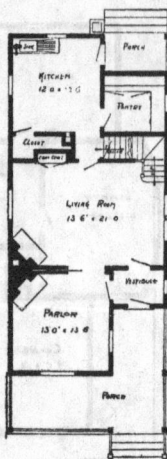

First Floor Plan   Second Floor Plan

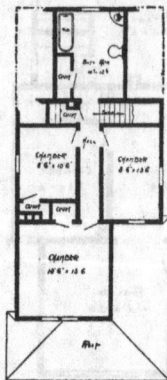

Size: Width, 22 feet; length, 45 feet, exclusive of porch

Blue prints consist of cellar and foundation plan; first and second floor plans; front, rear, two side elevations; wall sections and all necessary interior details. Specifications consist of about fifteen pages of typewritten matter.

Full and complete working plans and specifications of this house will be furnished for $5.00. Cost of this house is from about $1,750.00 to about $1,900.00, according to the locality in which it is built.

Design No. 1061

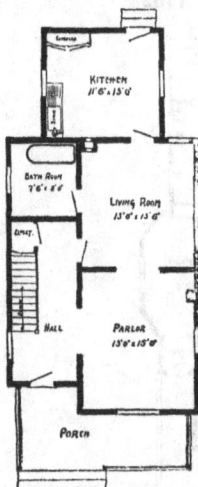

First Floor Plan

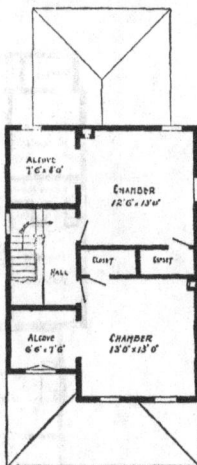

Second Floor Plan

Size: Width, 22 feet; length, 42 feet, exclusive of porch

Blue prints consist of foundation plan; first and second floor plans; front, rear, two side elevations; wall sections and all necessary interior details. Specifications consist of about fifteen pages of typewritten matter.

Full and complete working plans and specifications of this house will be furnished for $5.00. Cost of this house is from about $1,350.00 to about $1,500.00, according to the locality in which it is built.

Design No. 1162

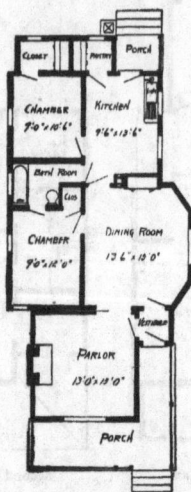

PRICE

of Blue
Prints, to-
gether with
a complete
set of type-
written
specifica-
tions is

ONLY

$5.00

We mail
Plans and
Specifica-
tions the
same day
order is re-
ceived.

Floor Plan

Size: Width, 24 feet; length, 48 feet 6 inches, exclusive of porch

Blue prints consist of foundation plan;
floor plan; front, rear, two side elevations; wall sections and all necessary
interior details. Specifications consist
of about twelve pages of typewritten
matter.

Full and complete working plans and
specifications of this house will be furnished for $5.00. Cost of this house is
from about $1,350.00 to about $1,500.00,
according to the locality in which it is
built.

Design No. 1074

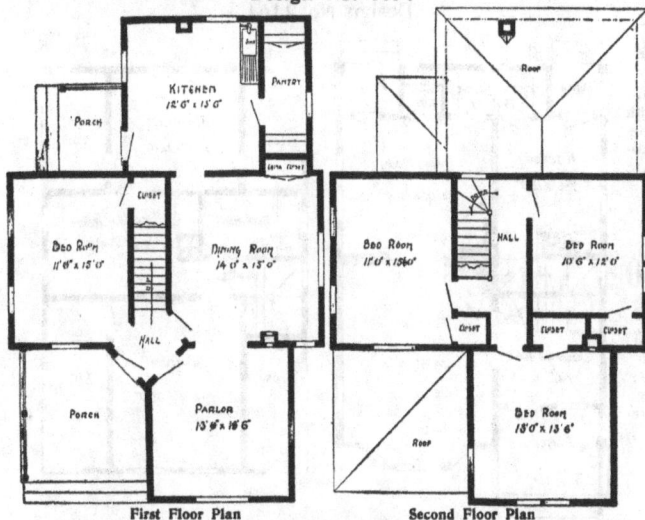

First Floor Plan

Second Floor Plan

Size: Width, 30 feet; length, 44 feet

Blue prints consist of foundation plan; first and second floor plans; front, rear, two side elevations; wall sections and all necessary interior details. Specifications consist of about twenty pages of typewritten matter.

Full and complete working plans and specifications of this house will be furnished for $5.00. Cost of this house is from about $1,650.00 to about $1,800.00, according to the locality in which it is built.

Design No. 1164

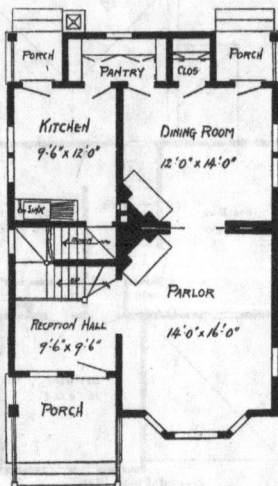

Porch

PANTRY   CLOS

Porch

KITCHEN
9'6" x 12'0"

DINING ROOM
12'0" x 14'0"

PARLOR
14'0" x 16'0"

RECEPTION HALL
9'6" x 9'6"

PORCH

CLOS

CHAMBER
10'6" x 12'0"

CLOS

BATH ROOM

SEWING ROOM
9'0" x 12'0"

CLOS

CHAMBER
12'0" x 15'0"

CLOS

First Floor Plan

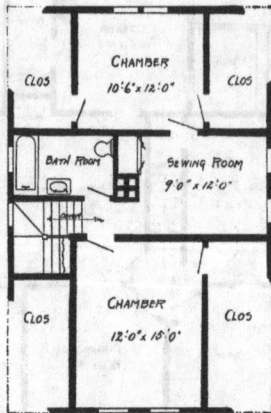

Second Floor Plan

Size: Width, 25 feet; length, 36 feet 6 inches, exclusive of porches

Blue prints consist of cellar and foundation plan; first and second floor plans;
front, rear, two side elevations; wall
sections and all necessary interior details. Specifications consist of about
fifteen pages of typewritten matter.

Full and complete working plans and
specifications of this house will be furnished for $5.00. Cost of this house is
from about $1,550.00 to about $1,700.00,
according to the locality in which it is
built.

www.ingramcontent.com/pod-product-compliance
Lightning Source LLC
Chambersburg PA
CBHW011801190326
41518CB00017B/2558